An Introduction to Spectroscopic Methods for the Identification of Organic Compounds

VOLUME 2

Mass Spectrometry, Ultraviolet Spectroscopy, Electron Spin Resonance Spectroscopy,
Nuclear Magnetic Resonance Spectroscopy (Recent Developments),
Use of Various Spectral Methods Together, and
Documentation of Molecular Spectra

EDITED BY
F. SCHEINMANN

Senior Lecturer, University of Salford, England

PERGAMON PRESS

Oxford · New York · Toronto
Sydney · Braunschweig

Pergamon Press Ltd., Headington Hill Hall, Oxford

Pergamon Press Inc., Maxwell House, Fairview Park, Elmsford, New York 10523

Pergamon of Canada Ltd., 207 Queen's Quay West, Toronto 1

Pergamon Press (Aust.) Pty. Ltd., 19a Boundary Street, Rushcutters Bay, N.S.W. 2011, Australia

Vieweg & Sohn GmbH, Burgplatz 1, Braunschweig

First edition 1973

Library of Congress Cataloging in Publication Data

Scheinmann, F   1933–

An introduction to spectroscopic methods for the identification of organic compounds.

Includes bibliographies.

CONTENTS: v. l. Nuclear magnetic resonance and infrared spectroscopy

1. Spectrum analysis. 2. Chemistry, Organic.

I. Title

QD271.S343   547'.34'085   76-99991

ISBN 0-08-006662-3 (v. l)

ISBN 0 08 016720 9 (flexicover)

ISBN 0 08 016719 5 (hard cover)

Printed in Hungary

An Introduction to
Spectroscopic Methods for the
Identification of Organic Compounds

VOLUME 2

Contents

Contents

Contents of Volume 1

Preface

VOLUME 2 of *An Introduction to Spectroscopic Methods for the Identification of Organic Compounds* follows largely the style adopted for Volume 1. Thus adequate theory, a study of some applications, seminar problems, and worked answers have been provided for:

(a) Mass spectrometry.

(b) Ultraviolet spectroscopy.

(c) Electron spin resonance spectroscopy.

(d) The combined use of data from the various spectral probes for elucidating organic structures.

There are also two additional chapters provided to keep pace with developments in organic spectroscopy and to serve the requirements of graduate students and research workers. These deal with:

(i) Recent developments in nuclear magnetic resonance spectroscopy which includes such techniques as the nuclear Overhauser effect, lanthanide shift reagents, and recent progress involving the use of isotope ^{13}C.

(ii) Documentation of molecular spectra which guides the chemists to the various collections of reference spectra (or data) now available in science libraries.

Many of the problems and answers given in this book have been used at Salford in the seminars given to undergraduates and M.Sc. and Ph.D. students in organic chemistry. The problems are arranged so that they become progressively more difficult with the latter problems in each case being suitable for graduates.

We thank the students and the various referees of the chapters in both Volumes 1 and 2 for their critical and helpful comments.

In finalizing Volume 2 we were very much encouraged by the favourable reviews accorded to Volume 1. It was pleasing to see the book being recommended by universities and colleges, especially The Open University (U.K.) where students receive their lectures and study guide by television, radio and correspondence, and study largely at home. The most serious criticism of Volume 1 was concerned with the absence of the companion chapters and seminar problems which now appear in Volume 2. While it is regretted that both volumes did not appear concurrently, we hope that the inclusion of some recent advances in organic spectroscopy, which have not been previously covered in a textbook at this level, will eventually compensate for the delay and any previous omissions.

It is again a pleasure to thank authors and publishers for permission to reproduce spectra and data where acknowledgements have been given.

F. SCHEINMANN

List of Contributors

Dr. L. A. Cort
Senior Lecturer in Organic Chemistry,
University of Surrey, England.

Professor J. A. Elvidge
Professor of Organic Chemistry,
University of Surrey, England.

Dr. B. J. Hopkins
Former Research Fellow,
University of Salford, England.

Dr. D. H. Maass
Lecturer in Organic Chemistry,
University of Salford, England.

Dr. R. W. A. Oliver
Lecturer in Physical Chemistry,
University of Salford, England.

Dr. F. Scheinmann
Senior Lecturer in Organic Chemistry,
University of Salford, England.

Dr. H. W. Wardale
Lecturer in Inorganic Chemistry,
University of Salford, England.

Dr. W. A. Wolstenholme
Products Manager, A.E.I. Inc.,
White Plains, New York, U.S.A.

An Introduction to Mass Spectrometry

L. A. CORT

Department of Chemistry, University of Surrey

Introduction

Whilst for the elucidation of unknown structures the whole range of available spectroscopic techniques is used whenever possible, one may sometimes be forced to use a limited number in a particular case. For determination of structure of the naturally occurring alkaloid aspidofractinine, the small quantity isolated dictated the methods of analysis, and the molecular formula and structure (I) were elucidated[1,2] by low resolution mass spectrometry in conjunction with i.r. absorption spectroscopy.

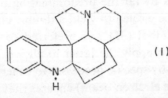

(I)

This single example is cited to emphasize the power of mass spectrometry, and illustrates what is possible provided that the processes giving rise to a spectrum are understood and adequate model compounds are available with which to make comparisons. With this sort of precise structural information evident from the mass spectrum, it can perhaps be appreciated that even with a minimum of knowledge of the processes giving rise to the spectrum, quite significant deductions may be made. This is a method of structure elucidation particularly suited to computer manipulation, in the acquisition, presentation, and analysis of data, and with such total equipment very complicated structures can be completely solved with certainty and rapidity,[3] but it is the purpose in this chapter to outline the origin of the spectrum and to introduce and exemplify manual methods of analysis.

In mass spectrometry the compound is induced to ionize in the vapour phase, and the resulting ions are then sorted according to their mass/charge (m/e) ratios. A *mass spectrum* is a record of the masses and relative abundances of the ions produced. These are characteristic of the compound concerned, and the identification of the ions enables deductions to be made about the parent molecule, both about the elements present and the method of their linkage. The great advantage of this spectroscopic technique, which utilizes less than one milligram of sample, is that by using a high resolution mass spectrometer the unique molecular formula can readily be determined, as can the unique compositions of the fragment ions; low resolution instruments do not yield this information with the same degree of certainty. Since the same atomic groupings in molecules give rise to ions by the same processes, molecules of the same chemical classes give similar spectra, with the differences being largely predictable.

Instrumentation

The generalized block diagram for a mass spectrometer is shown in Fig. 1. Organic mass spectrometry for structural analysis is most usually performed using a high energy electron beam (10–70 eV) to excite molecules from the sample in the vapour state, at elevated tem-

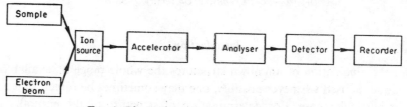

FIG. 1. Block diagram of a mass spectrometer.

peratures if necessary, in a very high vacuum (10^{-5}–10^{-8} torr). The electron beam generates ions from molecules in the *ion source*. There is ejection of an electron to give the *molecular ion* $(M)^+$ (or *parent ion*); this is by far the predominating process, but others also occur, in particular two electrons can be ejected to yield a doubly charged ion $(M)^{2+}$.

[The term molecular ion, symbol $(M)^+$, is preferred to the term parent ion, symbol $(P)^+$. There is now a tendency to apply the latter to all ions, including the molecular ion, which can be shown to give rise by specific processes to other (daughter) ions.]

The energy acquired (from the electron beam) ensures that many molecular ions undergo fragmentation to give ions of smaller mass together with neutral fragments. All the positive ions leaving the ion source are accelerated by means of an electrostatic voltage V and then they enter a magnetic field (in the analyser) H applied at right angles to their line of flight. The deflection they now experience (radius of curvature r) depends upon the m/e ratio and is given by the relationship

$$r^2 = k \; m/e \frac{V}{H^2} \, ,$$

where k is a constant. That is, all the ions with the same m/e ratio are deflected by equal amounts; this is the sorting process, and a series of collectors (one for every m/e value) would enable the relative abundances of each sort of ion to be measured by measuring the amount of charge brought to each collector in the same time. It is more convenient, of course, to use one fixed collector–detector and to bring the different sorts of ions successively to it by varying V or H in a known regular manner. The very low pressure employed ensures that each ion can reach the collector without collision with neutral fragments or molecules. For a full treatment of ion optics and modes of action of different mass spectrometers, see ref. 4a.

The spectrum is most conveniently recorded using mirror galvanometers because of their rapid response. The ion current detected for each m/e value activates the galvanometer and a beam of light is deflected by the mirror on to light-sensitive paper. Simultaneous recording at sensitivities of $\times 1$, $\times 10$, and $\times 100$, ensures that for each ion one of the traces will be optimum for measurement purposes. For a complete unknown of molecular weight 350 the spectrum may take 10 min to obtain and will cover a piece of paper 4 m long by 15 cm wide. Part of a spectrum is shown in Fig. 2. The peaks have width at the base because in practice, in the ion source, the generated ions have finite small, different

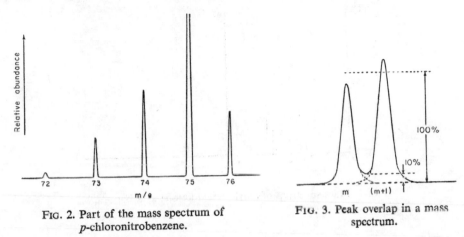

FIG. 2. Part of the mass spectrum of *p*-chloronitrobenzene.

FIG. 3. Peak overlap in a mass spectrum.

velocities even before acceleration. The m/e scale is not linear, but decreases exponentially towards the higher m/e values, and consequently at high m/e values the peaks overlap at the base. This is shown schematically in Fig. 3.

Depending upon the particular mass spectrometer used, at some point in the m/e scale it will become impossible to distinguish between two ions one mass unit apart, and to be able to differentiate between such ions is an important function of the instrument. As a useful working figure it is taken that when the height of the valley exceeds 10% of the average height of the two peaks (see Fig. 3), then the two peaks will not be separately resolved. The maximum m/e value where this valley contribution does not exceed 10% is termed the *resolution* of the instrument. (Note that some manufacturers give different definitions of resolution.)

Mass spectrometers are classed as high or low resolution instruments. The former class have resolutions exceeding 10,000; the latter have resolutions less than about 3000, and are correspondingly less expensive. A low resolution mass spectrum does not yield all the information obtainable from a high resolution spectrum, but, as will be seen from the following arguments, it is quite adequate for much structural analysis.

PRESENTATION OF SPECTRA

It is evident that it is quite impracticable to reproduce mass spectra such as have been described, in the same way as, for example, n.m.r. or i.r. absorption spectra. The essential information, the masses of the ions together with their relative abundances, may be presented in two ways. One way is to summarize this information as a simple bar graph, e.g. Fig. 4 shows the mass spectrum of ethyl methyl ketone. The molecular ion peak (M) is at m/e 72.

This is particularly suitable for making visual comparison with other spectra, but ions of low relative abundance are not readily shown on the same scale and these may be of

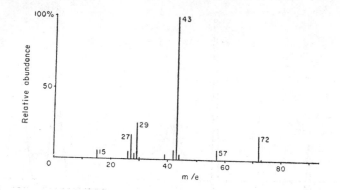

FIG. 4. Mass spectrum of ethyl methyl ketone as a bar graph.

importance in interpretation, e.g. there are such ions from ethyl methyl ketone at m/e 73 and 74 [$(M+1)$ and $(M+2)$ respectively].

In a spectrum one ion is of greatest relative abundance (here with m/e 43), and the response from it is said to give the *base peak* (B) in the spectrum. The alternative method of presentation is to draw up a table or list of ions (m/e values) and to show in parentheses the abundances relative to that of the base peak ion (arbitrarily taken as 100%). Even so, for the important low abundance ions in the vicinity of the molecular ion it may be preferable to report these relative to the molecular ion (taken as 100%). Thus the spectrum could equally be recorded: m/e 72 (17·0%) M, 57 (6·1), 44 (2·5), 43 (100) B, 42 (5·2), 39 (2·2), 29 (24·5), 28 (2·9), 27 (15·7), 26 (5·0), and 15 (5·2), with m/e 74 (0·3%) $(M+2)$, 73 (4·7) $(M+1)$, and 72 (100) (M).

(An equivalent method, used on an increasing scale, is to record the abundance of any particular ion as a percentage of the total ion abundance. This has the effect of mitigating the undue weighting otherwise given to the single measurement of the base peak ion.)

Both these presentations have their uses; the former is more like the original spectrum in appearance, but both still omit important information [see below, Metastable ion peaks (pp. 12, 57). It should be realized that manual transfer of the original data to either form is tedious and very time consuming, and, except for limited analysis, undesirable. It is possible to couple the detector in a mass spectrometer to other sorts of recorder,[4b] but the mirror galvanometer is universally used. The most useful development is simultaneous on-line computer recording with a digital computer collecting and processing the data as it is produced,[5, 6] thus eliminating laborious examination of the spectrum. A plotter, driven by the computer, can subsequently reproduce the spectrum as a bar graph if this is required.[7]

In any analysis of spectra it is essential to know several parameters if the analysis is to be meaningful; e.g. for an n.m.r. spectrum the solvent must be known. Similarly, the appearance of the mass spectrum is dependent on a number of variables, and comparisons can only be meaningfully made when conditions are the same.

Low-boiling liquids are handled by conventional vacuum line techniques. High-boiling liquids and solids are handled by a direct inlet system whereby they are introduced on a probe into the ion source. If the internal pressure is too high, secondary, unwanted reactions may occur. If the temperature is too high, thermally induced reactions may occur. Both pressure and temperature should be the lowest possible consistent with the production of a spectrum suitable for interpretation; it is usual to record the temperature.

Most low resolution spectra in the literature have been determined for an electron beam energy of 70 eV. A beam of lower energy, apart from giving fewer peaks overall, can bring about marked local changes in appearance because fragmentation processes can change. The energy of the ionizing electron beam should be recorded together with the total ion current.

There are other ways of bringing about the ionization of the compound; these are not as yet fully exploited, but developments may be expected. Methods such as field ionization[8] or chemical ionization[9] produce spectra of quite different appearance since the overall processes for ion production are not the same.

Analysis of Spectra

DETERMINATION OF MOLECULAR FORMULAE

For combinations of C, H, N, and O

For ethyl methyl ketone, C_4H_8O, using the whole number atomic weights ^{12}C, ^1H, ^{14}N, and ^{16}O, the calculated molecular weight is 72; but there are other combinations of atoms which give the same whole number total [of atomic mass units (a.m.u.)], and some of these are shown in Table 1. If the accurate, fractional values are used for the atomic weights, totals are obtained that differ in the fourth or fifth significant figure.

Now if the accurate mass of an unknown molecular ion, m/e 72, could be measured, the molecular formula could be uniquely decided. The most difficult case in the table would be to distinguish between 72·057511 and 72·068745. That is, to do this the accuracy would have to be better than 11 parts in 72,000, or 1 part in 7000. This is the same as saying that it would be necessary to count precisely and distinguish between two peaks of m/e 7000 and 7001, and this could be done if the mass spectrometer had a resolving power of at least 7000. To determine a unique molecular formula by accurate mass measurement when the molecular weight is much higher requires correspondingly increased resolution, double focusing, or the equivalent, is necessary to achieve this,[4a] and mass spectrometers are routinely designed in this way. In practice the accurate mass can be measured essentially by comparing the values of the spectrometer controls for the focusing operation with those used for focusing an ion of known mass used as a standard; an ion of mass comparable to that of the unknown is always selected.

To facilitate translation of accurate masses into unique atomic combinations there are published tables, based on either the 1959 standard ($^{16}O = 16 \cdot 000000$)[4c] or on the 1962 standard ($^{12}C = 12 \cdot 000000$)[10] of atomic weights. Each forms a self-consistent set and

TABLE 1

ACCURATE WEIGHTS FOR SOME COMBINATIONS OF C, H, N, AND O ATOMS WHERE THE TOTAL MASS IS NOMINALLY 72 A.M.U., BASED ON THE 1962 ATOMIC WEIGHTS ($^{12}C = 12 \cdot 000000$).

C_4H_8O	72·057511
$C_2H_4N_2O$	72·032360
$C_3H_4O_2$	72·021127
C_3H_6NO	72·045107
$C_3H_8N_2$	72·068745
C_5H_{12}	72·093896

either may be used (the former lists combinations up to 250 and the latter up to 500 a.m.u.). For accurate mass measurement work the reference standard against which the unknown mass is measured must, of course, be appropriate to the set of tables used. Another equivalent compilation[11b] based on the 1962 standard offers an alternative means of using the experimental data. A further alternative is to use a digital computer to process the data; a programme can readily be written to cater for any specific limits which might usefully be made to the possibilities in the light of compound history and/or experience (see ref. 12 for a suitable programme in Fortran).

The measurement of accurate mass may be made whilst the unknown compound is being manipulated in the spectrometer (a process limiting the number of measurements that can be made), or the whole spectrum can be stored on tape[13] or photographic plate[14] in such a way that the measurements can be made subsequently. Although the unique composition of the molecular ion is of greatest interest, the exact compositions of fragment ions are also important, since this greatly aids structure determination [see below, Saturated Aliphatic Hydrocarbons (p.32), and Nitriles (p.48)]. Sometimes a fragment ion so investigated proves to be inhomogeneous. Thus in the actual spectrum of ethyl methyl ketone the ion peak, m/e 43, is readily shown to be actually a doublet. The two ions of this nominal mass are $C_2H_3O^+$ and $C_3H_7^+$ (differing by 0·027 a.m.u.), and a resolution of 1600 will separate these. The latter ion is seen to contribute approximately 7% of the peak at m/e 43, and obviously a profound rearrangement gives rise to it. For the finer points of spectrum interpretation it is necessary to have such comprehensive knowledge about the ionic species contributing to each peak.

When the effective resolution is sufficiently high, determination of the ion composition in this way is obviously unambiguous, but as the effective resolution decreases (from one spectrometer to another), so more possibilities must be admitted for consideration. Thus for an unknown compound, if the resolution is 12,000 and the molecular weight is approximately 194, then accurate mass measurement of the molecular ion might indicate a value of $194 \cdot 227 \pm 0 \cdot 008$ based on $^{16}O = 16 \cdot 000000$. The possibilities indicated[4c] are $C_{11}H_{20}N_3$ $(194 \cdot 2278)$ and $C_{13}H_{22}O$ $(194 \cdot 2292)$. (With a resolution of 2,000 then the value determined might be $194 \cdot 25 \pm 0 \cdot 10$; this gives seventeen possibilities.) Fortunately there is an aid available by which possibilities for a molecular formula may be limited. This is termed the nitrogen rule.

The nitrogen rule

It happens that, with one exception, for all the common elements encountered in compounds of which mass spectra are obtainable, the common isotopes of even mass number have even valencies and common isotopes of odd mass number have odd valencies. For example, ^{12}C, ^{16}O, and ^{32}S have even valencies, ^{1}H, ^{35}Cl, and ^{31}P have odd valencies. The one exception is ^{14}N which has even mass but odd valency. This has an important consequence which may be stated as the *nitrogen rule*: molecules of odd molecular weight must contain an odd number of nitrogen atoms; molecules of even molecular weight must contain either an even number of nitrogen atoms or no nitrogen atoms.

An examination of genuine molecular formulae will convince the reader that this is so, and this rule can be most useful in molecular formula assignment.

The tables of combinations of C, H, N, and O atoms (e.g. refs. 4c, 10) contain under any particular a.m.u. total all the combinations (with certain exceptions noted)[†] of atoms giving this gross mass and not just combinations that are capable of being translated into complete molecules (e.g. C_3H_6NO for 72, and $C_{11}H_{20}N_3$ for 194 a.m.u.).

Thus, for instance, in the case of the compound above, of molecular weight 194, the nitrogen rule enables rejection of $C_{11}H_{20}N_3$ as a possible molecular formula, and the experimental value for the accurate molecular weight, when the resolution is 12,000, then gives $C_{13}H_{22}O$ as the unique molecular formula. (In this case, when the resolution is 2000, the nitrogen rule enables six of the seventeen possibilities to be rejected.)

Obviously it is advantageous to use the highest resolution possible; if only a low resolution instrument is available, then it may be possible to determine the molecular weight to the nearest whole number only. In this event another technique must be adopted to obtain the unique molecular formula from the available data. This involves the principle of isotope abundance analysis.

---

† For example, under 262 a.m.u. will not be found $C_{10}H_{14}O_8$, it being reasoned that this is a molecular formula that the user will encounter most infrequently, if at all. There are similar omissions throughout, and the user is referred to the introductory explanations to the individual tables.

ISOTOPE ABUNDANCE ANALYSIS

For combinations of C, H, N, and O

Using the whole number atomic weights ^{12}C, ^{1}H, ^{14}N, and ^{16}O, the calculated molecular weight of ethyl methyl ketone is 72. This is the figure obtained from the spectrum (see Fig. 4) by recognition of the molecular ion peak. But natural carbon atoms consist not only of ^{12}C but of ^{13}C, and the relative abundances are 98·89 : 1·11. Thus one in every hundred carbon atoms is not ^{12}C but ^{13}C, so that in a bulk sample of ethyl methyl ketone, for every twenty-four molecules containing ^{12}C only there is one molecule containing three ^{12}C atoms and one ^{13}C atom. The molecular weight of this molecule is 73.

Similarly, natural hydrogen, nitrogen, and oxygen contain the heavier isotopes ^{2}H, ^{15}N, ^{17}O, and ^{18}O, and other molecules will have a molecular weight of 73 if they contain one ^{15}N atom in place of ^{14}N, etc., and a molecule that contains one ^{18}O atom in place of ^{16}O, or one ^{13}C and one ^{2}H atom in place of ^{12}C and ^{1}H, etc., will have a molecular weight of 74. These give the ions producing what are termed the *first* and *second isotope peaks* $(M+1)$ and $(M+2)$ respectively.

Now from known natural relative abundances of these heavier isotopes it is possible to calculate[4d, 10] relative to the molecular ion the expected abundances of the ions giving the first and second isotope peaks. That is, the expected relative heights of the (M), $(M+1)$ and $(M+2)$ peaks in the spectrum can be calculated. For some combinations of atoms of 72 a.m.u. these figures appear in Table 2 together with the ratio of heights $(M+1)/(M+2)$.

TABLE 2
CALCULATED RELATIVE HEIGHTS OF THE FIRST AND
SECOND ISOTOPE PEAKS FOR COMBINATIONS OF C, H, N
AND O ATOMS OF 72 A.M.U.

	(M)	$(M+1)$	$(M+2)$	$(M+1)/(M+2)$
C_4H_8O	100	4·49	0·28	16
$C_2H_4N_2O$	100	3·03	0·23	13·2
$C_3H_4O_2$	100	3·38	0·44	7·7
C_3H_6NO	100	3·76	0·25	15
$C_3H_8N_2$	100	4·13	0·07	59
C_5H_{12}	100	5·59	0·13	43

In practice one determines these quantities from the spectrum (obviously not from the bar graph presentation), and searches for a "best fit" with the calculated values. If ethyl methyl ketone were used as an unknown compound, typical experimental figures might be respectively 100, 4·7, 0·3, and 15·7, and these would indicate clearly a molecular formula of C_4H_8O. Suitable tables of calculated isotope abundance ratios exist for up to 250 a.m.u.[4c, 15] and up to 500 a.m.u.[10]

Unfortunately there are circumstances encountered, resulting from the nature of the particular compound and the operating conditions, when the method fails. The most common effect is when the first and second isotope peaks are inflated by unknown, variable amounts, and the only indication that an incorrect molecular formula has been selected may be inconsistencies in the subsequent interpretation of the fragmentation pattern. Compounds which are readily protonated are particularly prone to give misleading experimental figures. [Experimental measurements are also affected by any unduly high background to the spectrum; see below, under The molecular ion peak (p. 14).] In any event, relative peak heights should be averaged from several spectra to eliminate fortuitous errors in a single spectrum.

[Even without recourse to tables it is possible to make simply a useful deduction about the molecular formula. For compounds not unduly rich in nitrogen, by far the largest contribution to the first isotope peak comes from the ions with ^{13}C in place of ^{12}C. Thus if the height of this peak (relative to that of the molecular ion peak as 100%) is divided by 1·1 and the result rounded off to the next lowest whole number, then this integer almost always gives the number of carbon atoms in the molecule.]

For combinations other than C, H, N, and O only

It is clear that whatever elements comprise the compound, the isotope distribution pattern will be manifest in the spectrum in the vicinity of the molecular ion peak. Fortunately in many cases these patterns are readily recognized and distinguished one from another, and this enables easy identification of such elements.

For example, Fig. 5 shows the low resolution spectrum of an unknown compound (II) of molecular weight 157. Whilst there is nothing untoward about the relative height of the first isotope peak (m/e 158), simple inspection shows that the height of the second isotope peak (m/e 159) is approximately one-third of the height of the molecular ion peak [measured relative abundances are m/e 157 (100), 158 (7·2), 159 (33·2), 160 (2·4), and 161 (0·2)]. This

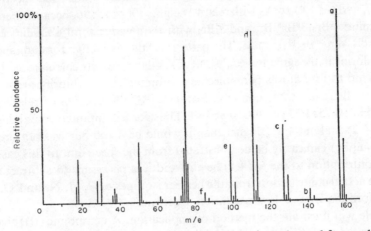

FIG. 5. Mass spectrum of compound (II), mol. wt. 157. Peaks at a, b, c, d, e and f are each accompanied by peaks at 2 atomic mass units higher having approximately one-third relative intensity.

clearly represents the isotope distribution pattern of an element other than C, H, N, and O. The molecule of compound (II) must contain one atom of an element where there are two isotopes, differing by 2 a.m.u., and the heavier is one-third as abundant as the lighter. It is apparent that this element must be chlorine [$^{35}Cl:^{37}Cl$: 75·53 : 24·47 (= 100 : 32·4)].

In order to complete the molecular formula for compound (II) from the low resolution spectrum, use is made of the $(M+1)$ peak height. Clearly, the combination of ^{12}C, ^{1}H, ^{14}N, and ^{16}O atoms for which this is relevant is the molecular weight less ^{35}Cl, that is $157-35 = 122$. Under 122 a.m.u. there is a combination, $C_6H_4NO_2$, for which the calculated relative height of the first isotope peak is 7·01; this compares well with the experimental value of 7·2. The molecular formula of compound (II) is thus $C_6H_4ClNO_2$ (this formula obeys the nitrogen rule). In such cases it is useless to attempt to extract the genuine contribution to the $(M+2)$ peak from the heavy isotopes of C, H, N, and O only.

If accurate mass measurement of the molecular ion (m/e 157) of compound (II) were possible, the experimental figure might be 157·040±0·007 a.m.u. on the $^{16}O = 16·000000$ standard. The appropriate ^{35}Cl mass to be subtracted is 34·980 a.m.u., and this leads[4c] to $C_6H_4NO_2$ as the unique composition of the remainder of the molecule. Tables of accurate combination masses for up to sixteen elements have been published,[16] but it is usual to combine the practice of accurate mass measurement with the principle of isotope abundance analysis, as shown, in order to compute unique ion compositions. A more recent publication[17] offers unique formula determination from accurate mass measurement by utilizing a slightly different method of manipulating the data, and appropriate tables are given.

The principle of isotope abundance analysis can be extended whenever the pattern in the vicinity of the molecular ion peak differs from that expected for combinations of C, H, N, and O only, and enables recognition of the presence of such elements as bromine, sulphur, silicon, mercury, etc.

With bromine, for example, the relative abundances of $^{79}Br:^{81}Br$ are 50·52 : 49·48. For a molecule containing one bromine atom, the $(M+2)$ peak will be at least 98% of the height of (M). Further, for a compound containing two bromine atoms, the spectrum will show (M), $(M+2)$, and $(M+4)$ peaks with relative heights 100 : 195 : 96, because there will be molecules containing $^{79}Br_2$, $^{79}Br^{81}Br$, and $^{81}Br_2$ with the relative total abundances readily calculated[4d] from the $^{79}Br:^{81}Br$ ratio. The pattern is still distinctive for combinations of bromine with chlorine in the same molecule and is, similarly, readily calculated.[4d] The patterns expected for up to four atoms per molecule of bromine and/or chlorine are shown in Fig. 6.

For sulphur there are three common isotopes, ^{32}S, ^{33}S, and ^{34}S of relative abundances (variable) 95·02 : 0·75 : 4·23 (= 100 : 0·8 : 4·4). Thus for a compound containing one atom of sulphur, $(M+2)$ is about 4·5% higher than it would be if sulphur was not present, and the presence of sulphur can easily be demonstrated from the spectrum. In this case ^{33}S makes a significant contribution to the $(M+1)$ height, and it is necessary to subtract 0·8 from this to obtain the actual contribution from the heavier isotopes of C, H, N, and O, when determining the formula from a low resolution spectrum.

An example will illustrate the method of application. A compound (III) gives the following figures for the peak heights in the vicinity of the molecular ion peak: m/e 154 (4·9%) $(M+2)$, 153 (9·4) $(M+1)$, and 152 (100) (M). What is the molecular formula?

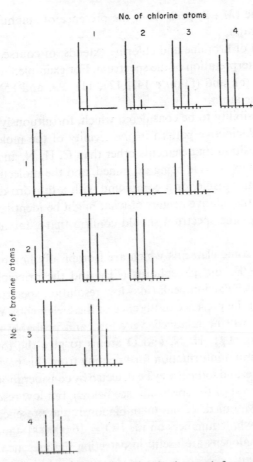

FIG. 6. Patterns expected in the vicinity of the molecular ion peak for molecules containing combinations of bromine and/or chlorine atoms [(M), $(M+2)$, $(M+4)$, $(M+6)$, etc., peaks].

The $(M+2)$ height indicates one sulphur atom in the molecule, since otherwise the maximum expected height for $(M+2)$ is 0·9 (tables). The contribution to the $(M+1)$ height less that from ^{33}S is 9·4−0·8 = 8·6. For a total of 120 a.m.u. (= 152−32) this as a first isotope contribution indicates a fragment $C_7H_8N_2$ (8·46 calculated). The formula is thus $C_7H_8N_2S$.

When actually using experimentally determined peak heights to give a molecular formula, some latitude should be allowed on the measured values; peaks from any background spectrum will inflate those from the unknown compound, and the second isotope peak is very susceptible to such influence. Experience will quickly show what allowance should be made in practice. This has the effect of permitting other formulae for consideration after the first choice. It may become evident, from other internal evidence in the mass spectrum or from the integral for the n.m.r. spectrum, for example, that the first choice is not, in fact, correct.

It should be noted that presence or absence of bromine and/or chlorine is immediately obvious merely by inspection of the spectrum, even as a bar graph presentation, whereas the smaller effect on the $(M+2)$ peak from the presence of sulphur is not always readily obvious from the bar graph.

This easy recognition of bromine and chlorine extends, of course, to fragment ions, and may aid considerably interpretation of the spectrum. For example, in Fig. 5 clearly there are five peaks (b), (c), (d), (e), and (f) (m/e 141, 127, 111, 99, and 85) from ions which still each contain the chlorine atom.

[There is another possibility to be considered which, fortuituously, might lead to a spectrum where there is a distinctive pattern in the vicinity of the molecular ion peak which would lead to the supposition that elements other than C, H, N, and O were present. This is when a *mixture* of compounds is being examined, and the molecular weights differ by 2 a.m.u., e.g. *m*-xylene with anisole, or a compound with a dihydro-derivative. The relative amounts will determine the relative heights of what might be identified as (M) and $(M+2)$. In such a case the rest of the spectrum should confirm that a mixture was, in fact, being examined.]

There are, of course, some elements which are isotopically homogeneous, in particular fluorine (^{19}F), iodine (^{127}I), and phosphorus (^{31}P), and the principle of isotope analysis in the vicinity of the molecular ion peak (in a low resolution spectrum) cannot be applied to the spectrum to detect the presence or absence of these elements. When iodine is present the $(M+1)$ peak height will be relatively very low, and probably lower than calculated heights for combinations of C, H, N, and O atoms to give the total molecular weight. This represents the maximum information forthcoming from this part of the spectrum.

The presence of fluorine, and iodine, may be detected by consideration of the fragmentation pattern evident from the rest of the spectrum (see below), but low resolution spectra unfortunately do not offer a way of detecting unambiguously the presence of phosphorus. The mass deficiencies (below whole numbers on the $^{16}O = 16 \cdot 000000$ standard) of the atoms of phosphorus and all the halogens are useful in detecting the presence of these elements by accurate mass measurement.

Metastable ions and peaks therefrom

Examination of an actual spectrum will often reveal sharp peaks at half mass numbers; for example, there are two such peaks at m/e 121·5 and 122·5 in the spectrum of triphenylamine. These must, of course, derive from ions of mass 243 and 245 respectively, each of which carries a positive charge of two units (instead of the customary one).

In addition to such peaks there are many others, with broad bases, in spectra, that appear at fractional as well as at whole number m/e values. These arise from ions (*metastable ions*) that are undergoing a fragmentation process after they leave the ion source and as they are being accelerated and focused before detection. The peaks from metastable ions are referred to as "metastable peaks", and these take all forms from simple Gaussian, through flat-topped peaks, to ones with a double maximum. None of the "metastable peaks" are shown in bar graph presentations.

The spectrum of compound (II), $C_6H_4ClNO_2$, is rich in such peaks. Two of them, with centres at m/e 102·7 and 104·7, are shown in the part of the spectrum given in Fig. 7.

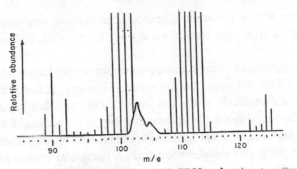

FIG. 7. Part of the spectrum of compound (II), $C_6H_4ClNO_2$, showing two "metastable peaks".

When a parent ion (mass m_1) decomposes in this way to give a neutral fragment and a daughter ion (mass m_2) and there is a corresponding "metastable peak" (apparent mass m^*), then a simple mathematical relationship exists[4e] between m_1, m_2, and m^* in the form $m^* = m_2^2/m_1$.

Thus the presence of a metastable ion can be used to identify the parent of a particular daughter ion, or to identify the daughter of a particular parent.

For compound (II) (spectrum Fig. 5) there are two ions, m/e 157 and 127, such that $127^2/157 = 102.73$. The "metastable peak" in the spectrum is thus good evidence that the mass loss of $157 - 127 = 30$ a.m.u. *occurs in one stage*. [Note that the transition m/e 159 → 129 is similarly supported by the experimental $m^* = 104.7$ (104.66 calculated), and that the relative abundances of the two metastable ions are in agreement with the single halogen atom content of both parents and daughters; $m^* = 102.7$ is for ions containing only ^{35}Cl and $m^* = 104.7$ for ions containing only ^{37}Cl.]

To obviate laborious calculations for each and every case, there are comprehensive tables[18] of calculated metastable transitions and helpful nomograms.[4f]

By detecting metastable ions one is really detecting fragmentations which take an appreciable time, that is for redistribution of electrons within the ion compared with, say, the initial process of ionization. In the case of compound (II) the loss of 30 a.m.u. in one stage from the molecular ion is quite useful knowledge; the only reasonable one step loss is of NO or CH_4N. Each of these could arise by scission of a single bond, but in the particular case of aromatic nitro compounds we find that NO is ejected from the nitro group, in a process which must involve rearrangement, and that this process always manifests itself by the occurrence of the appropriate metastable ion [see below (pp. 52, 57)]. Thus we have the immediate suggestion that compound (II) contains an aromatic nitro group; this is not readily apparent from the otherwise unsupported observation that there is an ion (among others) in the spectrum, 30 a.m.u. less than the molecular ion.

Again, one fragmentation of the molecular ion of benzene $[C_6H_6]^+$, m/e 78, is to give $[C_4H_4]^+$, m/e 52, ($+C_2H_2$) in one stage; the appropriate metastable ion (m/e 34.7) gives a peak in the spectrum. For mono-substituted benzenes there is the equivalent process for the derived phenyl ion, $C_6H_5^+$, m/e 77 → $C_4H_3^+$, m/e 51 ($+C_2H_2$); the metastable ion here has m/e 33.8 and is visible in the spectrum. If, then, ions are detected with m/e 77 and 51, together with a metastable ion, m/e 33.8, the compound is probably a mono-substituted benzene which gives a phenyl cation. Considered alone, the one metastable ion with m/e 33·8 might also arise from the transitions $80 \rightarrow 52$, $96 \rightarrow 57$, $339 \rightarrow 107$, or $378 \rightarrow 113$, for example.

The fact that a metastable ion is not detected for a particular process does not mean that such a process does not occur but only that if it occurs it is comparatively very rapid (and may take place exclusively in the ion source). On the other hand, simple scissions of single bonds may be accompanied by metastable ions, e.g. the loss of terminal methyl from a long-chain compound. The great usefulness of metastable ions is that they demonstrate the formation of a particular daughter ion from a particular parent as a one-step process.

It should be mentioned that a good spectrum from the point of view of using it for molecular structure determination and comparison with other spectra is almost certainly not the best for observing the metastable ion peaks; to get the best for the latter operation requires individual tuning of the various spectrometer controls for the particular compound concerned. Nevertheless, an ordinary spectrum will show several useful metastable ion peaks, and it is the presence of these (as in Fig. 7) which helps to make simultaneous on-line computer recording and processing of the spectral data a difficult operation.

The molecular ion peak

It will be realized that in analysis of spectra much depends upon the correct identification of the molecular ion peak. It is not unknown for the molecular ion to be less than 0·5% as abundant as the base peak ion, and it is easy for the peak from such a low abundance ion to be lost in the background (this is especially the case when there is automatic computer recording and processing of the data, since it would be reasonable initially to programme the recording device to ignore all ions of less than, say, 1·0% relative abundance, in order to yield a manageable print-out). Some saturated alcohols, amines, mono-, and polyhalides are particularly prone to give low abundance molecular ions.

The problem is aggravated by the fact that there are always background signals and, according to circumstances, some of these may be as large as genuine signals from the compound under investigation. Sometimes it is obvious when signals are entirely background, for instance a regular, weak, repeating pattern with corresponding peaks 14 a.m.u. apart may well derive, under certain operating conditions, from long-chain hydrocarbons present in the oil of the vacuum system. When in doubt it is often useful to compare the spectrum with an immediately previous "blank" or with the spectrum of the previous compound handled.

Two further circumstances may cause confusion and complications, particularly when the molecular formula has to be determined using a low resolution instrument.

(1) If a hydrogen radical is very readily lost from the molecular ion, then the M and $(M+1)$ peaks contain contributions from the first and second isotope peaks respectively of the $(M-H)^+$ ion, and it is not easy to work out the quantitative contributions. Further, if the M peak is small, then the $(M-H)$ peak may be mistaken for the molecular ion peak.

(2) Some compounds, particularly carbonyl compounds, ethers, and amines will react with protons generated in the ion source to form stable $(M+H)^+$ ions [e.g. $(CH_3)_2C \overset{+}{=\!\!=} \overset{+}{O}—H$

ions]. In this case the $(M+1)$ peak is too great by an amount depending upon the (unknown) extent of the ion–molecule reaction, and quantitative estimation of peak heights may lead to a quite incorrect molecular formula. However, the ion–molecule reaction is dependent upon the internal pressure, and if such a state of affairs is suspected it may be confirmed by obtaining spectra of the compound with a range of internal pressures; if decreasing the pressure results in a decreased relative height of the $(M+1)$ peak, then this complicating reaction is indicated.

It is appropriate to record the checks, a selection of which may be employed according to circumstances, for ensuring correct identification of the molecular ion.

(a) It is self-evident that a true molecular ion is homogeneous, and that if the mass spectrometer resolves an ion peak into two or more components, then it follows that none of these can be the molecular ion, and that this must be searched for at higher m/e ratios.

Thus in the spectrum of *N*-acetylleucine ethyl ester (mol. wt. 201) run as an unknown compound, the very weak peaks above m/e 158 are likely to be overlooked. Accurate mass measurement for the m/e 158 ion shows that this comprises two ions, differing by 0·037 a.m.u. (they arise by loss of $CH_3CO^{\cdot}$, and of $(CH_3)_2CH^{\cdot}$ from the molecular ion, and are $C_8H_{16}NO_2^+$, and $C_7H_{12}NO_3^+$). Therefore it is clear that the molecular weight is not 158.

(b) Fragmentation must always proceed by loss of reasonable fragments. It is unreasonable for the first loss from the molecular ion to be an improbable combination of atoms. In particular, combinations of atoms to give totals of 6–14 and 21–24 a.m.u. are of very low probability (what would be the combinations?).

If the first peak below the apparent molecular ion peak represents a loss of, say, 8 a.m.u., then clearly it is likely that the apparent molecular ion peak is itself a fragment of an undetected parent.

(c) For all transitions supported by metastable ions, all daughters must have parents. That is, there must be two ions in the spectrum at m/e values higher than every m^* value, and the $m^* = m_2^2/m_1$ relationship must apply.

Thus if the top end of a spectrum shows peaks at m/e 172 and 187 only (with appropriate first and second isotope peaks) with a metastable ion peak at m/e 170·6 then clearly the latter does not support the transition m/e 187 → 172 (m^* calculated 158·2), but it indicates an undetected parent ion of m/e 205 for m^* at 170·6 supports the transition m/e 205 → 187 (m^* calculated 170·6).

This check obviously can only be applied to metastable ions of relatively high m/e values, and it should be pointed out that if the spectrum is rich in doubly charged ions then metastable ions may be evident for processes involving these, and the simple relationship above may not be applicable for these ions.

(d) A doubly charged ion cannot have more than half the mass of the molecular ion. If there is a doubly charged ion with an m/e value of $(x+0·5)$, then the mass of the molecular ion must be at least $(2x+1)$ a.m.u.

(e) No fragment ion may contain more atoms of any one sort than does the molecular ion.

Suppose that the apparent molecular ion peak was at m/e 129 and that high or low resolution measurements lead to a molecular formula of $C_4H_3NO_2S$. If there was in the spectrum a prominent ion at m/e 41, then it is apparent that of the four atomic combinations possible for this, the most probable as a prominent ion is $C_3H_5^+$, and hence the ion at m/e 129 is probably, in fact, a fragment ion from a parent which contains at least five hydrogen atoms; this must be the case if accurate mass measurement confirms the composition of the m/e 41 ion (or, alternatively, this may be evidence for the presence of impurities in the sample).

In the case of many alcohols there is ready loss of H_2O from the molecular ion leading to an olefin, and the molecular ion may not be observed. Other fragmentations can lead to ions still containing the oxygen atom, the presence of which is apparent from accurate mass measurement on these ions. It is possible in such cases to deduce a molecular ion peak which is lost in the background.

A neutral fragment eliminated from the molecular ion in the ion source may there itself become a positive ion, which will appear in the spectrum, and the identification of this ion will aid establishment of the total molecular formula. For instance, t-butyl bromide shows no molecular ion, but there are ions in the spectrum at m/e 79 and 80, which obviously derive from Br^+ and $HBr^{\ddagger}$ since the isotope pattern for one bromine atom is evident in each case. The largest carbon-containing ion is at m/e 57 (B), so the molecular weight is at least 136 for the compound giving this spectrum. As the spectrum can be very conveniently scanned for the presence of Cl^+, $HCl^{\ddagger}$, Br^+, and $HBr^{\ddagger}$, it is easy to infer the presence of these halogens even if the molecular ion peak is not discernible or if the ions at the top end of the spectrum do not show the requisite isotope peak pattern. (Note the complications arising when quaternary ammonium halides or amine salts of halogen acids are examined by mass spectrometry.)

Since fluorine and iodine are isotopically homogeneous, there are no isotope patterns to help identify ions containing these elements, but fortunately ions of m/e 19, 20, or 127, 128 are sufficiently uncommon, so that if these are present fluorine (F^+, $HF^{\ddagger}$) or iodine (I^+, $HI^{\ddagger}$) can be suspected. If the molecular ion peaks are actually detectable, then the appropriate ions $(M-X)^+$, $(M-HX)^{\ddagger}$ will, of course, help to confirm these elements.

(f) As explained below, fragment ions may be either odd- or even-electron ions, whereas molecular ions are always odd-electron ions. For compounds which are largely saturated, the majority of ions in the spectrum will be even-electron ions. Thus if, in the spectrum, the chief components of each group of peaks arise from ions of even mass, then these each probably contain an odd number of nitrogen atoms (e.g. one). An apparent molecular ion of even mass would be suspect as tending to be inconsistent.

Fragmentation Processes

Fragmentation processes can be actually very simple, and they can be exceedingly complicated; some of the latter are deceptively simple. Thus, for instance, one fragmentation of benzene is to give C_2H_2, and the use of labelled benzenes has revealed that prior to this there is complete carbon scrambling,[19] as well as hydrogen scrambling,[20] in the molecular ion, so that the C_2H_2 fragment is not formed exclusively from two adjacent carbon atoms in the original molecule plus the hydrogen atoms attached to them.

Nevertheless, such a deceptively simple fragmentation can be very useful in structure deduction, and for these and for actually simple fragmentations, identification of the fragments suggests very strongly the partial parent structure(s) giving rise to them.

SYMBOLISM

It is as well to emphasize here that the precise structure (electronic or nuclear) of the great majority of individual ions in the mass spectrometer cannot be established, but it is a very useful concept for the purposes of rationalization of the spectrum and structure deduction to consider that the very rapid ejection of an electron to give the molecular ion occurs without change in nuclear structure. The molecule loses a single electron from a full complement of paired electrons and so becomes a radical ion.

When the charge and radical site are conveniently considered non-localized, the symbolism is as in structures (IV) and (V) (molecular ions from butadiene and benzene); often all except the top right-hand corner of the square bracket is omitted, giving structure (VI) for the molecular ion from benzene. When the charge and radical site are conveniently considered to be localized, the symbolism is as in structures (VII), (VIII), and (IX) (molecular ions from butadiene, again, ethyl methyl ketone, and pyrrole). The concept of initial localization of charge on a hetero-atom is a particularly fruitful one.

Structures (IV) and (VII) are different ways of writing the same molecular ion. If charge and radical site are not to be located on the same atom, then obviously for open-chain structures the two atoms concerned must initially either be joined by a multiple bond or be connected by conjugated multiple bonds, otherwise the symbolic scission of a single bond will result in fragmentation; for cyclic structures, symbolic scission of such a single bond will not result in mass loss.

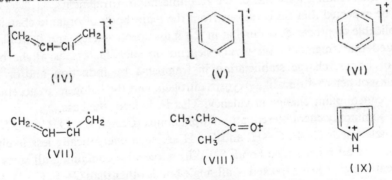

In the mass spectrometer scission of single bonds may be heterolytic or homolytic, and the conventional symbols for two- and one-electron shifts are used; for the homolytic process it is also usual to draw one only of the arrows concerned with the shift of two electrons, as this indicates unambiguously the overall process.[21a, 22] Thus the formation of the n-butyl carbonium ion and a bromine radical from the molecular ion of 1-bromobutane is symbolized as (X) (heterolytic scission), the formation of the ethyl radical and an ion, $CH_3 \cdot C = O^+$, from the molecular ion of ethyl methyl ketone is symbolized as (XII)

(homolytic scission), and the formation of a hydrogen radical and a resonance-stabilized carbonium ion from the molecular ion of propene is symbolized as (XIII) (also homolytic scission).

(X) $CH_3-CH_2-CH_2-CH_2-\overset{+}{Br} \cdot$ ⟶ $CH_3-CH_2-CH_2-CH_2^+ + Br\cdot$

                                          (XI)

(XII) $\underset{CH_3}{\overset{CH_3-CH_2}{{\Large\diagdown}\;{\Large\diagup}}}C{=}\overset{+}{O}\cdot$ ⟶ $CH_3-CH_2^\cdot + CH_3-C{\equiv}O^+$

(XIII) $H{-}CH_2{-}\overset{+}{C}H{-}\overset{\cdot}{C}H_2$ ⟶ $H\cdot + CH_2{\cdots}\underset{CH}{\overset{+}{}}{\cdots}CH_2$

In addition, it is convenient to adopt some convention as shown (XIV) to indicate the separate total masses of the atomic combinations at either end of a designated bond.

$$\overset{\displaystyle 122 \quad\;\; 107 \quad\; 93 \quad\;\; 79}{\underset{\displaystyle 15 \quad\; 29 \quad\; 43 \quad\; 57}{CH_3 \;\}CH_2\}CH_2\}CH_2\}Br}} \qquad (XIV)$$

Fragmentation processes may be divided into two categories, scission only, and scission with rearrangement. These will be dealt with in order, and the important concept introduced of even- and odd-electron ions.

When the molecular ion undergoes fragmentation, the factors which determine which bonds will break and which ions will be formed are the relative bond strengths and the stabilities of the resulting ions and of the resulting neutral fragments.

It might be expected that observations from the main body of organic chemistry would also be applicable to processes occurring in the mass spectrometer, and it would therefore be anticipated that fragmentation would depend on such factors as allylic or benzylic activation of bonds, charge stabilization in fragments by induction and/or resonance, and the ability of hetero-atoms like oxygen, nitrogen, and the halogens to accept a positive charge with concomitant change in valency. This is, indeed, the case.

Of bonds commonly encountered, all multiple bonds ($C{\equiv}N$, $C{=}O$, $C{=}C$, $N{=}O$, etc.) and three single bonds C—F, C—H, and O—H are each energetically less likely to suffer cleavage than are all other single bonds.[23] With molecules containing all sorts of bonds, then, preferential cleavage is expected of all single bonds other than C—F, C—H, and O—H, and this is true generally; some C—H and O—H bonds will cleave readily.

Scission only

Ions arising by loss of H· radicals are generally not prominent in spectra unless the ions produced are stabilized by resonance [as is the ion from propene (XIII)] or by a hetero-atom changing its valency state [as in the ion $C_6H_{12}N^+$ (XV) from 4-methylpiperidine].

It is unusual for molecular ions to lose more than two H⋅ radicals, although, for example, this occurs in polycyclic aromatics where the nuclear structure is particularly favourable for the sustaining of a positive charge or charges.

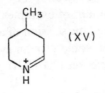

$$(XV)$$

Predictions concerning the formation and stability of, say, carbonium ions are useful in mass spectrometry, as they are in the main body of organic chemistry. Note, however, that much of our knowledge of carbonium ions is derived from solution chemistry, and heterolytic scission of the C—Br bond in 1-bromobutane yields two ions. In solution, reaction of analogous parent compounds may depend on such factors as solvation and equilibria, whereas in the mass spectrometer heterolytic scission (X) of ions is a unimolecular process, as is homolytic scission. Therefore the analogy between reactions (electron shifts) from the main body of organic chemistry and those in the mass spectrometer is not quantitative by any means, but is usefully qualitative.

This is simply illustrated, for example, by the spectra in Figs. 8 and 9, which are of 1-bromopropane and 2-bromopropane respectively. The more stable carbonium ion $C_3H_7^+$ (m/e 43), and hence relatively the more abundant by fragmentation of the molecular ion,

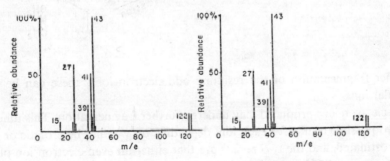

FIG. 8. Spectrum of 1-bromopropane. FIG. 9. Spectrum of 2-bromopropane.

should be given by 2-bromopropane, and this is the case as may be seen by comparing peak heights relative to the molecular ion peak height in the two cases (relative abundances of the $C_3H_7^+$ ion and the molecular ion are 690 : 100 and 910 : 100 respectively). A further illustration of the relative ease of fragmentation to give carbonium ions from straight- and branched-chain structures is provided by the spectra[24a] of the eighteen isomeric octanes. It may be emphasized here that to make such comparisons meaningfully implies that conditions for obtaining each set of spectra are quite rigorously standardized.

Even- and odd-electron ions

All molecular ions are odd-electron ions. When scission of single bonds produces mass loss there are two cases only for consideration: (a) scission of open-chain and exocyclic bonds of the original structure, and (b) scission of cyclic bonds of the original structure. [In this consideration the bare minimum of examples is used.]

(a) In this case, heterolytic or homolytic scission of any bond would lead in theory to the same general result, namely that this would give an even-electron ion and an odd-electron neutral fragment (radical) [for examples see (X), (XII), and (XIII)], and this is what happens in practice.

(b) In this case, for mass loss to occur, it is necessary to postulate that any two cyclic bonds undergo scission (one of them perhaps in formation of the equivalent molecular ion), and this time the result in theory for either heterolytic or homolytic scission of the second bond is that it would give an odd-eletcron ion and a neutral fragment [e.g. from the molecular ion (XVI) of cyclohexanol]; and this is what happens. [In practice other scissions occur in addition to those shown.]

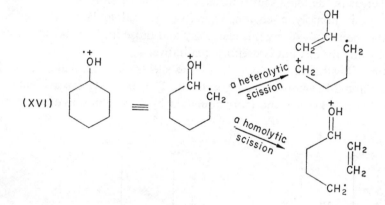

For further fragmentation of the resulting odd-electron ions, these can be treated as new molecular ions.

The even-electron ions produced may undergo further fragmentation. This time in theory there are two possible results for each ion, depending on whether heterolytic or homolytic scission is postulated, and the two results are that either an even-electron ion plus a neutral fragment, or an odd-electron ion plus a neutral fragment would be produced [see, for example, the two scissions for the even-electron ion (XI) from 1-bromobutane].

$$(XI)\quad CH_3\!-\!CH_2\!-\!CH_2\!-\!CH_2^+\ \xrightarrow[\substack{a\ homolytic\\ scission}]{\substack{a\ heterolytic\\ scission}}\ \begin{array}{l} CH_3\!-\!CH_2^+\ +\ CH_2\!=\!CH_2 \\[4pt] CH_3\!-\!\dot{C}H_2\ +\ \dot{C}H_2\!-\!CH_2^+ \end{array}$$

Of the two possible general results, it is found that the formation of an odd-electron ion is by far the less-preferred alternative, and that even-electron ions fragment in practice almost exclusively to give even-electron ions. Odd-electron ions can be recognized since they will be of even mass unless they contain an odd number of nitrogen atoms, when they will have odd mass.

The majority of ions in a spectrum, certainly of open-chain compounds, are even-electron ions (see Figs. 4, 8, and 9), but it is usually the odd-electron ions which are of greatest use in structure deduction through postulation of possible precursors.

There is a fairly prominent odd-electron ion, m/e 42, $[C_3H_6]^{\ddagger}$, in the spectrum (see Fig. 8) of 1-bromopropane. This does not arise by loss of a hydrogen radical from $C_3H_7^+$ (the less-preferred alternative, as has been said); this follows from two pieces of evidence, one of which may be summarized by saying that when the n-$C_3H_7^+$ ion is generated in the mass spectrometer from other sources, then the $[C_3H_6]^{\ddagger}$ ion is not always present, and even when it is then its abundance relative to $C_3H_7^+$ is not constant.

If there is not consecutive loss of bromine, then hydrogen (or hydrogen, then bromine), it follows that these are lost simultaneously, and there is positive evidence that this is so (from the presence of metastable ions). That is, there is effective loss of hydrogen bromide in one stage, a process involving *concerted cleavage of two bonds*. In order to obtain positive evidence for the specific origin of particular atoms in fragments it is customary to use compounds labelled isotopically (e.g. with deuterium) in known positions, and by such means it has been established, as might be expected, that the bromine and hydrogen are not lost from the same carbon atom. For 1-bromopropane the process, then, may be pictured as in (XVII), but it is by no means universal in analogous cases for the entities to be lost from adjacent carbon atoms.

$$(XVII) \quad CH_3-\overset{\overset{H}{|}}{CH}-\overset{\overset{Br^+}{|}}{CH_2} \quad \longrightarrow \quad CH_3-\overset{\cdot}{CH}-\overset{+}{CH_2} \;+\; HBr$$
$$m/e\ 122 \qquad\qquad\qquad m/e\ 42$$

Consecutive scission of open-chain and exocyclic bonds leads to even-electron ion fragments and neutral entities. The concerted scission of two such bonds leads to an odd-electron ion fragment and ejection of a neutral entity. For cyclic bonds the scission of two of these also leads to an odd-electron ion fragment and ejection of a neutral entity; these scissions may be concerted or consecutive, and for the majority of such fragmentations this point cannot be established.

Scission with rearrangement

It will be seen that the dividing line between "scission only" and "scission with rearrangement" is not very sharply defined. The reaction symbolized in (X) is clearly scission only, and that symbolized in (XVII) is clearly scission with rearrangement, but it is not easy to classify the change in going from the molecular ion of cyclohexanol (XVI) to the equivalent open-chain form.

In the concerted double scission with rearrangement for open-chain and exocyclic bonds, which is quite common, generally one of the scissions involves hydrogen. Note that the symbolism here, and subsequently, concerned with the breaking of an X–H bond and the formation of an H–Y bond, does not imply the transfer of a *proton* (H$^+$). The H and Y moieties in the original molecule may be separated by one, two, three, or four atoms. In diethyl sulphide, for example [molecular ion (XVIII)], they are again separated by two atoms, and the second most prominent ion in the spectrum has *m/e* 62 and is an odd-electron ion. This arises as shown; this time the transferred hydrogen radical appears in the ion.

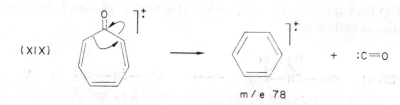

The spectrum of tropone shows a molecular ion (XIX) peak at *m/e* 106, and the first significant peak below this is at *m/e* 78 (an odd-electron ion). The remainder of the spectrum is remarkably like that from benzene. The loss of 28 a.m.u. represents, then, CO (which is confirmed by accurate mass measurement) and this scission of two cyclic bonds to expel the neutral fragment is probably a concerted process with simultaneous formation of a new bond to give benzene.[25]

(XIX)

m/e 78

In the light of what has been said, it will be anticipated that the effect of a total of three scissions, with or without rearrangement, of the type described above will be to give even-electron ions. Two examples will suffice—ethyl isobutyl ether[24b] and cyclohexylamine[24c] [molecular ions (XX) and (XXI) respectively]. An ion from the former, *m/e* 31 CH$_3$O$^+$, is the second most abundant, and one from the latter, *m/e* 56 C$_3$H$_6$N$^+$, is the most abundant, in each spectrum, and these probably arise as shown.

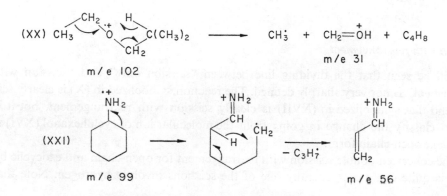

However, in addition to such as these, there are two types of concerted three-bond scissions with rearrangement which lead to odd-electron ions, and both of these are found to be of extraordinarily wide occurrence. When recognized they are very useful in structure deduction. These are (a) the retro-Diels–Alder fragmentation, and (b) the McLafferty rearrangement.

Retro-Diels–Alder fragmentation

This is shown by six-membered cyclic olefins. Taking cyclohexene [molecular ion (XXII)], the fragmentation gives ethylene and butadiene, with the carbon atoms of the original olefinic bond appearing in the 2,3-positions of the butadiene. Either fragment may carry the positive charge according to the mode of fragmentation, here illustrated.

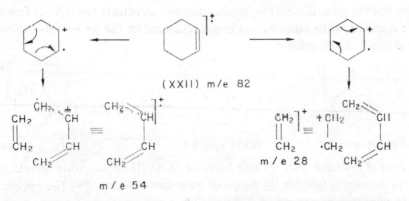

(XXII) m/e 82

m/e 54

m/e 28

The three bonds that have undergone scission are two single bonds and one-half of the double bond. Substituted cyclohexanes behave similarly, and the kind of substitution will determine the relative abundances of the monoene and diene ions, which will be of differing stabilities. The original cyclic olefinic bond may be part of another, aromatic, ring. The base peak in the spectrum of tetralin [molecular ion (XXIII)] arises from a retro-Diels–Alder fragmentation.

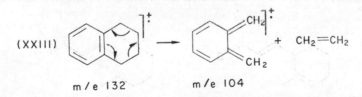

(XXIII)   m/e 132   m/e 104

It is instructive to compare the spectra of norbornane (bicyclo[2,2,1]heptane) and norbornene (bicyclo[2,2,1]hept-2-ene) (Figs. 10 and 11 respectively). The saturated hydrocarbon [molecular ion (XXIV)] contains no hetero-atom or unsaturation which might be involved in the types of scissions leading to odd-electron ions, so the ions expected should be predominantly even-electron ions, which they are. The unsaturated hydrocarbon [molecular ion (XXV)], on the other hand, contains one double bond in the correct location to promote a retro-Diels–Alder fragmentation. Fragmentations like those in the saturated hydrocarbon

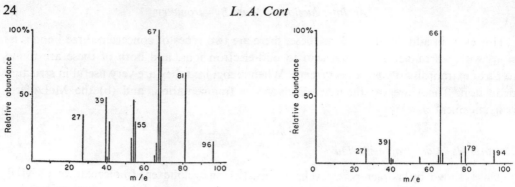

FIG. 10. Spectrum of norbornane.   FIG. 11. Spectrum of norbornene.

(From S. R. Shrader, *Introductory Mass Spectrometry*, Copyright 1971 by Allyn and Bacon, Inc., Boston, Mass. Used by permission of the publisher.)

will occur, but the great distinction between the two structures (as regards fragmentation) should be manifest in the odd-electron ion(s) generated by the retro-Diels–Alder fragmentation, and this is clearly evident.

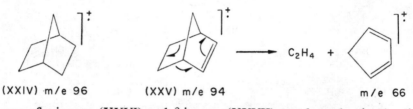

(XXIV) m/e 96   (XXV) m/e 94   m/e 66

In the case of α-ionone (XXVI) and β-ionone (XXVII) an elegantly simple distinction between the isomers is possible by means of mass spectrometry.[26a] The spectra are quite different in appearance, one major difference being the presence of a very prominent m/e 136 peak in one case and the presence of a very prominent m/e 177 peak in the other.

The retro-Diels–Alder fragmentation should give ions, m/e 136 in the case of α-ionone (with neutral fragment C_4H_8) and m/e 164 in the case of β-ionone (with neutral fragment C_2H_4). However, for β-ionone there is a competing fragmentation, the loss of $CH_3^{\cdot}$, because of allylic activation of the appropriate bond by the cyclic unsaturation, and this accounts for the prominence of the m/e 177 peak and the weakness of the m/e 164 peak from this isomer.

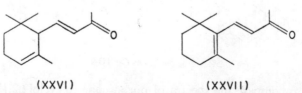

(XXVI)   (XXVII)

If another ring is fused to the cyclohexene in the 3,4-positions, then the retro-Diels–Alder fragmentation can occur without mass loss, and in this event its occurrence may only be apparent if subsequent fragmentations occur.

For the purposes of rationalization of the appearance of spectra it may be noted that apparently the purely olefinic double bond of cyclohexene may be replaced by a double bond to a charged nitrogen atom, and a fragmentation analogous to the retro-Diels–Alder postulated. Thus 4-methylpiperidine [molecular ion (**XXVIII**)] gives the base peak ion, m/e 98 (**XV**) by loss of a hydrogen radical, and it is known[24d] that this ion fragments to give an ion, m/e 56 $C_3H_6N^+$. This fragmentation can be pictured as shown; such a fragmentation in this and similar cases correctly predicts the size of ions formed, but proof that this is the pathway has yet to be adduced.

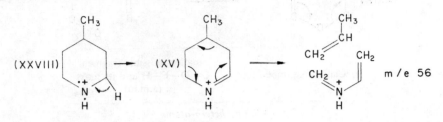

McLafferty rearrangement

This rearrangement is shown by a large number of classes of compounds, and involves the transfer of a hydrogen radical in a six-membered transition state, to an atom which must be joined to the adjacent atom of the six by a double or triple bond. The general case is illustrated by (**XXIX**).

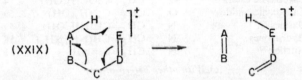

The number of classes is large since all the atoms A, B, C, D, and E may be carbon atoms, or atoms A, C and/or E, or D may be hetero-atoms, and even this does not exhaust all the possibilities. The fragmentation is facile and frequently produces the base peak; the charge usually resides on the C=D—E—H fragment. In Table 3 appear some of the classes of compounds (a) when all the atoms are carbon atoms, (b) when atom E alone is a hetero-atom, and (c) when atom C, and atoms C and E are hetero-atoms. The nature and mass (a.m.u.) of the smallest possible fragment C=D—E—H is given for each case.

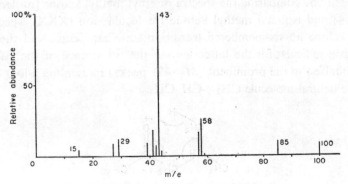

Fig. 12. Spectrum of isobutyl methyl ketone.

TABLE 3

SOME CLASSES OF COMPOUNDS, OF GENERAL FORMULA
(XXIX), SHOWING THE McLAFFERTY REARRANGEMENT

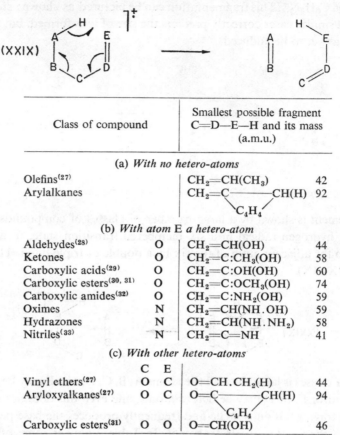

Class of compound	Smallest possible fragment C=D—E—H and its mass (a.m.u.)
(a) *With no hetero-atoms*	
Olefins[27]	$CH_2{=}CH(CH_3)$ 42
Arylalkanes	$CH_2{=}C\!-\!\!-\!\!-\!CH(H)$ 92 C_4H_4
(b) *With atom E a hetero-atom*	
Aldehydes[28] O	$CH_2{=}CH(OH)$ 44
Ketones O	$CH_2{=}C{:}CH_3(OH)$ 58
Carboxylic acids[29] O	$CH_2{=}C{:}OH(OH)$ 60
Carboxylic esters[30, 31] O	$CH_2{=}C{:}OCH_3(OH)$ 74
Carboxylic amides[32] O	$CH_2{=}C{:}NH_2(OH)$ 59
Oximes N	$CH_2{=}CH(NH.OH)$ 59
Hydrazones N	$CH_2{=}CH(NH.NH_2)$ 58
Nitriles[33] N	$CH_2{=}C{=}NH$ 41
(c) *With other hetero-atoms* C E	
Vinyl ethers[27] O C	$O{=}CH.CH_2(H)$ 44
Aryloxyalkanes[27] O C	$O{=}C\!-\!\!-\!\!-\!CH(H)$ 94 C_4H_4
Carboxylic esters[31] O O	$O{=}CH(OH)$ 46

Where the course of the rearrangement has been established unequivocally, it has been done so by isotopic labelling (e.g. in the case of esters[30]); the six-membered transition state is preferred above all others. The effect of the occurrence of this type of fragmentation can be seen quite easily by comparing the spectra of ethyl methyl ketone [molecular ion (VIII), spectrum, Fig. 4] and isobutyl methyl ketone [molecular ion (XXX), spectrum, Fig. 12]. In the former ketone no six-membered transition state can occur, and the molecular ion loses odd-electron radicals; for the latter ketone the occurrence of the McLafferty rearrangement is manifest in the prominent $(M-42)$ peak, representing a loss from the molecular ion of the neutral molecule $CH_2{=}CH.CH_3$.

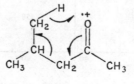

(XXX)

Carboxylic esters of methanol show this rearrangement when the acid chain is long enough. If the alcohol is ethanol or higher, then the rearrangement can occur through this group. Thus not only does β-phenylethyl isovalerate [molecular ion (XXXI)] show the rearrangement, but the occurrence of the base peak at m/e 104 shows clearly that (a) the C—H bond involved in the transfer of the hydrogen radical is preferentially one of a methylene group subject to benzylic activation rather than one of a methyl group without activation, and (b) the charge here resides on the highly conjugated styrene fragment.

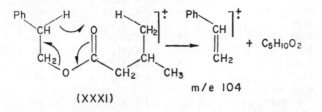

(XXXI)

m/e 104

Alkyl phenyl ethers where the alkyl group is ethyl or higher similarly show the rearrangement,[27] as do esters of phenols other than formates, aromatic carboxylic esters,[31] and arylalkanes with at least an n-propyl side-chain.[27] In these cases transfer of a hydrogen radical is to a carbon atom of the aromatic ring, e.g. with phenyl acetate [molecular ion (XXXII)] where the base peak, m/e 94, arises in this way. When both *ortho-* positions of the aromatic nucleus are substituted, this rearrangement is found not to occur.

It is not practicable to make Table 3 completely comprehensive. The reader will be able to visualize other cases of the rearrangement, e.g. obviously with organic carbonates, ketimines, and aldehyde semicarbazones, and appreciate the extension to include such classes as organic sulphites, epoxides[21b] (XXXIII), and organic phosphates[34] (XXXIV). There is also an obvious resemblance to the McLafferty rearrangement of a rearrangement undergone by such compounds as *o*-toluic acid[31] [molecular ion (XXXV)].

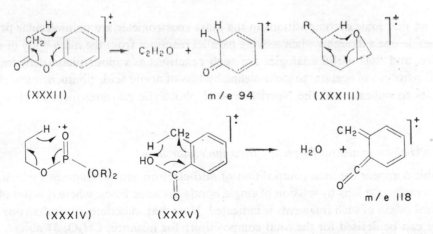

(XXXII) m/e 94   (XXXIII)

(XXXIV)   (XXXV) m/e 118

In all the rearrangements discussed there are many instances of ejection of even-electron neutral fragments. The most frequently encountered of these are shown in Table 4.

TABLE 4

TOTAL COMPOSITION OF EVEN-ELECTRON NEUTRAL
FRAGMENTS MOST FREQUENTLY EJECTED ON FRAG-
MENTATION WITH REARRANGEMENT

Fragment	Mass (a.m.u.)	Fragment	Mass (a.m.u.)
H_2	2	CO_2	44
CH_2	14	C_2H_7N	45
O	16	C_2H_6O	46
H_3N	17	CH_4S	48
H_2O	18	C_4H_6	54
HF	20	C_4H_8	56
C_2H_2	26	C_3H_4O	56
HCN	27	C_3H_6O	58
C_2H_4 and homologues	28, 42, ...	C_3H_9N	59
CO	28	C_3H_8O	60
$[NO]$	30	$C_2H_4O_2$	60
CH_5N	31	C_2H_6S	62
CH_4O	32	SO_2	64
H_2S	34	$C_3H_6O_2$	74
HCl	36	C_6H_6	78
C_2H_2O	42	HBr	80
$CHNO$	43	HI	128
C_2H_4O	44		

The scissions with and without rearrangement of ions in the mass spectrometer, such as have been described and as will be outlined for various classes of organic compounds below, often appear somewhat bewildering on first acquaintance with their courses difficult to rationalize. Perhaps the easier in this respect are the retro-Diels–Alder fragmentation and the McLafferty rearrangement, with their obvious parallels in classical organic chemistry.

The fact that ionic decompositions in the mass spectrometer are unimolecular processes should guide one's thoughts when seeking parallel reactions from the main body of organic chemistry, and thus useful analogies are such reactions as vapour phase dehydration of alcohols, pyrolysis of acetates to yield olefins by loss of acetic acid, photochemical cleavage of ketones to radicals, and the Norrish type II photolytic rearrangement of olefins from ketones.

Loss of odd-electron neutral fragments from molecular ions

In Table 5 appear the total composition of odd-electron neutral fragments which can be lost from molecular ions by scission of single bonds. In some cases, where it is not obvious, the general origin of such fragments is indicated. Note that sometimes more than one radical structure can be devised for the total composition; for instance, CH_3O, 31 a.m.u., can be CH_3O· deriving from methyl esters or ethers, and it can be ·CH_2OH deriving from primary alcohols.

TABLE 5

ODD-ELECTRON NEUTRAL FRAGMENTS LOST FROM MOLECULAR IONS, WITH AN INDICATION OF SOME COMMON PRECURSOR STRUCTURES

[Note that there can be precursors other than those listed.]

Fragment	Origin	Mass (a.m.u.)	Fragment	Origin	Mass (a.m.u.)
H	Aldehydes, acetals, phenols, methylaromatics, aliphatic amines	1	NO_2		46
			CH_3S	Methyl sulphides, thiolacid esters, primary thiols	47
CH_3		15	C_3H_4N	Nitriles	54
H_2N	Amides	16	C_4H_7		55
HO	Alcohols, acids, *ortho*-disubstituted aromatics where one function carries oxygen and the other carries hydrogen (e.g. *o*-nitrotoluene)	17	C_4H_9		57
			C_3H_5O	Ethyl ketones	57
			C_3H_8N		58
			C_2H_4NO	Amides	58
			CNS		58
			C_3H_7O	Dimethylcarbinols	59
			$C_2H_3O_2$	Methyl esters, acetates, acids	59
F		19	C_2H_5S	Primary thiols, methyl sulphides	61
CN		26			
C_2H_3		27			
CH_2N		28	C_4H_6N	Nitriles	68
C_2H_5		29	C_5H_9		69
CHO	Aldehydes, phenols	29	C_5H_{11}		71
CH_4N	Primary aliphatic amines	30	C_4H_7O	Ketones	71
			$C_3H_5O_2$		73
NO		30	C_3H_7S	Ethyl sulphides	75
CH_3O	Methyl esters, ethers, primary alcohols	31	C_6H_5		77
			Br		79
			C_6H_{11}		83
H_2NO		32	C_6H_{13}		85
HS		33	C_5H_9O	Ketones	85
Cl		35	$C_4H_7O_2$	Esters	87
C_3H_3		39	C_7H_7		91
C_2H_2N	Nitriles	40	C_7H_{13}		97
C_3H_5		41	C_7H_{15}		99
C_2H_4N		42	$C_6H_{14}N$	Primary aliphatic amines	100
CNO		42			
C_3H_7		43	C_8H_9	Phenylalkyl compounds	105
C_2H_3O		43			
C_2H_6N		44	C_7H_5O	Benzoyl compounds	105
CH_2NO	Primary amides	44	I		127
C_2H_5O	Ethyl esters, ethers, methylcarbinols	45			
CHO_2	Acids	45			

TABLE 6

CHARACTERISTIC LOW-MASS FRAGMENT IONS NOT ALREADY LISTED IN TABLES 3, 4, AND 5, WITH AN INDICATION OF SOME COMMON PRECURSOR STRUCTURES

Fragment	Origin	Mass (a.m.u.)	Fragment	Origin	Mass (a.m.u.)
H_4N		18	C_5H_4NO	Pyrrolyl-CO- compounds	94
N_2		28			
C_4H_2	Phenyl or pyridyl compounds	50	C_6H_4F		95
			$C_5H_3O_2$	Furyl-CO-compounds	95
C_4H_3	Phenyl or pyridyl compounds	51	C_5H_5S	Thienyl-CH_2- compounds	
C_4H_4	Phenyl or pyridyl compounds	52	$C_5H_6O_2$	Esters of aliphatic dibasic carboxylic acids higher than pimelic	98
C_3H_3O	Cyclohexanones	55			
C_3H_5N	Nitriles with α-methyl	55			
C_3H_6N	Cyclohexylamines	56	$C_5H_7O_2$	δ-Lactones	99
C_2H_4S	Cyclic sulphides	60	C_8H_7	Styrenes, cinnamates	103
CH_2NO_2		60	C_7H_7O	Methoxyphenyl, benzyloxy compounds	107
C_5H_8	Cyclohexenes, cyclohexanols	68			
C_4H_8N	Pyrrolidines	70	C_6H_6NO	N-Methylpyrrolyl-CO- compounds	108
C_4H_7O	Tetrahydrofurfuryl compounds	71			
			C_6H_4Cl		111
C_4H_8O	Aldehydes with α-ethyl	72	C_5H_3OS	Thienyl-CO- compounds	111
C_3H_7NO	Amides with α-methyl	73			
C_3H_6S	Cyclic sulphides	74	C_9H_{11}	Phenylalkyl, tolylalkyl compounds	119
C_5H_4N	Pyridines	78			
C_5H_6N	Pyrrolyl-CH_2- compounds	80	C_8H_7O	Toluoyl compounds	119
			$C_7H_4O_2$	Salicylates	120
C_5H_5O	Furyl-CH_2-compounds	81	$C_7H_5O_2$	Hydroxybenzoyl compounds (salicylates)	121
C_4H_3S	Thienyl compounds	81			
$C_4H_4O_2$	Esters of aliphatic dibasic carboxylic acids higher than pimelic	84			
			$C_6H_4NO_2$		122
			C_7H_4FO	Fluorobenzoyl compounds	123
$C_5H_{10}N$	Piperidines	84	$C_7H_{10}O_2$	Esters of aliphatic dibasic carboxylic acids higher than pimelic	126
$C_4H_5O_2$	γ-Lactones	85			
$C_4H_8O_2$	Methyl esters with α-methyl	88			
C_6H_4O	Salicylates	92	$C_{10}H_7$	Naphthyl compounds	127
C_6H_6N	Pyridyl-CH_2- compounds	92	C_9H_7O	Cinnamates	131
			$C_8H_7O_2$	Methoxybenzoyl compounds	135
C_6H_5O	Phenyl ethers (esters)	93	C_7H_4ClO	Chlorobenzoyl compounds	139
			$C_8H_5O_3$	Dialkyl phthalates	149
			$C_7H_4NO_3$	Nitrobenzoyl compounds	150
			C_6H_4Br		155

Characteristic low-mass fragment ions

Most of the odd-electron fragments listed in Table 5 can, of course, appear as even-electron ions. Neutral fragments ejected by fragmentation and rearrangement are listed in Table 4; many of these can appear as odd-electron ions. Some odd-electron ions (fragments) that can arise by McLafferty rearrangement are listed in Table 3.

In addition to all these ions there are others often encountered in spectra, and those not already given in Tables 3, 4, and 5, are listed in Table 6.

Double bond and/or ring equivalents implied from a formula

When a molecular ion is selected a formula can be suggested, either with or without the use of tables. When considering potential structures, a useful precept to bear in mind is the one of the number of double bond and/or ring equivalents deducible from the simple formula (see also p. 239).

The general case is expressed by the formula $C_aH_bN_dO_c$, and the number of double bond and/or ring equivalents (DBE) is given by

$$DBE = 1+(2a-b+d)/2.$$

The elements are assumed to be in their lowest valency states; if other elements are present these may be equated with those in the general formula of the same valencies.

Thus for ethyl methyl ketone, C_4H_8O, DBE $= 1+(8-8)/2 - 1$ (the double bond in $C=O$), and for 2-bromo-4-nitroaniline, $C_6H_5BrN_2O_2$, DBE $= 1+(12-6+2)/2 - 5$ (four double bond equivalents for the benzene ring, and one for the nitro-group as the nitrogen atom in this group is actually in the higher valency state).

For ions, the same calculation can be used, but note that the result will be either an integer (for odd-electron ions) or not (for even-electron ions); in the latter case it is necessary to subtract finally 0·5 to give the true DBE value of a radical.

Fragmentations of the Chief Classes of Organic Compounds

Below are summarized the fragmentations of the more important classes of organic compounds. For more comprehensive treatments, with full rationalizations of the origins of the various fragments together with many spectra presented as bar graphs, the reader is referred to texts by Beynon *et al.*[24] and by Budzikiewicz *et al.*[21]

Simultaneous, different fragmentation pathways, directed by even two functional groups, will produce effectively overlapping spectra which are not easy to perceive in isolation. For a full, collected treatment of the mass spectra of complicated and polyfunctional molecules the reader is referred to texts on natural products by Budzikiewicz *et al.*[11]

In the following discussion, heights of peaks are relative to that of the base peak in each instance.

SATURATED ALIPHATIC HYDROCARBONS

If high resolution spectroscopy is employed, it will quickly be established that all the individual fragment ions are homogeneous, i.e. the hydrocarbon nature of the compound will be apparent [otherwise, in particular, if oxygen is present, inhomogeneous ions of the type $(C_{n+1}H_{2n+3}^+ + C_nH_{2n-1}O^+)$ are to be expected, and this possibility makes it difficult to decide for a hydrocarbon structure alone from a low resolution spectrum].

(i) *Open-chain*

For isomers, the relative height of the molecular ion peak is greatest for the straight-chain compounds and decreases as the degree of branching increases.

For an homologous series, the relative height of the molecular ion peak decreases with increasing molecular weight.

Cleavage takes place preferentially at chain branches; the more branched the more likely the cleavage. (This is in accordance with the relative stability of the types of carbonium ions.)

The molecular ion peak is visible (on the original spectrum, but not always on a bar graph) for straight-chain compounds up to about C_{45}, but is not detectable for highly branched compounds of comparatively low molecular weight. (For the octanes, where at least one carbon atom carries no hydrogen, the molecular ion is less than 0·04% as abundant as the base peak ion.[24a])

For fragment ions, the abundance of the $(M-15)^+$ ions is a minimum for straight-chain compounds; a strongly abundant $(M-15)^+$ ion implies a branched-chain methyl group. The fragmentation of unbranched chains produces prominent peaks 14 a.m.u. (CH$_2$) apart, and the major peaks above m/e 57 decrease progressively in height towards the high mass end of the spectrum. Peaks at m/e 43 $(C_3H_7)^+$ and 57 $(C_4H_9)^+$ are always large from paraffins, but these and others in the C_3 and C_4 regions are generally not characteristic of the particular compound under investigation.

The majority of ions are formed by cleavage of C—C bonds in the molecular ion, giving

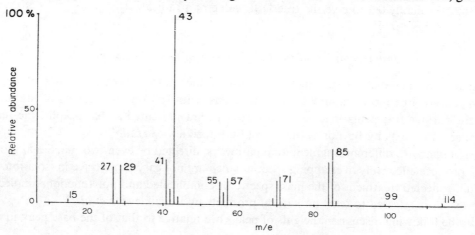

FIG. 13. Mass spectrum of 3-ethylhexane (XXXVI)

ions of odd mass in the series m/e 29, 43, 57, 71, 85, 99, etc.; associated with each of the very prominent peaks is another of lower abundance two mass units lower.

The spectrum[24a] of 3-ethylhexane (XXXVI) is shown in Fig. 13. Straight scissions give prominent ions including the base peak ion. The ion of m/e 57 must arise by rearrangement, and this is confirmed by the presence of a metastable ion, m/e 38·2, indicating that the parent has m/e 85; a probable process is that suggested for the ion (XXXVII). It is impossible to say, without further evidence from labelling experiments or the presence of metastable ions, whether the ion of m/e 84 arises by expulsion of a methyl radical from an ion of m/e 99 or by expulsion of a hydrogen radical from an ion of m/e 85; in the former case a product such as the ion (XXXVIII) (the molecular ion of a cyclobutane) could readily account for the other ions in the spectrum, of m/e 69, 55, and 56, with the last of these subsequently yielding the m/e 41 ion by loss of CH_3 (scission of a bond β- to olefinic unsaturation).

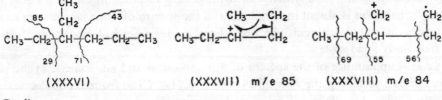

(XXXVI)   (XXXVII) m/e 85   (XXXVIII) m/e 84

(ii) *Cyclic*

Cyclic structures stabilize the molecular ion, and such compounds give more prominent molecular ion peaks than do the open-chain compounds of the same carbon content; fusion of rings increases the effect. Cleavage of the bond connecting the ring to the remainder of the molecule is favoured (α-cleavage).

Cleavage of the ring bonds also occurs, and the expulsion of one- and two-carbon fragments takes place. The fragments are 15 (CH_3), 28 (C_2H_4), and 29 a.m.u. (C_2H_5) from unsubstituted atoms; the second is not found as a loss from the molecular ion of open-chain compounds.

This is illustrated by the behaviour of norbornane (spectrum, Fig. 10). The molecular ion (XXIV) probably suffers ring opening, to (XXXIX), which can rearrange as indicated with elimination of an ethyl radical to give a cyclopentenyl cation. Alternative charge distribution on ring opening of (XXIV) would enable ethylene to be eliminated subsequently or a methyl radical to be eliminated giving a cyclopentenylmethyl cation (XL). Loss of C_2H_2 or C_2H_4 from (XL) accounts for the ions of m/e 55 and 53, and loss of these fragments from the cyclopentenyl cation accounts for the ions of m/e 41 and 39. It is a matter of speculation that the two remaining ions, m/e 54 and 27, both arise from (XL) also.

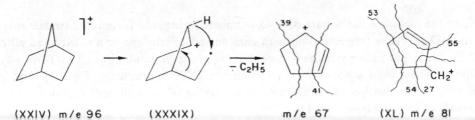

(XXIV) m/e 96   (XXXIX) m/e 67   (XL) m/e 81

OLEFINS

(i) Open-chain[27]

Double bonds stabilize the molecular ion and promote allylic cleavage as the predominating fragmentation. The molecular ion peak of an olefin containing less than six carbon atoms is stronger than that from the corresponding paraffin, and the base peak frequently arises by allylic cleavage with the charge remaining on the unsaturated fragment (resonance stabilized allylic carbonium ion). There are usually prominent peaks in the series m/e 27, 41, 55, 69, 83, 97, etc. Rearrangement ions are quite common, including those from McLafferty rearrangement[27] and hydrogen rearrangement resulting in migration of a radical site along a chain (this is absent in paraffins), and the spectra of some isomers are so similar that it is very difficult to distinguish between them; simple empirical prediction of abundant ions is frequently impossible.

The general appearance of the spectra of a mono-olefin and an isomeric cyclic paraffin are very similar, including the loss of 28 a.m.u. ($CH_2{=}CH_2$) from the molecular ion of open-chain olefins, but olefins of small molecular weight do not lose a terminal methyl group to any significant extent.

It is similarly difficult to make unambiguous identification of acetylenes.[35]

To aid location of unsaturation in a given compound recourse is often made to formation of suitable derivatives with subsequent mass spectra determination of these. For example, after epoxidation of olefins, reaction with dimethylamine affords a mixture of two (dimethylamino) alcohols, which cleave in the mass spectrometer between the carbon atoms of the original olefinic bond with the charge remaining on the nitrogen. Identification of the amine fragments automatically identifies the olefin. Olefins can also be converted into the vicinal diol from which an isopropylidene derivative (a dimethyldioxolan) can be prepared, and acetylenes can be hydrated in the presence of ethylene glycol, which process also gives a dioxolan, the fragmentations of which are characteristic.[21c] See also trimethylsilyl ethers (p. 38).

(ii) Cyclic

When the olefinic bond is part of a six-membered ring, the favoured cleavage involves the retro-Diels–Alder fragmentation with elimination of ethylene or a substituted ethylene (m/e 28, 42, 56, etc.). Compare the fragmentations of the molecular ion (XXII) from cyclohexene and that from norbornene [molecular ion (XXV), spectrum, Fig. 11]. For other rings there is elimination of (substituted) ethylene or of (substituted) acetylene. For the behaviour of terpenoids, see ref. 11.

AROMATIC HYDROCARBONS[36a]

(i) *Alkylbenzenes*

The presence of an aromatic nucleus stabilizes the molecular ion even more than do olefinic bonds, and the molecular ion peak is almost always detectable. The favoured cleavages preserve the aromatic ring, and benzylic cleavage is particularly likely (β-cleavage to the ring).

Where there is no branching at the α-carbon atom, the expected ion, $C_7H_7^+$, is formed (as from ethylbenzene), but this is almost certainly not the benzyl cation but the tropylium ion formed by rearrangement.[36a, 37] (This ion is often formed even when the α-carbon atom of a monosubstituted benzene carries only one hydrogen, e.g. it provides the base peak ion from styrene oxide.) When the side-chain is long enough, McLafferty rearrangement takes place (see Table 3).

Ions from the breakdown of an aromatic nucleus are of relatively low abundance, but are characteristic [m/e 39, 50, 51, 65, and (76), 77, (78)]; the metastable ion peak at m/e 33·8 (77 → 51) for the fragmentation of the aromatic nucleus is detectable. Thus, although there is complete carbon and hydrogen scrambling before fragment ejection [see the intro-

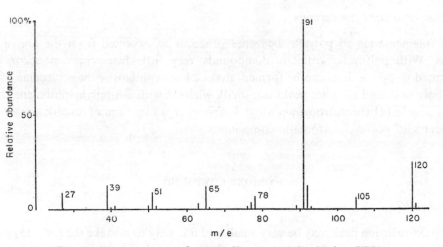

FIG. 14. Mass spectrum of n-propylbenzene, molecular ion (XLI).

duction to Fragmentation processes (p. 16)], if no attempt is made to identify particular atoms in fragments, then the total composition of fragments is a useful guide as to their origin.

Figure 14 shows the spectrum of n-propylbenzene [molecular ion (XLI), m/e 120], the principal fragment ions from which arise as indicated in Scheme 1.

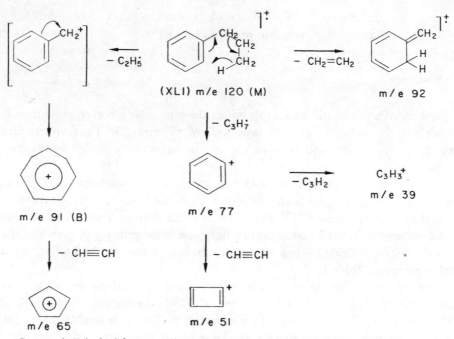

SCHEME 1. Principal fragmentations of the molecular ion (XLI) of n-propylbenzene.

(ii) *Other aromatic hydrocarbons*

The fragmentation of polyalkylbenzenes proceeds as expected from the above observations. With polycyclic aromatic compounds very little fragmentation occurs unless substituted tropylium ions can be formed, that is, for naphthalene the molecular ion peak is the base peak and all other peaks are small, whereas with 2-n-butylnaphthalene the base peak is at m/e 141 (benzotropylium ion); doubly charged ions are of notable occurrence in the spectra of polycyclic aromatic compounds.

HYDROXY COMPOUNDS

(i) *Alcohols*[38]

The molecular ion peak may be very small and it is easy to mistake the $(M-18)$ peak for this. The molecular ion is formed by removal of one of the non-bonding electrons on the oxygen,[39] the larger substituent on the carbon carrying the oxygen is then expelled preferentially to give an oxonium ion in the series m/e 31, 45, 59, 73, etc. An $(M-1)^+$ ion is formed by expulsion of hydrogen from the carbon carrying the hydroxyl group.

A minor process generates cyclic oxonium ions (same series as above, starting at m/e 45) by scission of a bond to the β-, γ-, or δ-carbon atoms (m^* often observed).

Loss of water from the molecular ion of long-chain alcohols involves hydrogen from the carbon atom γ- or δ- from the hydroxyl group, with formation of a cyclic hydrocarbon

(m^* often observed); the spectra of such hydrocarbons are therefore to be seen superimposed on any other fragmentation patterns. Alcohols higher than butanol show substantial ($M-46$) (or corresponding) peaks; this represents loss of water plus ethylene (or substituted ethylene) from the molecular ion;[38] this may be pictured conveniently as involving a cyclic six-membered transition state, but no metastable ion can be detected for this as a concerted process.

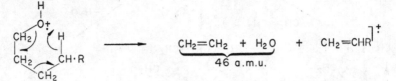

When the molecular ion peak is not readily detected, accurate mass measurement will reveal the presence of oxygen in fragment ions. Even if one is limited to low resolution spectroscopy, ions in the homologous series begining at m/e 31 point to the presence of oxygen, and confirmation can often be obtained by the presence of significant peaks at m/e 18 $(H_2O)^+$ and 17 $(OH)^+$.

4-Methyl-4-heptanol [molecular ion (XLII), m/e 130] shows no molecular ion peak in the spectrum (Fig. 15), but there are abundant ions characteristic of oxygen-containing fragments (m/e 45, 59, 73, 87, and 115). The largest of these could not be the molecular

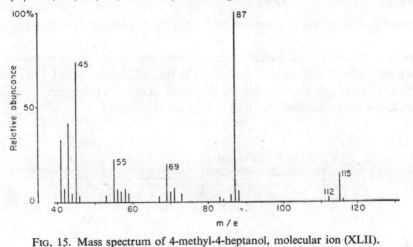

FIG. 15. Mass spectrum of 4-methyl-4-heptanol, molecular ion (XLII).
(From S. R. Shrader, *Introductory Mass Spectrometry*. Copyright 1971 by Allyn and Bacon, Inc., Boston, Mass. Used by permission of the publisher.)

ion as the first mass loss from it would then be only 3 a.m.u. The major fragmentations are as indicated in Scheme 2. Most difficult to rationalize is the prominent ion of m/e 45, which clearly has the composition $(C_2H_5O)^+$, and extensive rearrangement must give rise to it. The origin of this ion is not at present known; it may well derive from one of the ions of m/e 115.

In the predominating fragmentation of cyclic alcohols[40] the first step is scission of a ring bond to give an oxonium ion, and this is followed by transfer of a hydrogen radical

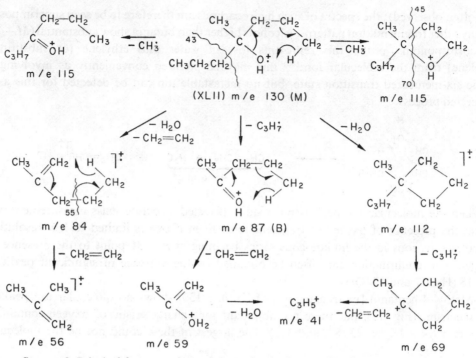

SCHEME 2. Principal fragmentations of the molecular ion (XLII) of 4-methyl-4-heptanol.

to the created radical site and simultaneous formation of an olefin–oxonium ion; the transition state for hydrogen radical transfer can be five- or six-membered. For cyclohexanol [molecular ion (XVI)] this accounts for the base peak at m/e 57. [Compare the origin of the m/e 56 ion from the molecular ion (XXI) of cyclohexylamine.]

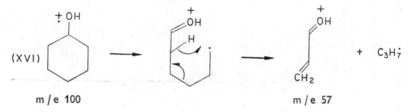

Minor fragmentations, as a cyclic hydrocarbon, also occur.

A useful derivative for alcohols, from the point of view of both gas chromatography and mass spectrometry, is the trimethylsilyl ether.[41] Preparation is very simple and nearly quantitative, and by-products are readily volatile. An alcohol R.OH becomes the more-volatile ether $R—O—Si(CH_3)_3$. The molecular ion peak of the derivative often is visible (silicon isotopes enable easy recognition), and instead of eliminations involving the —OH group the preferred fragmentations of the ether are scission of bonds β- to the ethereal oxygen to give oxonium ions, all of which are relatively of high abundance.

Thus the trimethylsilyl ether (XLIII) of 3-hexanol gives a molecular ion, m/e 174, and three prominent fragments of m/e 159, 145, and 131 (each containing the silicon).

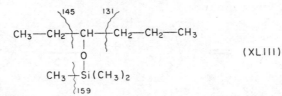

(XLIII)

This can obviously be extended to the preparation of bistrimethylsilyl ethers from the glycols obtained from olefins, with a view to locating the original double bond.

Aromatic alcohols[42, 43] like benzyl alcohol show strong molecular ion peaks, and fragmentations as expected from the foregoing (both the hetero-atom and the π-system compete for the location of the positive charge). In addition the $(M-1)^+$ ion loses CO (m^* observed) and the resulting ion loses H_2 (m^* observed) to produce effectively the aryl cation.

(ii) *Phenols*[42, 43]

The molecular ion peak of phenols is strong; where a suitable side-chain can lead to a hydroxy-tropylium ion, this ion may be more abundant than the molecular ion [e.g. the $(M-1)^+$ ion for the cresols]. The molecular ion loses 28 (CO) and 29 a.m.u. (CHO) (m^* observed in each case), the former probably from a cyclohexadienone intermediate giving a cyclopentadiene cation; the composition of the lost fragments has been confirmed by accurate mass measurement.

The spectrum of 2-ethyl-4-methylphenol [molecular ion (XLIV), m/e 136] is shown in Fig. 16, and the principal fragmentations are as indicated in Scheme 3; fragmentation both

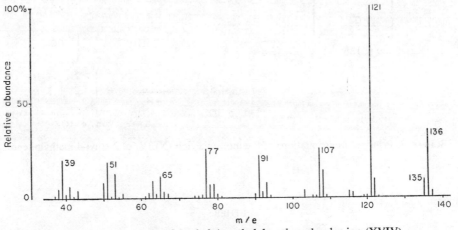

FIG. 16. Mass spectrum of 2-ethyl-4-methylphenol, molecular ion (XVIV).

as a phenol and as an alkylbenzene is evident. Note that the process depicted as loss of H_2 is the same as that mentioned above in the fragmentation of an ion from benzyl alcohol, and that, for example, the ion of m/e 91 undergoes the further fragmentation expected by considering it as the $C_7H_7^+$ ion from benzyl compounds. It is impossible, using this single spectrum, to decide the orientation of the phenol as an unknown compound.

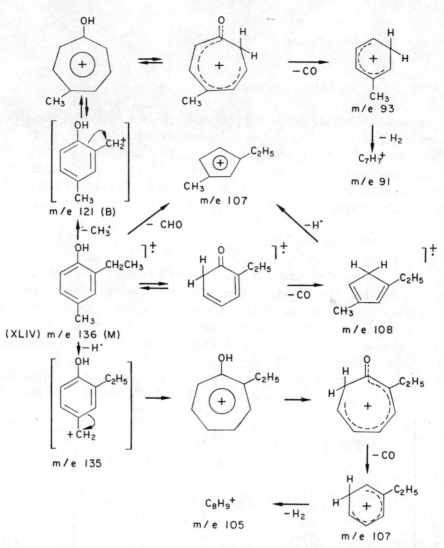

SCHEME 3. Principal fragmentations of the molecular ion (XLIV) of 2-ethyl-4-methylphenol.

ETHERS

(i) *Aliphatic*[48]

As with alcohols, cleavage of a bond β- to the oxygen occurs with formation of an oxonium ion in the series m/e 31, 45, 59, 73, etc., and one of these may be the base peak. Such an oxonium ion loses ethylene or a substituted ethylene in a rearrangement involving a four-membered transition state [for the decomposition of the molecular ion (XX) of ethyl isobutyl ether these fragmentations have been depicted (p. 22) as a concerted process].

The scission of a bond from the oxygen, with rearrangement of a hydrogen radical from the β-carbon atom, can also occur for the molecular ion and either fragment (an alcohol or an olefin) may carry the positive charge. Another fragmentation is simple scission of a bond from the oxygen, with the charge remaining on the oxygen-free fragment (especially if this is highly branched at the site of cleavage).

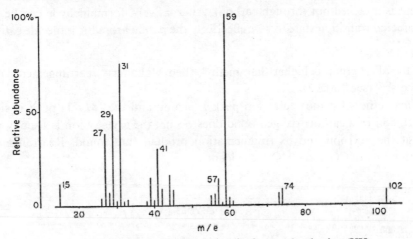

FIG. 17. Mass spectrum of ethyl isobutyl ether, molecular ion (XX).

The spectrum of ethyl isobutyl ether [molecular ion (XX), m/e 102] is shown in Fig. 17 and the chief fragmentations are as indicated in Scheme 4. Note that the fragmentation with elimination of ethylene can also give an ion, m/e 28, when the charge does not remain with the oxygen, and that the generated molecular ions of isobutanol and isobutene also produce their own (superimposed) spectra.

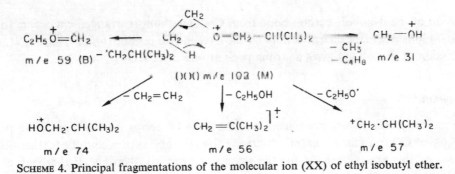

SCHEME 4. Principal fragmentations of the molecular ion (XX) of ethyl isobutyl ether.

In distinction to alcohols, $(M-17)^+$ and $(M-18)^+$ ions do not occur, and $(M-1)^+$ ions are strongly abundant for cyclic ethers and acetals (open-chain and cyclic) but are not for open-chain ethers or alcohols (except benzyl alcohols). The molecular ion peak of aliphatic ethers is small.

(ii) *Aromatic*

The molecular ion peak of aromatic ethers is large. The molecular ion of methyl ethers loses CH_3^- and the resulting ion loses CO (m^* observed) to give the cyclopentadienyl cation (m/e 65, and encountered generally in the spectra of benzenoid compounds when the substituent is attached not through carbon). Alternatively, formaldehyde can be lost via a four-membered transition state to give effectively the parent aromatic molecule cation ($C_6H_6^{+\cdot}$ from $C_6H_5\overset{\cdot+}{O}CH_3$).

When the alkyl group is higher than methyl, then McLafferty rearrangement is a prominent process[27] (see Table 3).

For diaryl ethers the molecular ion peak is strong and the ($M-1$) peak is significant. Simple scission of a carbon–oxygen bond does occur; the aryloxy ion is usually not significant, but the aryl ion and its fragmentation products are found. Rearrangement ions representing loss of CO and CHO are evident.

ALDEHYDES

(i) *Aliphatic*[28]

The most facile ionization process for carbonyl compounds corresponds to the removal of one of the lone pair electrons. The molecular ion peak and the ($M-1$) peak are both usually pronounced; losses of 18 (H_2O) and 28 a.m.u. (CO) also occur from the molecular ion.

When the chain is long enough the base peak ion usually derives from McLafferty rearrangement with the charge remaining on the oxygen, giving a peak in the series m/e 44, 58, 72, etc. (see Table 3); the ion arising when the charge remains on the olefin is also prominent.

Scission of the β-carbon–carbon bond from C=O without rearrangement occurs to give significant ions where the charge resides on the oxygen-free residue (m/e 29, 43, 57, 71, etc.), and scission of the α-bond gives a strong peak at m/e 29 (CHO^+).

(ii) *Aromatic*[44]

The molecular ion is very prominent, and the ($M-1$)$^+$ ion is usually even more prominent. Loss of CO from the latter (m^* observed) gives the aryl cation, and the further decompositions of this are evident (as for the same ion from alkylbenzenes); the peak at m/e 29 (CHO^+) is usually insignificant.

KETONES

(i) *Aliphatic*[45]

The molecular ion peak is pronounced; the base peak comes from scission of a C—C bond α- to C=O, with the charge remaining on the oxygen, the loss of the larger alkyl group is more favoured.

McLafferty rearrangement gives prominent peaks in the series m/e 58, 72, 86, etc. {see Table 3, and the spectrum (Fig. 12) of isobutyl methyl ketone [molecular ion (XXX)]}.

Note that if the alkyl groups in the ketone are each propyl or higher, then the oxygen-containing fragment after McLafferty rearrangement can itself undergo a further McLafferty rearrangement as an olefin.

For cyclic ketones[21d] one prominent fragmentation is analogous to that shown above for the molecular ion (XVI) of cyclohexanol, and leads to the base peak ion, m/e 55, from cyclohexanone. Loss of CO is not a favoured process (but loss of $CH_2=CH_2$ occurs, and these are not readily differentiated except by accurate mass measurement), but H_2O is lost from the molecular ion.

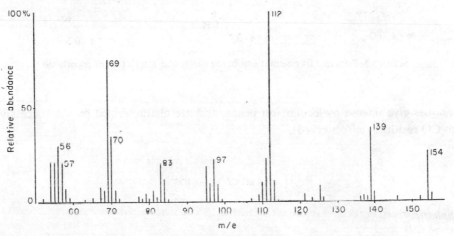

FIG. 18. Part of the mass spectrum of menthone, molecular ion (XLV).

Figure 18 shows part of the spectrum of the cyclic ketone menthone [molecular ion (XLV), m/e 154]. The major fragments arise as indicated[46] in Scheme 5; most interesting is that McLafferty rearrangement gives an ion in which a retro-Diels–Alder fragmentation is possible, and this is the predominating mode of decomposition of this ion.

(ii) *Aromatic*[47]

The molecular ion peak is pronounced, and the expected α-scissions occur. For mixed ketones, the positive charge resides preferentially on fragments containing the aromatic nucleus, these further break down as indicated above for aromatic aldehydes. If McLafferty rearrangement occurs it involves C=O and not a double bond of the ring.

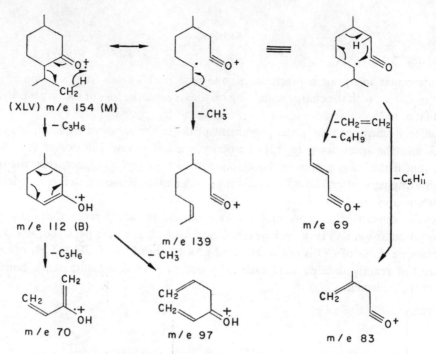

SCHEME 5. Principal fragmentations of the molecular ion (XLV) of menthone.

Quinones give intense molecular ion peaks, and are characterized by the successive loss of two CO residues (m^* observed).

CARBOXYLIC ACIDS

(i) Aliphatic[29]

The molecular ion peak of the normal saturated acids is readily discernible. The base peak usually arises by McLafferty rearrangement (see Table 3); the characteristic peak for acids unbranched at the α-carbon is at m/e 60.

For the lower acids, significant peaks occur at $(M-17)$, $(M-18)$, and $(M-45)$. A peak at m/e 45 (CO_2H^+) occurs for many acids.

(ii) Aromatic[44]

The molecular ion peak is strong, as are peaks at $(M-17)$ and $(M-45)$. $(M-18)$ is insignificant unless there is a suitable side-chain in the *ortho*- position, as with *o*-toluic acid [molecular ion (XXXV)] (p. 27).[31]

CARBOXYLIC ESTERS

(i) *Aliphatic*[30, 36b]

The molecular ion peak is discernible. The four possible ions by scission of bonds α-to $C{=}O$ are all obtained, prominent ones being $R.\overset{+}{C}{=}O$ and $\overset{+}{C}{=}O.OR'$. McLafferty rearrangement of methyl esters gives a prominent peak (at m/e 74 when the α-carbon atom is unbranched), as does similar rearrangement of ethyl esters (m/e 88).

A rearrangement involving the transfer of two hydrogen radicals is quite common for esters, typically ethyl esters, and gives diagnostically useful peaks in the series m/e 61, 75, 89, 103, etc.

$$R{\cdot}\underset{\underset{{\cdot}+}{\overset{\|}{O}}}{C}{\cdot}O{-}(C_nH_{2n-1})-H_2 \longrightarrow R{\cdot}\underset{\underset{+}{\overset{\|}{OH}}}{C}{\cdot}OH + C_nH_{2n-1}^{\cdot}$$

Symmetrical esters of dibasic acids behave as expected from the foregoing, very prominent ions being $(M-OR)^+$ and $(M-CO_2R)^+$ from straight scission. McLafferty rearrangement occurs when possible, and also the rearrangement above with the transfer of two hydrogen radicals. The ion produced in the latter case, for esters of higher than pimelic acid, fragments to give ions $\overset{+}{\cdot}(C_nH_{2n}).C({=}OH)OH$ in the series m/e 84, 98, etc., which are sufficiently unusual to be diagnostic.

(ii) *Aromatic*[31]

The esters of benzoic acid give strong molecular ion peaks unless the ester is from an aliphatic alcohol higher than C_5. The major fragmentations are α to the CO group with the charge remaining on the aromatic moiety; the base peak usually arises from the scission that is also β- to the ring.

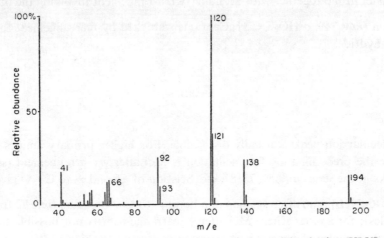

Fig. 19. Part of the mass spectrum of n-butyl salicylate, molecular ion (XLVI).

Rearrangement fragments are prominent, of the type considered above and also those arising from the presence of a substituent *ortho-* to the carboxyl and having an α-hydrogen atom. Salicylates are particular examples of the latter type of compound, and Fig. 19 shows the spectrum of n-butyl salicylate [molecular ion (XLVI), *m/e* 194]. The principal fragmentations are as indicated in Scheme 6; note that one of the ions (*m/e* 138) is the molecular ion of salicylic acid, and all the expected fragments of this are observed, as are all the smaller fragments of protonated salicylic acid (*m/e* 139).

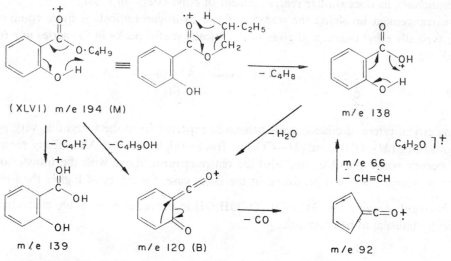

SCHEME 6. Principal fragmentations of the molecular ion (XLVI) of n-butyl salicylate.

Whilst esters of isophthalic and terephthalic acids, and dimethyl phthalate, behave[31] like esters of benzoic acid, higher esters of phthalic acid give[31] a very prominent peak at *m/e* 149 (the base peak for many esters). This ion probably arises by loss of an ·OR radical from one ester group together with McLafferty rearrangement involving the other, and the resulting ion (*m/e* 149, $o\text{-}HO_2C.C_6H_4.C{=\!\!=}O$) is stabilized by rearrangement to protonated phthalic anhydride.

AMIDES

(i) *Aliphatic*[32]

The molecular ion peak is usually discernible. For higher primary amides higher than butyramide the predominating fragmentation is McLafferty rearrangement (see Table 3), giving peaks in the series *m/e* 59, 73, 87, etc. Scission of a bond α- to $C{=\!\!=}O$ gives a peak at *m/e* 44 ($C{=\!\!=}O.NH_2$), and often scission of a bond γ- to $C{=\!\!=}O$ occurs (*m/e* 72 from straight-chain amides); for amides where McLafferty rearrangement is not possible these become the important fragmentations.

For secondary and tertiary amides additional processes are apparent when there is the possibility of scission of bonds β- or γ- to the nitrogen in the amine residue. The first of these gives an ion $R.CO.\overset{+}{N}R'{=}CH_2$, which subsequently fragments with rearrangement of a hydrogen radical from the acid residue to give an ion, $\overset{+}{H}NR'{=}CH_2$ (m/e 30 from secondary amides). In the second case the initial scission is accompanied by rearrangement to give an ethyleneimine, which then undergoes the hydrogen radical transfer to give an ion,

$$HR'\overset{+}{N}\underset{CH_2}{\overset{CH_2}{\diagdown \diagup}}|\quad (m/e\ 44\ \text{from secondary amides}).$$

In the specific case of acetamides, there is a prominent peak at m/e 43 $\left(CH_3.\overset{+}{C}{=}\overset{+}{O}\right)$, and a peak at m/e 60 (protonated acetamide) which arises after extensive rearrangement.

The spectrum[32] of *NN*-diethylacetamide (Fig. 20) can be explained on the basis of the foregoing.

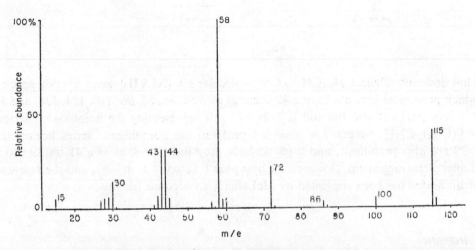

FIG. 20. Mass spectrum of *NN*-diethylacetamide.

N-Arylamides (other than formanilide) undergo the McLafferty rearrangement (giving an ion of m/e 93 from anilides), and also fragment with rearrangement to give the amine molecular ion.

(ii) *Aromatic*

The spectra of substituted benzamides are comparatively simple. The molecular ion peak is reasonably prominent, and the base peak arises from the $Ar.\overset{+}{C}{=}O$ ion by ejection of the amine radical. The expected fragments from the base peak ion largely account for the rest of the spectrum.

NITRILES

(i) *Aliphatic*[33]

The molecular ion peak is of very low intensity. Loss of a hydrogen α- to C≡N gives an $(M-1)$ peak of only slightly higher intensity. An $(M+1)$ peak can be obtained especially when the sample pressure is high. McLafferty rearrangement occurs[33] (see Table 3) giving peaks in the series m/e 41, 55, 69, etc., but note that ions $C_nH_{2n-1}N^{\ddot{+}}$ have the same gross mass as ions $C_{n+1}H_{2n+1}^{+}$ so that it is very difficult in low resolution spectrometry to be sure that a given ion peak represents a single species, and rearrangements are very common with nitriles; this a clear case where high resolution spectroscopy is advantageous.

Long-chain nitriles give prominent peaks arising from expulsion of olefinic (ethylene) molecules from the molecular ion; many processes are responsible for this.[48]

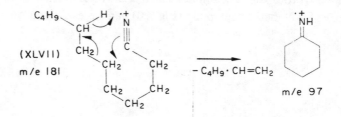

Thus dodecanonitrile, $CH_3.(CH_2)_{10}.CN$ [molecular ion (XLVII)], gives a mass spectrum in which prominent ions above m/e 40 occur at m/e 54, 68, 82, 96, 110, 124, 138, and 152, with minor peaks at m/e 166 and 180 $(M-1)$, this representing the homologous series of ions $[\cdot(CH_2)_n.CN]^{\ddot{+}}$. Several low mass ion peaks in the homologous series beginning at m/e 29 are also prominent, and a particularly prominent peak at m/e 41 testifies to the McLafferty rearrangement. However, the base peak is actually at m/e 97, and the rearrangement indicated has been suggested by McLafferty to account for this.

(ii) *Aromatic*

The molecular ion peak is very intense and is often the base peak. Loss of CN from the molecular ion is a very minor event, and the major fragmentation is loss of HCN (27 a.m.u.) (m^* observed); loss of H_2CN also occurs as a minor event.

AMINES

(i) *Aliphatic*

The molecular ion peak is weak; removal of one of the lone pair electrons of the hetero-atom is easier than with alcohols (the ionization potential of ethylamine is lower than that of ethanol); this directs fragmentation, which occurs more readily with amines.

The process producing the most abundant ions in the spectra of primary, secondary, and tertiary n-alkylamines[49] is expulsion of a neutral radical by scission of a bond β- to the nitrogen, the largest alkyl group being lost preferentially. This gives ions in the series

m/e 30 ($\overset{+}{CH_2}=NH_2$), 44, 58, 72, etc. Hydrogen is not lost from the α-carbon atom unless there is no alternative. As with alcohols, when a chain is long enough, loss of a radical from it to give a cyclic ion is also favoured; the effect decreases progressively from the nitrogen except that a six-membered ring is apparently more stable than a five-membered.

When chain-branching occurs at an α-carbon atom, then a rearrangement process can become very important and can give rise to the base peak. After scission of a bond β- to the nitrogen, there is another, α-bond, scission with rearrangement of a hydrogen radical on to the nitrogen atom. This gives a peak in the same series as above (note that a very prominent peak at m/e 30 in the spectrum of an amine does not imply solely a primary amine containing $\cdot CH_2.NH_2$). The transferred hydrogen comes from the β-carbon atom exclusively only when the alkyl chain is not longer.

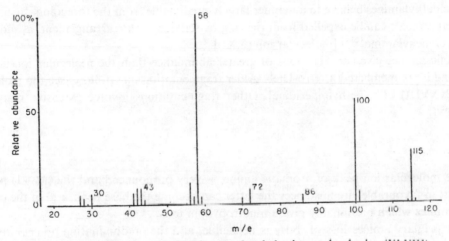

FIG. 21. Mass spectrum of *N N*-di-isopropylmethylamine, molecular ion (XLVIII).

Thus the major peaks in the spectrum[49] (Fig. 21) of *NN*-di-isopropylmethylamine [molecular ion (XLVIII), m/e 115] arise as shown. It is instructive to compare the spectrum with that (Fig. 20) of *NN*-diethylacetamide, which has the same molecular weight and also contains a single nitrogen atom; the most significant difference is the peak at m/e 44 in the amide spectrum, which must represent the ion $\overset{+}{H_2}N(CH_2)_2$ and this cannot arise by rearrangement from the amine.

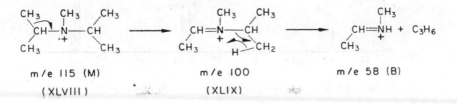

After scission of the β-bond, the $C{=}N$ double bond in the resulting ion may promote a McLafferty rearrangement if the remaining alkyl chains are long enough [i.e. in ions corresponding with (XLIX)].

With amines the same difficulty exists, referred to under fragmentation of nitriles (above), in deciding whether a given peak is from a purely hydrocarbon ion and/or a nitrogen-containing ion. Confirmation for the presence of nitrogen can be sought (in a low resolution spectrum) by examining the spectrum for the presence of peaks at m/e 18 and 17, and by comparing these peak heights. For amines the ratio of abundances of ions $\overset{+}{N}H_4\big/\overset{+\cdot}{N}H_3$ is high; if oxygen is present and these ions are due respectively to $\overset{\cdot+}{H_2O}$ and $\overset{+}{O}H$, then the ratio $\overset{\cdot+}{H_2O}\big/\overset{+}{O}H$ is much lower.

Ions representing the loss of NH_3 from the molecular ion are of no importance in amine spectra (compare the loss of H_2O and H_2S from alcohols and thiols).

Cycloalkylamines behave in a manner largely predictable from the foregoing. In addition neutral radicals can be expelled from the ring by scission with rearrangement, as indicated for cyclohexylamine [24 c] [molecular ion (XXI)].

Cyclic amines give $(M-1)^+$ ions of greater abundance than the molecular ions, and if the ring is six-membered a retro-Diels–Alder fragmentation may follow [see the molecular ion (XXVIII) of 4-methylpiperidine]. Other fragmentations involve extensive rearrangement.[21e]

(ii) *Aromatic*

The molecular ion peak of aromatic amines is very pronounced and the $(M-1)$ peak is strong; in favourable circumstances the latter can become the base peak, e.g. in the case of tolylamines which presumably give aminotropylium ions.

For primary amines loss of $\cdot NH_2$ is negligible, and the predominating process involves loss of HCN from the molecular ion (m^* observed) to give a cyclopentadiene ion which

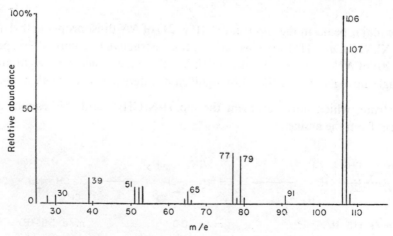

FIG. 22. Mass spectrum of *o*-toluidine, molecular ion (L).

then loses a hydrogen radical (*m** observed) to give a cyclopentadienyl cation. (Compare the loss of CO and CHO from phenol molecular ions.)

This is shown by the spectrum (Fig. 22) of *o*-toluidine [molecular ion (L), *m/e* 107], with the chief fragmentations being as indicated in Scheme 7.

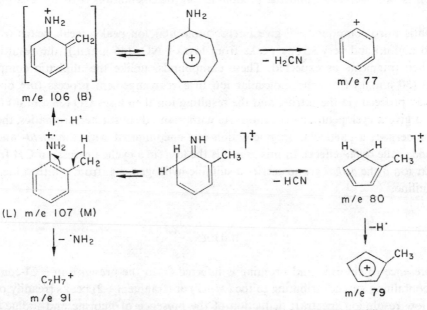

SCHEME 7. Principal fragmentations of the molecular ion (L) of *o*-toluidine.

With secondary and tertiary amines, when an alkyl chain is long enough McLafferty rearrangement (with elimination of an olefin) occurs to only a small extent, and the predominating fragmentation is scission of a bond β- to the nitrogen. [Note that this is a bond γ- to the aromatic ring, so that it is the presence of the heteroatom and not the aromatic ring which directs the initial fragmentation.]

Pyridines give very pronounced molecular ion peaks, and the fragmentation depends very much upon the position of substituents.[26b]

(iii) *Amine and quaternary ammonium salts*

These, of course, are non-volatile, but spectra will be produced from them if they are heated on a probe in a direct inlet system. Amine salts will decompose and give spectra which are the sum of those from the amine and the corresponding acid. The apparent molecular ion peak will then come from the amine, and if the salt is of a halogen acid the only indication of the nature of the acid will be the appropriate molecular ion peak from it, with isotope peaks from chlorine or bromine.

Quaternary ammonium halides both dissociate into the amine and the alkyl halide, and undergo the Hofmann degradation where appropriate.

NITRO COMPOUNDS

The nitro group is not a charge-stabilizing entity. Molecular ions of nitroalkanes eject the nitro group so readily that molecular ion peaks are scarcely discernible (the only exception is nitromethane), and the predominating fragmentation is of the charged alkyl residue.

Aromatic nitro compounds[50] give intense molecular ion peaks, weak peaks from loss of O (16 a.m.u.) and very strong peaks from loss of NO_2 (46 a.m.u.); the resulting aryl cation then fragments as expected. These compounds, unlike the aliphatic compounds, lose NO (30 a.m.u.) from the molecular ion in a rearrangement process (m^* observed) which may proceed via the nitrite, and the resulting ion then loses CO (28 a.m.u.) (m^* observed) to give a cyclopentadienyl cation. As with many disubstituted aromatics, the *ortho*-isomer undergoes a particular fragmentation not encountered with the *meta*- and *para*-compounds, (the *ortho*- effect). In this case it is the loss (in a cyclic process) of OH from the molecular ion if the *ortho*- group carries a suitable hydrogen, e.g. from *o*-nitrotoluene and *o*-nitroaniline.

HALIDES

The presence of chlorine and bromine is inferred from the presence of ^{37}Cl-containing or ^{81}Br-containing ions contributing to the $(M+2)$ or (fragment$+2$) peaks (readily obvious even in low resolution spectra); deduction of the presence of fluorine and iodine cannot rely on the presence of isotope peak clusters [see above, under recognition of the molecular ion peak (e) (p. 14)].

(i) *Aliphatic*[51]

Elimination of a halogen radical or of hydrogen halide occurs and is progressively more easy from fluorides to iodides. Thus low molecular weight bromides and iodides give a molecular ion peak that is very weak (not discernible for the t-butyl compounds).

In the case of alcohols, cyclic oxonium ions are of minor importance, but in the case of halides cyclic ions containing five atoms in the ring including positively charged chlorine or bromine are of very prominent abundance (smallest possible is $C_4H_8X^+$); such ions are not important when the halogen is fluorine or iodine.

These fragmentations occur for saturated alkyl halides; when other groups are present the pattern may be considerably modified, e.g. whilst the halogen directs the course of fragmentation for *endo*-5-bromo-2-norbornane[21f] (base peak m/e 95, from $C_7H_{11}^+$), it is the retro-Diels–Alder fragmentation directed by the olefinic bond that predominates for *endo*-5-chloro-2-norbornene[21f] (base peak at m/e 66, from the cyclopentadiene cation). Halogen attached directly to an olefinic or aromatic carbon atom is not ejected readily from molecular ions in comparison with ejection from saturated alkyl halides.

(ii) *Aromatic*[52]

Aryl halides give strong molecular ion peaks. Ejection of fluorine does not seem to occur, but in aryl chlorides, bromides, and iodides the most abundant fragment is due to loss of the halogen unless substituents larger than methyl are present.

SULPHUR-CONTAINING COMPOUNDS

The presence of sulphur is usually inferred from the contribution from ^{34}S-containing ions to the $(M+2)$ or (fragment$+2$) peaks, and a low resolution spectrum will reveal this.

The molecular ion peak for a given compound is stronger than that for the oxygen analogue (noticeably for thiols-alcohols), and the fragmentations parallel, of course, those of the corresponding oxygen-containing compounds, with the noteworthy exception that thiophenols also lose ·SH radicals from the molecular ion.

Disulphides give strong molecular ion peaks and, where it is possible, strong peaks from the rearrangement to give R—S—S—H $^{\cdot+}$ ions with expulsion of an olefin (CH$_2$=CHR'). This rearrangement is repeated to give H—S—S—H $^{\cdot+}$ ions (m/e 66). Otherwise, straight scission of R—S and S—S bonds occurs.[53]

HETEROCYCLIC COMPOUNDS

Considerations of space do not permit a treatment of the fragmentations of heterocyclic compounds, and the reader is referred to two recent texts[54 55] on the subject, of which the first is very detailed. Reference 21 also includes comprehensive coverage of heterocyclic substances, ref. 24 includes many of the commoner heterocycles, and ref. 11 includes heterocyclic natural products.

Fragmentations of Compounds of Designated Structure

It should be possible, using the information above, to predict the general appearance of the spectrum of a compound of designated structure. This can be a very valuable approach to deciding whether a given spectrum may arise from a deduced structure and whether isomeric compounds would give rise to significant differences in the spectra.

As an example, consider the spectrum to be expected from the ethyl ester (LI) of the amino-acid methionine.

$$\underset{\substack{\\ \\ (130)(116)NH_279}}{\overset{\substack{47\quad 61\quad 75\quad 104\quad 132\\ }}{CH_3-S-CH_2-CH_2-CH-CO-OC_2H_5}} \qquad (LI)$$

102

Although open-chain and aliphatic, the large number of heteroatoms should confer stability on the molecular ion, and the peak from this should be quite visible at m/e 177. The more obvious fragmentations to be expected are summarized.

(i) As an ethyl ester, ions expected are of m/e 132 (M—OC_2H_5), 45 (OC_2H_5), 104 (M—$CO_2C_2H_5$), 73 ($CO_2C_2H_5$), 148 (M—C_2H_5), 29 (C_2H_5), 103 from McLafferty rearrangement (with the charge remaining on oxygen), 74 (from the same rearrangement with the charge remaining on the olefinic thioether), and 150 [by the double rearrangement giving (M—C_2H_3)].

(ii) As an amine, ions expected are of m/e 102 and 104 (β-cleavage with the charge remaining on nitrogen), 75 and 73 (if the charge remains on the other fragments from β-cleavage). From further breakdown of the ion of m/e 104 there would be expected ions of m/e 30 (CH_2=NH_2, with the charge remaining on nitrogen), 74 (charge remaining on the other fragment), 43 ($CH_2.CH$=NH_2), 57 ($CH_2.CN_2.CH$=NH_2), and 56 (CH_2=$CH.CH$=NH_2).

(iii) As a thioether, ions expected are of m/e 47 (CH_3S), 48 (CH_3SH), 129 (M—CH_3SH), 61 ($CH_3.S$=CH_2), and 116 (M—$CH_3.S$=CH_2).

It is to be noted that because of the nature of this particular molecule, the great majority of fragmentations leave at least one heteroatom in each fragment, and that therefore both fragments may be expected to appear as ions. Some fragmentations can be expected to be particularly favourable, these are scission β- to the sulphur to give m/e 61, and scission β- to nitrogen to give m/e 102 and 104 (especially the latter which is also a scission α- to C=O); the ion of m/e 56 would be well stabilized by resonance. Similarly, the ion of

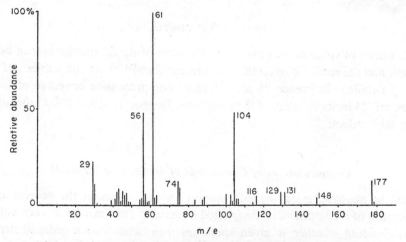

FIG. 23. Spectrum of the ethyl ester (LI) of methionine. (From K. Biemann, *Mass Spectrometry: Organic Chemical Applications*. Copyright 1962 by McGraw-Hill Book Company. Used with permission of McGraw-Hill Book Company.)

m/e 150 would be probably of low abundance since other scissions would be more favourable.

The actual spectrum[26c] is given as a bar graph in Fig. 23 and accords well with prediction. Fourteen of the seventeen predicted ions are seen to occur (two others, m/e 132 and 150, are also present, but each of less than 1% relative abundance) and the three most abundant ions are not unexpected. It is perhaps surprising that the ion of m/e 73 does not apparently appear at all; evidently in this scission the charge resides exclusively on the nitrogen-containing fragment, and the preferred alternative is McLafferty rearrangement.

Interpretation of Spectra

STRUCTURE ANALYSIS FROM A SINGLE SPECTRUM

When no information other than the isolated mass spectrum is available, then structure derivation from this alone can be quite difficult if the structure is complicated. It is clear that there is so much data to be derived from the mass spectrum that one should be systematic about its acquisition. However, the sheer bulk of data means that for manual manipulation one cannot proceed logically and systematically through all of it, taking every structural possibility into account at every stage and accepting or rejecting a possibility as a result of close reasoning.

Usually one proceeds as in the case of similar analysis from other isolated spectra, like i.r. or n.m.r. That is, one considers the whole accumulated data, when observations about the type of compound (? aliphatic, aromatic) and functional groups can be made. Starting from an obvious deduction, the basis of which will vary from case to case, confirmation is sought from the remainder of the data, and it usually quickly becomes apparent whether one is thinking on the right lines or not. As some idea of the nature of the compound becomes evident it is a customary and obvious practice to compare spectra of authentic compounds with that of the unknown to confirm that the general fragmentation pattern is applicable to the particular case in hand or to consult the literature for the known ways in which compounds of exactly the same type undergo fragmentation. In the event that some correspondence is found, then identity of the unknown may be established by matching spectra (*these must be obtained under the same experimental conditions*); it may well be impossible to decide between isomers by using only the spectrum of the unknown. No attempt is made, of course, to account for the origin of every ion in a spectrum.

Summary of data acquisition from the mass spectrum:

(i) Examine the high-mass end for a likely molecular ion, applying checks for internal self-consistency if necessary [see above, The molecular ion peak (p. 14)].

(ii) In the vicinity of the molecular ion peak, look for peak clusters resembling the characteristic isotope abundance patterns for elements other than C, H, N, and O.

(iii) Determine the molecular formula. Accurate mass measurement with a high resolution instrument renders this a simple matter; if a low resolution spectrum only is available, and the M, $(M+1)$, and $(M+2)$ relative heights can be ascertained with some accuracy, determine the molecular formula from these. The molecular formula must obey the nitrogen rule. Consider first the most likely possibilities.

When accurate mass measurement is possible, determine also the formulae of fragment ions, and note particularly whether any prominent ions are inhomogeneous.

(iv) Note the metastable ion peaks (impossible from a bar graph alone) and identify the transitions that these support by their presence. The most important of these initially are those involving parent ions at the high-mass end.

(v) Note, if possible, the odd-electron ions in the spectrum, especially those of high relative abundance.

(vi) Note the small fragments lost from the molecular ion (Table 4 for even-electron neutral fragments and Tables 3 and 5 for odd-electron neutral fragments).

(vii) Look for characteristic low-mass fragment ions above m/e 28 (Table 6, and ions deriving from fragments in Tables 3, 4, and 5).

Of course, in any particular case circumstances will dictate whether all these seven suggestions can be followed and, if so, how completely.

With as much data acquired from the spectrum as possible, the general way of using it is outlined below. There follow some examples of the method of analysis for specific compounds. Further instances of the ways in which spectra can be analysed appear in the Chapter on Problems in Mass Spectrometry (p. 67), where the relevant analyses are fully explained.

In the light of the data, partial structures to accommodate the fragments are considered. When a definite anomaly is apparent it is necessary to go back to the previous steps and make alternative postulates. For example, one may have to consider the second most likely molecular formula, etc. The problem here is really that given fragment weights (a.m.u.) can often be accommodated by more than one combination of atoms (a loss of 29 a.m.u. might not be CHO, but C_2H_5 or CH_3N), and the problem is more difficult the larger the fragment. Obviously, unique fragment composition can be determined using accurate mass measurement, but if this is not possible then there are useful correlation tables[56] which list total composition of fragments, their masses (a.m.u.), and the relative probabilities of finding the fragment as an ion from given precursor structures. For instance, $C_2H_3O^+$ (m/e 43) may arise as $[CH_3 . \overset{.}{C}{=}O]^+$ from methyl ketones, when the relative probabilities of finding this ion as the most abundant, second most abundant, and third most abundant in the spectrum are 30 : 4 : 14 respectively, and it may also arise as $[CH_2{=}C(H)O \cdot]^+$ from vinyl ethers, when the corresponding relative probabilities are 0 : 2 : 3. Thus a prominent ion, m/e 43, is by no means always from methyl ketones only, but it is more likely that if this is the base peak ion then a methyl ketone is more probable than is a vinyl ether.

In order to evaluate a given deduced structure it is customary to consider isomers, to postulate the spectra from these (see the example above, under fragmentations of compounds of designated structure), and to suggest how the spectra might differ from the spectrum under consideration. It must be emphasized that some isomers cannot readily be distinguished by mass spectrometry; other spectroscopic techniques are better suited for this. For example, it is difficult to decide the exact nature of a multibranched alkyl chain or to assign unequivocal orientation to some disubstituted benzenes.

A literature search may provide authentic individual spectra for comparisons, and there are four accessible collections of data: (i) the *Index of Mass Spectral Data*[57] (3200 spectra abbreviated to the molecular ion and the four most abundant ions in each spectrum);

(ii) the *MSDC Mass Spectral Data Sheets*[58] (now over 10,000 spectra reported fully including the metastable ions in many cases); (iii) the *Compilation of Mass Spectral Data*[59] (some 7300 spectra giving the ten most abundant ions and cross-indexed in four ways); and (iv) the *Eight Peak Index of Mass Spectra*[60] (some 17,000 spectra giving the eight most abundant ions and cross-indexed in three ways).

STRUCTURE ANALYSIS: EXAMPLES

1. As the first example let us consider the compound (II) giving the (low resolution) spectrum in Fig. 5 (p. 9). We have the following data:

(i) The most likely molecular ion is at m/e 157; the first mass loss is 16 a.m.u. which is quite reasonable.

(ii) The isotope pattern for one atom of chlorine is apparent in the vicinity of the molecular ion peak [see the calculation, above (p. 13)]; moreover the chlorine atom is still present in the ions of m/e 141, 127, 111, 99, and 85.

(iii) Measurement of the relative heights of the molecular ion peak and the first and second isotope peaks leads to a probable molecular formula of $C_6H_4ClNO_2$ [see the calculation, above (p. 9)].

(iv) Metastable ions m^* are present with m/e 102·7, 87·3, 77·1, 53·8, 50·8, 33·0, and 29·5. Calculation shows readily that these support the following transitions (calculated m^* values in parentheses): m/e 157 → 127 (102·7), 141 → 111 (87·4), 127 → 99 (77·2), 99 → 73 (53·8), 111 → 75 (50·7), 76 → 50 (32·9), and 85 → 50 (29·5). [Where parent ions contain ^{35}Cl, there are present in addition metastable ions for the corresponding ^{37}Cl-containing parents.]

(v) Odd-electron ions. Since the molecular ion contains an odd number of nitrogen atoms, namely one, odd-electron ions will have even mass if they do not contain the nitrogen atom and odd mass if they contain the nitrogen atom. Simple inspection does not therefore reveal the odd-electron ions, although the ion of m/e 50 is one such, for it cannot really be anything except $[C_4H_2]^+$. We may note that all major ions have odd mass.

(vi) Small fragments lost from the molecular ion. The first five ions below the molecular ion represent losses of, respectively, 16, 30, 46, 58, and 72 a.m.u. The first three of these, with loss of 30 a.m.u. occurring in one stage [see above (p. 13)], indicates clearly an aromatic nitro group, and the most reasonable total losses will be O, NO, NO_2, CNO_2, and $C_2H_2NO_2$ respectively.

(vii) Characteristic low-mass fragment ions. These occur in groups around m/e 38, 50, and 75. These are not likely to contain the nitrogen atom (ions below m/e 111 probably represent fragments where the nitrogen has been ejected in the nitro group), and probable compositions are C_3H_2, C_4H_2, and C_6H_3 respectively.

These particular relatively low abundance ions, together with the occurrence of high abundance ions in the higher-mass region only, suggest a stable, aromatic type of nucleus. The persistence of the chlorine atom through many fragmentations confirms this, and further suggests that the chlorine is directly attached to the nucleus. The molecular formula will accommodate a chloronitrobenzene. Of the possible isomers, for the *ortho-* and *para-* compounds (but not for the *meta-*) it is possible to write additional resonance structures for some ions by placing the positive charge on the halogen. (This could be rationalization of the high relative abundance of these ions, but the spectra of all three isomers are required before this hypothesis can be tested.)

Taking the *para-* isomer, the fragmentation can be envisaged as in Scheme 8 (transitions supported by the presence of metastable ions, *m**, are indicated).

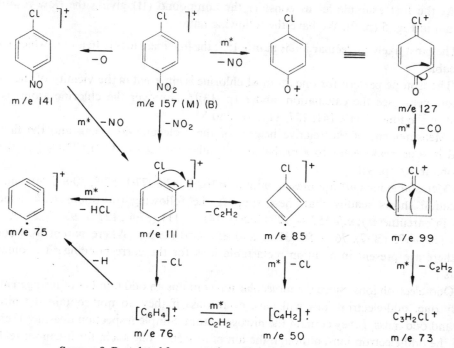

SCHEME 8. Postulated fragmentation of ions from *p*-chloronitrobenzene.

Perhaps it is worth mentioning that the ion of m/e 73 is now seen to contain the chlorine atom. This is not evident from inspection of the (low resolution) spectrum because the m/e 75 peak contains also a contribution from a second ion of m/e 75 ($C_6H_3^+$) arising from another fragmentation process.

It is clear that the requirements of the spectrum can be met by a chloronitrobenzene structure for compound (II), but without further evidence the orientation cannot be decided. [Compound (II) is, in fact, *p*-chloronitrobenzene.]

Of course, as soon as the molecular formula is apparent as $C_6H_4ClNO_2$ it is inevitable that a chloronitrobenzene structure will be immediately considered, but note how the spectrum provides positive confirmation. If an alternative formulation as a chloropyridinecarboxylic acid is made, then in spite of the similarity as a chloroaromatic compound, the predicted spectrum for this will be very different at the high-mass end.

2. For the second example, consider structure elucidation of compound (LII) for which the only information available is the mass spectrum in the form of the bar graph shown in Fig. 24.

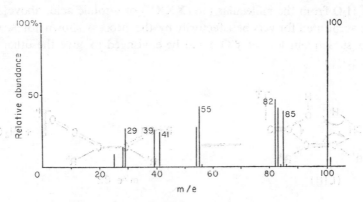

FIG. 24. Mass spectrum of compound (LII).

The analysis may be carried out in the following way.

(i) The apparent molecular ion peak is at m/e 100. The first loss would then be of 15 a.m.u., which could be possible.

(ii) No elements other than C, H, N, and O can be present since the height of the $(M+2)$ peak is too small to show on this scale.

(iii) Accurate measurement of the $M, (M+1),$ and $(M+2)$ relative heights is not possible, so nothing can be said of the molecular formula except that it contains no, or an even number of, nitrogen atoms.

(iv) No metastable ions are reported.

(v) If nitrogen is present, mere inspection does not reveal the odd-electron ions, but we note that the majority of the ions in the spectrum are of odd mass. Considering the even mass molecular weight this suggests that we are dealing with a compound containing no nitrogen. Assuming in the first instance that this is so, then the two ions of m/e 82 and 54 are odd-electron ions, and the latter can be only $[C_3H_2O]^{+}$ or $[C_6H_4]^{+}$.

(vi) Small fragments lost from the molecular ion. The first five are of 15, 17, 18, 45, and 46 a.m.u.; if no nitrogen is present the first three must be $CH_3^{.}$, $HO^{.}$, and H_2O, and, taken in conjunction with the loss of $HO^{.}$, the loss of 45 a.m.u. is almost certainly $HO^{.} + CO$, i.e. compound (LII) is a carboxylic acid.

(vii) Low-mass fragment ions are of m/e 29, 39, 41, 54, and 55. With nitrogen absent, the second must be $C_3H_3^{+}$ (and it is very likely that the third is $C_3H_5^{+}$).

It would seem that the molecule contains a carboxyl and a methyl group (it may be necessary to consider later that the latter may arise from a cyclic structure in which there are initially no methyl groups, but the presence of such a group is the first suggestion). To account for the remainder of the molecule (40 a.m.u.), the only two possible residues are C_2O or C_3H_4, and obviously the second is much more likely.

There is no odd-electron ion corresponding with that expected from McLafferty rearrangement (in the series m/e 60, 74, 88, etc.), but there is a prominent odd-electron ion corresponding with loss of H_2O, and this suggests an acid with $\alpha\beta$-unsaturation, [compare the loss of H_2O from the molecular ion (XXXV) of *o*-toluic acid, above], and the ion, m/e 82, can be accounted for very satisfactorily by the process shown for structure (LIII). Moreover, the subsequent loss of CO could be envisaged to give the other odd-electron ion, m/e 54 ($C_4H_6^{+}$).

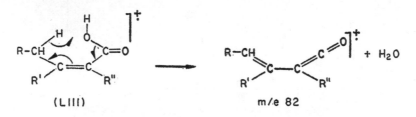

(LIII)   m/e 82

To decide the pattern of substitution in the olefin is not easy. Simple olefins always show prominent ions arising by extensive rearrangement, and it is dangerous to assign structures on the mere presence or absence of certain ions. Thus *trans*-hex-2-ene (LIV), 2,3-dimethylbut-2-ene (LV), and *trans*-but-2-ene (LVI) each show prominent $(M-15)^+$ and $(M-29)^+$ ions although the last two compounds do not contain ethyl groups.

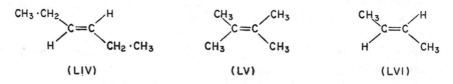

(LIV)   (LV)   (LVI)

However, the relative abundances of the $(M-29)^+$ ions are found to be far less than those of the corresponding $(M-15)^+$ ions for the two cases when no ethyl group is present, and far greater in the case when an ethyl group is present (see ref. 38). On this basis we can reject the pent-2-enoic acid formula for compound (LII). The compound is either 2-methylbut-2-enoic acid (with carboxyl *trans-* to hydrogen), or 3-methylbut-2-enoic acid (senecioic acid), but we cannot decide between these isomers. [The spectrum in Fig. 24 is, in fact, from the latter compound.]

3. For the third example, we have the analysis of the spectrum given in Fig. 25, which is that from a compound (LVII).

(i) The apparent molecular ion peak is at m/e 115, requiring a molecule with an odd number of nitrogen atoms. The first mass loss is $(115-101) = 14$ a.m.u., which is unlikely. Further, all the major ions in the spectrum have odd mass and this is a little unexpected since some of them must contain an odd number of nitrogen atoms. It must be suspected that the true molecular ion peak is not visible.

Rerunning the top end of the spectrum at higher ($\times$20) sensitivity shows that there is another, low abundance, ion at m/e 130 (Fig. 26). This could certainly be the molecular ion.

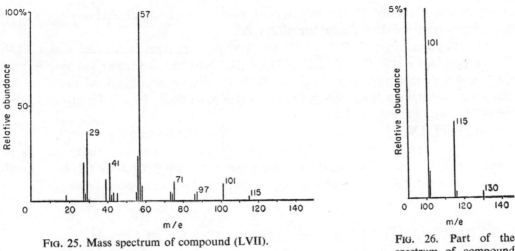

FIG. 25. Mass spectrum of compound (LVII).

FIG. 26. Part of the spectrum of compound (LVII) at high sensitivity.

(ii) No elements other than C, H, N, and O can be detected from the presence of isotope peak clusters.

(iii) The low abundance of the molecular ion renders the measurement of relative M, $(M+1)$, and $(M+2)$ peak heights not sufficiently accurate for molecular formula determination.

(iv) No metastable ions are reported.

(v) The even molecular weight and odd mass for all prominent ions is consistent with a molecule containing no nitrogen. Taking this as a first alternative, the only odd-electron ion of note is of m/e 56, $[C_2O_2^+$ (unlikely), $C_3H_4O^+$, or $C_4H_8^+$ only].

(vi) Small fragments lost from the molecular ion are of 15, 29, 43, 44, 55, 56, and 57 a.m.u. The series 15, 29, 43, and 57 could represent purely saturated hydrocarbon radicals, but we note that the prominent ions m/e 115 and 101 must contain oxygen [purely saturated alkyl radicals would have masses of 113 and 99 a.m.u., and unsaturated hydrocarbon radicals ($C_9H_7^-$ and $C_8H_5^-$) are unlikely]. Thus 29, 43, and 57 a.m.u. may represent CHO, C_2H_3O, and C_3H_5O.

(vii) The ions of low mass are of m/e 29, 41, 56, 57, 73, and 75. Although the molecular ion loses 43 a.m.u., the lost fragment does not appear as a significant ion. This suggests that the lost fragment is not C_2H_3O, but C_3H_7. In support of this, the prominent ion of m/e 41 could derive from $C_3H_7^+$ by loss of H_2, a known facile process since the resulting ion is stabilized by resonance.

The base peak ion, of m/e 57, however, could well contain oxygen and would be $C_3H_5O^+$. If this ion were $C_4H_9^+$ we would expect to see the derived $C_4H_7^+$ ion much more prominent than is the ion of m/e 55.

The situation is, then, that the molecular formula must contain oxygen (and no nitrogen). The molecular ion loses CH_3^-, $C_3H_7^-$, and $C_3H_5O^-$ If the molecule is of the form C_3H_7—Z—C_3H_5O, then Z can be only CH_2O, and the formula is $C_7H_{14}O_2$ (DBE = 1); an ester is immediately suspected. [If it is not an ester, then it must be an alcohol or ether, and no ions characteristic of these classes are observed.]

From the general formula R.CO.OR' we expect prominent peaks from R^+ and $R.CO^+$, thus a propionate is indicated, $C_2H_5.CO.OC_4H_9$. Now the molecular ion readily loses $CH_3^•$, and since this is not expected if the alcohol residue were n-butyl, we postulate that this residue contains a branched-chain methyl group, and only 2-methylpropyl can give a prominent ion of m/e 43 ($C_3H_7^+$).

Compound (LVII) is thus isobutyl propionate. The odd-electron ion, m/e 56, arises by McLafferty rearrangement [of (LVIII)], a process that also accounts for the ion of m/e 74. The ion, m/e 91, arising from the double rearrangement giving $R.C(OH)=\overset{+}{O}H$ is not in this instance detected; this reflects the very ready scission to lose the isopropyl group.

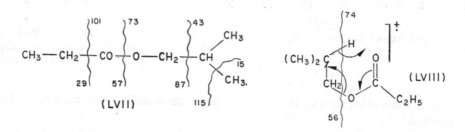

(LVII)   (LVIII)

This example will serve to illustrate one further method by which a check may be carried out on the progress of the analysis. It often happens that a particular odd-electron fragment ion is assigned a structure which is the molecular ion of a known compound (here the ion, m/e 56, is postulated as the molecular ion of isobutene, and there have been several other cases throughout this chapter).

If this is correct, there must be present in the spectrum all the fragment ions expected from this molecular ion in the correct relative abundances. Thus the spectrum can be searched for these ions; if they are not present, or if they are present but with very different relative abundances, then it is likely that the original odd-electron fragment ion does not have the structure postulated. Conversely, if they are all present, with relative abundances as expected, this is good confirmatory evidence that the fragmentation of the original compound proceeds in the manner deduced (and has the total structure postulated).

Note that there will be coincidences in that any given peak in the spectrum is not necessarily from a single ionic species (C_4H_8 has the same mass as C_3H_4O, for example, so that a particular peak height, especially in the low-mass end of the spectrum, may be the sum of more than one contribution, and allowances may have to be made. Even if a peak represents a single ionic species, the ions may arise by more than one process, and the suggested fragmentation with rearrangement may be only one of these.

Nevertheless, this constitutes a good method for confirming the identity of fragments, and may be extended to certain ions. Thus all benzyl compounds unbranched at the α-carbon, for instance, give the $C_7H_7^+$ (tropylium) ion,[37] and the fragmentation products of this all appear, so that there are well-defined peaks up to mass 91 quite characteristic of this ion (see, for example, the spectrum of n-propylbenzene, Fig. 14).

Concluding Comments

It is obvious from the preceding account that structure determination from mass spectra is a very worthwhile proposition. As with other spectroscopic methods, practice using compounds of known structure is necessary for the acquiring of experience. However, since mass spectrometry is such a sensitive method for probing structure, it is quite essential to ensure that any compound actually submitted to such analysis is of very high purity. Trace impurities will give their own spectra superimposed upon that of the major component. If an impurity is of lower molecular weight, apparent fragment ions will be present, which may be very difficult to rationalize or which may lead to incorrect structure deduction, and if an impurity is of higher molecular weight (and high molecular weight impurities are not easily removed from solids by crystallisation alone), then an ion from it may be erroneously identified as the molecular ion of the compound.

In this connection it should be borne in mind that, for example, an investigation of something like a drug metabolite may yield only a very small quantity of the material sought after, and consequently small quantities of impurities, including artefacts, may be mistaken for the genuine metabolite. The storage of biological fluids and reagents in plastic containers often results in the leaching out of minute quantities of plasticisers, and after considerable concentration of non-volatile material these may be isolated and genuinely confused with true metabolites.

Biemann has drawn attention to this aspect and has given appropriate spectra.[26d] The artefacts include tri-n-butyl orthophosphate and di-n-octyl phthalate (from plastic containers and tubing, and the latter also from diffusion pump oil of the inlet system to the spectrometer), silicones (from the stationary phase in gas chromatography), and high molecular weight hydrocarbons (from tap grease). Experience plus unrelenting suspicion are the only safeguards against making erroneous conclusions from the identification of compounds, even though the identification may be correct.

References

1. C. DJERASSI, H. BUDZIKIEWICZ, R. J. OWELLEN, J. M. WILSON, W. G. KUMP, D. J. LeCOUNT, A. R. BATTERSBY, and II. SCHMID, *Helv. chim. Acta*, 1963, **46**, 742.
2. G. SPITELLER, *Z. analyt. Chem.*, 1963, **197**, 1.
3. R. VENKATARAGHAVEN, F. W. McLAFFERTY, and G. E. VAN LEAR, *Org. Mass Spectrometry*, 1969, **2**, 1.
4. J. H. BEYNON, *Mass Spectrometry and its Applications to Organic Chemistry*, Elsevier, London, 1960; (a) Chapter 1; (b) Chapter 6; (c) Appendix 1; (d) Chapter 8; (e) p. 252; (f) Appendix 2; (g) p. 362.
5. B. HEDFJALL, P. A. JANSSON, Y. MARDE, R. RYHAGE, and S. WIKSTROM, *J. Phys.* (E), 1969, **2**, 1031.
6. P. A. JANSSON, S. MELKERSSON, R. RYHAGE, and S. WIKSTROM, *Arkiv Kemi*, 1969, **31**, 565.
7. H.-Y. LI, J. WALDREN, R. SAUNDERS, D. SIMPSON, L. MILLS, K. KINNEBERG, and G. R. WALLER, *Proceedings of the 16th Annual Conference on Mass Spectrometry and Allied Topics*, ASTM E14 Committee, Pittsburgh, 1968, p. 26.
8. H. D. BECKEY, *Field Ionisation Mass Spectrometry*, Pergamon Press, Oxford, 1971.
9. H. M. FALES, G. W. A. MILNE, and M. L. VESTAL, *J. Amer. Chem. Soc.*, 1969, **91**, 3682.
10. J. H. BEYNON and A. E. WILLIAMS, *Mass and Abundance Tables for Use in Mass Spectrometry*, Elsevier, 1963.

11. (a) H. Budzikiewicz, C. Djerassi, and D. H. Williams, *Structure Elucidation of Natural Products by Mass Spectrometry*, Vol. I, *Alkaloids*, Vol. II, *Steroids, Terpenoids, Sugars, and Miscellaneous Classes*, Holden-Day, San Francisco, 1964; (b) J. Lederberg, Appendix in Vol. II.
12. S. R. Shrader, *Introductory Mass Spectrometry*, Allyn & Bacon Inc., Boston, 1971, p. 144 and Appendix C.
13. A. L. Burlingame, *Adv. Mass Spectrometry*, 1968, **4**, 15.
14. K. Biemann, P. Bommer, and D. M. Desiderio, *Tetrahedron Letters*, 1964, 1725.
15. R. M. Silverstein and G. C. Bassler, *Spectrometric Identification of Organic Compounds*, 2nd edn., John Wiley, London, 1967, Chapter 2, Appendix A.
16. D. D. Tunnycliff, P. A. Wadsworth, and D. O. Schissler, *Mass and Abundance Tables*, Shell Development Co., Emeryville, California, 1965, Vols. I–III.
17. R. Binks, J. S. Littler, and R. L. Cleaver, *Tables for Use in High Resolution Mass Spectrometry*, Heyden & Son Ltd., London, 1970.
18. J. H. Beynon, R. A. Saunders, and A. E. Williams, *Table of Metastable Transitions for Use in Mass Spectrometry*, Elsevier, 1965.
19. I. Hormon, A. N. H. Yeo, and D. H. Williams, *J. Amer. Chem. Soc.,* 1970, **92**, 2131.
20. W. O. Perry, J. H. Beynon, W. E. Baitinger, J. W. Amy, R. M. Caprioli, R. N. Renaud, L. C. Leitch, and S. Meyerson, *J. Amer. Chem. Soc.*, 1970, **92**, 7236.
21. H. Budzikiewicz, C. Djerassi, and D. H. Williams, *Mass Spectrometry of Organic Compounds*, Holden-Day, San Francisco, 1967; (a) Note to the Reader, and throughout; (b) p. 454; (c) p. 58; (d) p. 143; (e) p. 309; (f) p. 439.
22. J. S. Shannon, *Proc. Roy. Austral. Chem. Inst.*, 1964, **31**, 323.
23. J. D. Roberts and M. C. Caserio, *Basic Principles of Organic Chemistry*, W. A. Benjamin Inc., New York, 1965, Table 3.5.
24. J. H. Beynon, R. A. Saunders, and A. E. Williams, *The Mass Spectra of Organic Molecules*, Elsevier, 1968; (a) p. 98; (b) p. 164; (c) p. 267; (d) p. 284.
25. J. M. Wilson, M. Ohashi, H. Budzikiewicz, C. Djerassi, Shô Itô, and T. Nozoe, *Tetrahedron*, 1963, **19**, 2247.
26. K. Biemann, *Mass Spectrometry: Organic Chemical Applications*, McGraw-Hill, New York, 1962; (a) p. 102; (b) pp. 130, 134; (c) p. 274.
27. F. W. McLafferty, *Analyt. Chem.*, 1959, **31**, 2072.
28. J. A. Gilpin and F. W. McLafferty, *Analyt. Chem.*, 1957, **29**, 990.
29. G. P. Happ and D. W. Stewart, *J. Amer. Chem. Soc.*, 1952, **74**, 4404.
30. D. H. Williams, H. Budzikiewicz, and C. Djerassi, *J. Amer. Chem. Soc.*, 1964, **86**, 284.
31. F. W. McLafferty and R. S. Gohlke, *Analyt. Chem.*, 1959, **31**, 2076.
32. J. A. Gilpin, *Analyt. Chem.*, 1959, **31**, 935.
33. F. W. McLafferty, *Analyt. Chem.*, 1962, **34**, 26.
34. F. W. McLafferty, *Analyt. Chem.*, 1956, **28**, 306.
35. Z. Dolejšek, V. Hanuš, and K. Vokač, *Adv. Mass Spectrometry*, 1966, **3**, 503.
36. F. W. McLafferty (ed.), *Mass Spectrometry of Organic Ions*, Academic Press, New York, 1963; (a) Chapter 10; (b) Chapter 9.
37. S. Meyerson and P. N. Rylander, *J. Chem. Phys.*, 1957, **27**, 901.
38. R. A. Friedel, J. L. Shultz, and A. G. Sharkey, *Analyt. Chem.*, 1956, **28**, 926.
39. C. S. Cummings and W. Bleakney, *Phys. Rev.*, 1940, **58**, 787.
40. P. Natalis, *Bull. Soc. roy. Sci. Liège*, 1962, **31**, 790.
41. A. E. Pierce, *Silylation of Organic Compounds*, Pierce Chemical Co., Rockford, Illinois, 1968.
42. J. S. Shannon, *Austral. J. Chem.*, 1962, **15**, 265.
43. T. Aczél and H. E. Lumpkin, *Analyt. Chem.*, 1960, **32**, 1819.
44. T. Aczél and H. E. Lumpkin, *Analyt. Chem.*, 1961, **33**, 386.
45. A. G. Sharkey, J. L. Shultz, and R. A. Friedel, *Analyt. Chem.*, 1956, **28**, 934.
46. B. Willhalm and A. F. Thomas, *J. Chem. Soc.*, 1965, 6478.
47. J. H. Beynon, *Adv. Mass Spectrometry*, 1959, **1**, 355.
48. R. Beugelmans, D. H. Williams, H. Budzikiewicz, and C. Djerassi, *J. Amer. Chem. Soc.*, 1964, **86**, 1386.
49. R. S. Gohlke and F. W. McLafferty, *Analyt. Chem.*, 1962, **34**, 1281.
50. J. H. Beynon, R. A. Saunders, and A. E. Williams, *Ind. Chim. Belge*, 1964, **29**, 311.
51. F. W. McLafferty, *Analyt. Chem.*, 1962, **34**, 2.

52. F. W. McLafferty, *Analyt. Chem.*, 1962, **34**, 16.
53. J. H. Bowie, S.-O. Lawesson, J. Ø. Mads en, C. Nolde, G. Schroll, and D. H. Williams, *J. Chem. Soc.* (B), 1966, 946.
54. Q. N. Porter and J. Baldas, *Mass Spectrometry of Heterocyclic Compounds*, Wiley—Interscience, 1971.
55. G. Spiteller, Chapter 5 in *Physical Methods in Heterocyclic Chemistry*, Vol. III (ed. A. R. Katritzky), Academic Press, 1971.
56. F. W. McLafferty, *Mass Spectral Correlations*, American Chemical Society, Washington DC, 1963.
57. *Index of Mass Spectral Data*, Special Technical Publication No. 356, American Society for Testing and Materials, Philadelphia, 1963.
58. *MSDC Mass Spectral Data Sheets*, Mass Spectrometry Data Centre, AWRE, Aldermaston, Reading, 1970, with subsequent additions.
59. A. Cornu and R. Massot, *Compilation of Mass Spectral Data*, Heyden & Sons, Ltd., 1966, First Supplement 1967, Second Supplement 1971; more will be issued.
60. *Eight Peak Index of Mass Spectra*, Mass Spectrometry Data Centre, AWRE, Aldermaston, Reading, 1st edn. 1970, Vols. I and II.

Mass Spectrometry Seminar Problems and Answers

F. SCHEINMANN

University of Salford

and

W. A. WOLSTENHOLME

A.E.I., Scientific Apparatus Incorporated, White Plains, New York

A discussion of the answers is provided on page 82.

Problems

MASS SPECTRUM PROBLEM 1

Deduce the structure of the hydrocarbon $C_{12}H_{26}$.

There are metastable peaks at m^* 117, 94·8, 75·2, 57·6, 44·6, 39·7, 38·3, 32·8, 29·6, 28·8, and 25·6.

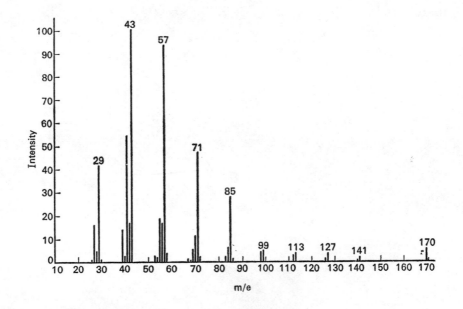

MASS SPECTRUM PROBLEM 2

Deduce the structure of the amine $C_4H_{11}N$.

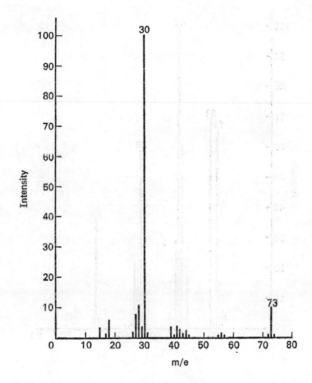

MASS SPECTRUM PROBLEM 3

Deduce the structure of the compound $C_5H_{12}O$. There is a metastable peak at m^* 43·3.

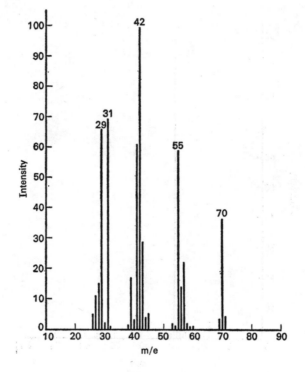

MASS SPECTRUM PROBLEM 4

Suggest a structure for the compound $C_8H_{18}O$.

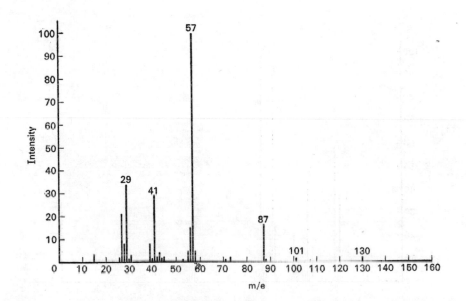

MASS SPECTRUM PROBLEM 5

Comment on the structure of the compound $C_8H_{18}S$. It is difficult to interpret the origin of the base peak and attempts should be made to analyse the other major fragment ions first.

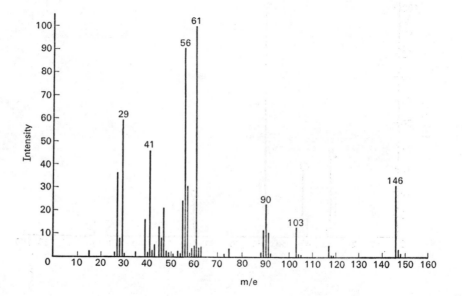

MASS SPECTRUM PROBLEM 6

Comment on the structure of the compound $C_6H_{12}O$. There is a metastable peak at m^* 72·2.

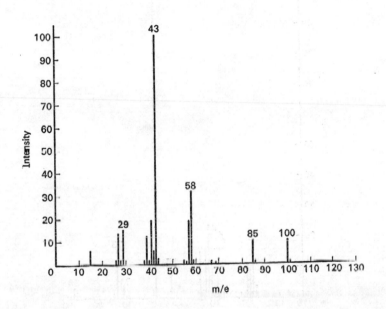

MASS SPECTRUM PROBLEM 7

Deduce the structure for the compound $C_4H_8O_2$. There is a metastable peak at m^* 41.

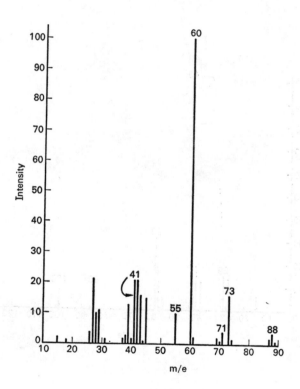

MASS SPECTRUM PROBLEM 8

Suggest the structure of the compound $C_8H_{16}O_2$. There is a metastable peak at m^* 53·7.

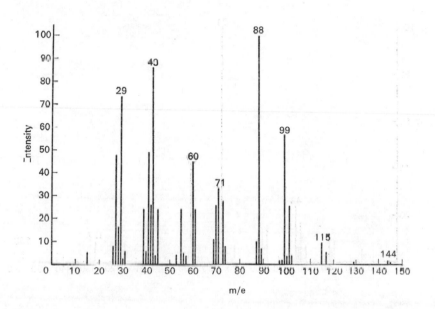

MASS SPECTRUM PROBLEM 9

Deduce the structure of the compound $C_9H_{18}O_2$. There is a metastable peak at m^* 34·7.

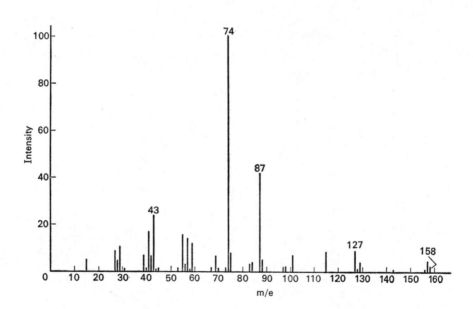

MASS SPECTRUM PROBLEM 10

Comment on the structure of the hydrocarbon C_9H_{18}.

There are metastable peaks at m^* 36·5 and 24·4.

Note: It is not possible to unambiguously assign the structure from the mass spectrum, but it is possible to deduce the nature of the double-bond equivalent indicated by the formula.

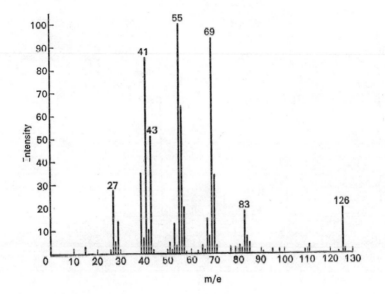

MASS SPECTRUM PROBLEM 11

Deduce the structure of the compound C_8H_{10}.

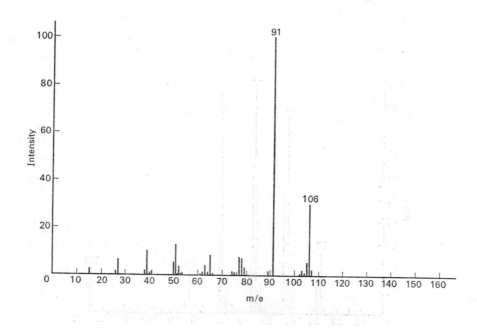

MASS SPECTRUM PROBLEM 12

Deduce the structure of the compound C_8H_7OBr. There is a metastable peak at m^* 56·5.

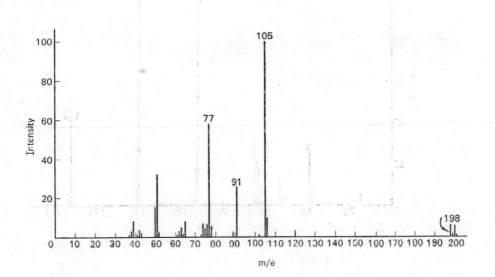

MASS SPECTRUM PROBLEM 13

The spectra 13a and 13b represent positional isomers of two methyl hydroxybenzoates. Analyse the spectra and suggest which is the *ortho* isomer.

Spectrum 13a has a metastable peak at m^* 94·7. Spectrum 13b has metastable peaks at m^* 96·3 and 71·5.

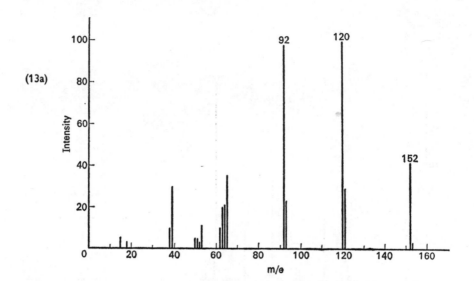

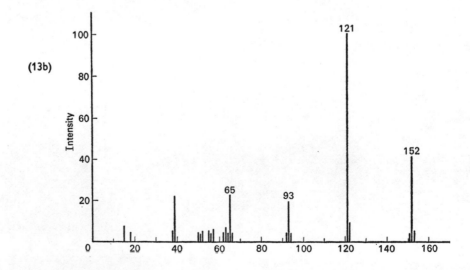

MASS SPECTRUM PROBLEM 14

Comment on the structure of the natural phenolic oxygen heterocyclic compound which has a molecular formula $C_{15}H_{12}O_6$. It is not possible to make a complete assignment with the given data.

There are metastable peaks at m^* 233 and 105.

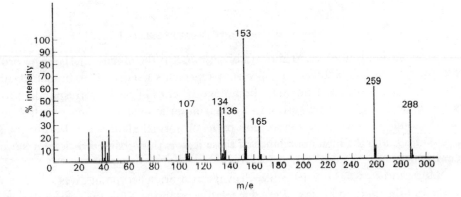

Answers

GENERAL COMMENTS

Although many of the fragment ions given in these answers are represented by formal structural formulae, it should be noted that this is done primarily as an aid to interpreting the fragmentation of the original molecule. It is necessary to remember that, since we are dealing with excited ions, these structures may not represent the true species involved.

For more exact information on the mode of fragmentation, deuterium labelling experiments are necessary, and the atomic compositions of the fragment ions should be verified by accurate mass measurements. No attempt has been made, or should be made, to interpret every peak in the mass spectrum.

The most intense peak in a mass spectrum is often referred to as the base peak, and this terminology has been used in the answers.

ANSWER TO MASS SPECTRUM PROBLEM 1

The molecular ion is at m/e 170 and can correspond to the alkane $C_{12}H_{26}$. The problem is therefore to establish whether the spectrum represents a normal or branched alkane. The main features of the spectrum are a homologous series of peaks corresponding to ions of the formulae $C_nH_{2n}+1$ which ascend to a maximum abundance when n is 3, 4, and 5, (m/e 43, 57, and 71). Several decomposition paths of normal alkanes can lead to the same lower mass ions, and the high abundance of these ions is thus characteristic of a saturated hydrocarbon.

Branching can be easily detected since cleavage at a branched point gives rise to a more stable ion and therefore to a peak of greater relative intensity. Thus the typical distribution

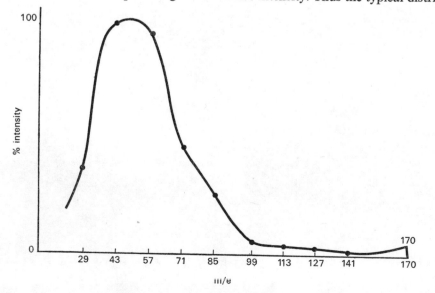

of peaks in n-alkanes (see Figure) would be disturbed and the smooth curve of decreasing intensities broken ·by preferred fragmentation at each branch.

The spectrum in this case shows no such breaks in the smooth curve and must therefore represent dodecane. The peaks at m/e 41 and 55 correspond to the loss of two hydrogen atoms from the propyl and butyl cations to yield the corresponding allyl carbonium ions. The metastable peaks (m^*) given by the relationship

$$m^* = \frac{(\text{daughter ion})^2}{\text{parent ion}}$$

support the following fragmentations:

$C_{12}H_{26}^+$ fragmentation to:
- $m^*\ 117$ → $C_{10}H_{21}^+$ m/e 141
- $m^*\ 94\cdot8$ → $C_9H_{19}^+$ m/e 127
- $m^*\ 75\cdot2$ → $C_8H_{17}^+$ m/e 113
- $m^*\ 57\cdot6$ → $C_7H_{15}^+$ m/e 99
- $m^*\ 29\cdot6$ → $C_5H_{11}^+$ m/e 71

$C_{10}H_{21}^+$ m/e 141 $\xrightarrow{\ m^*\ \ 35\cdot8\ }$ $C_5H_{11}^+$ m/e 71

$C_9H_{19}^+$ m/e 127 $\xrightarrow{\ m^*\ \ 39\cdot7\ }$ $C_5H_{11}^+$ m/e 71

$m^*\ 25\cdot6$ → $C_4H_9^+$ m/e 57

$C_8H_{17}^+$ m/e 113 $\xrightarrow{\ m^*\ \ 44\cdot6\ }$ $C_5H_{11}^+$ m/e 71

$m^*\ 28\cdot8$ → $C_4H_9^+$ m/e 57

$C_7H_{15}^+$ m/e 99 $\xrightarrow{\ m^*\ \ 32\cdot8\ }$ $C_4H_9^+$ m/e 57

$C_6H_{13}^+$ m/e 85 $\xrightarrow{\ m^*\ \ 38\cdot3\ }$ $C_4H_9^+$ m/e 57

ANSWER TO MASS SPECTRUM PROBLEM 2

The presence of an odd number of nitrogens in the amine leads to the molecular ion of odd mass number at m/e 73. This can correspond to the formula $C_4H_{11}N$ which represents a saturated amine. The problem is therefore to establish whether the spectrum represents a primary, secondary, or tertiary amine and to provide information about the alkyl portion. The most-favoured fission of amines is at the α-carbon atom and the base peak at m/e 30 is characteristic of primary amines.

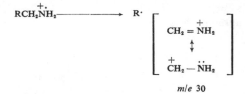

$$m/e\ 30$$

The peak at m/e 30 is a good example of cleavage where the charge is retained on the fragment with the heteroatom. Absence of other prominent fragmentations would suggest that the alkyl chain is not branched. Thus spectrum 2 represents n-butylamine since the branched isomers give significant $M-15$ peaks.

Note: That 2-methylpropylamine is an alternative is supported by the data recorded in *Index of Mass Spectral Data* published by The American Society For Testing and Materials, 1969.

ANSWER TO MASS SPECTRUM PROBLEM 3

The molecular ion from formula $C_5H_{12}O$ should appear at m/e 88 but the ion of highest significant mass is at m/e 70. The ready loss of 18 mass units suggests that water has been eliminated either in the inlet system or on electron impact and that compound $C_5H_{12}O$ is an alcohol. The base peak at m/e 42 is of even mass number and has therefore arisen from a rearrangement process.

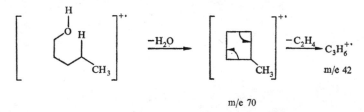

$$m/e\ 70$$

Cleavage at the C—C bond next to the oxygen atom gives rise to the ion at m/e 31 which, as expected, is more intense than the ion at m/e 57 which results from loss of 31 mass units from the molecule $C_5H_{12}O$. The peak at m/e 29 is attributed largely to fragment $(CHO)^+$.

The peaks at m/e 55 and m/e 41 are consistent with fragmentation after elimination of water. The metastable peak, m^* 43·2 ($55^2/70$), is consistent with the loss of a methyl group from the m/e 70 ion. The spectrum 3 represents n-pentyl alcohol, but 2- and 3-methyl-butanols could be alternative answers.

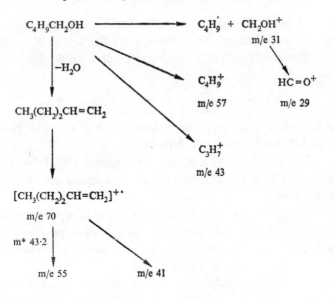

$$C_4H_9CH_2OH \longrightarrow C_4H_9^{\cdot} + CH_2OH^+$$
$$m/e\ 31$$

$$-H_2O$$

$$CH_3(CH_2)_2CH=CH_2$$

$$C_4H_9^+$$
$$m/e\ 57$$

$$HC=O^+$$
$$m/e\ 29$$

$$C_3H_7^+$$
$$m/e\ 43$$

$$[CH_3(CH_2)_2CH=CH_2]^{+\cdot}$$
$$m/e\ 70$$

$$m^*\ 43\cdot2$$

$$m/e\ 55 \qquad m/e\ 41$$

ANSWER TO MASS SPECTRUM PROBLEM 4

The molecular formula $C_8H_{18}O$ gives a molecular ion at m/e 130. Compounds with the general formula $C_nH_{2n}O$ are saturated and acyclic, and include alcohols and ethers. Since the molecular ion does not readily lose 18 mass units (corresponding to a loss of water) mass spectrum 3a is more likely to represent an ether.

The molecular ions of ether fragment in the following ways:

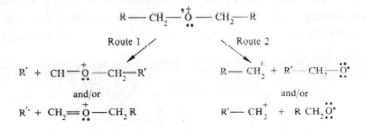

$$R-CH_2-\overset{+\cdot}{\underset{\cdot\cdot}{O}}-CH_2-R$$

Route 1   Route 2

$$R^{\cdot} + CH-\overset{+}{\underset{\cdot\cdot}{O}}-CH_2-R' \qquad R-\overset{+}{CH_2} + R'-CH_2-\overset{\cdot\cdot}{\underset{\cdot\cdot}{O}}{}^{\cdot}$$

and/or   and/or

$$R'^{\cdot} + CH_2=\overset{+}{\underset{\cdot\cdot}{O}}-CH_2R \qquad R'-\overset{+}{CH_2} + R\,CH_2\overset{\cdot\cdot}{\underset{\cdot\cdot}{O}}{}^{\cdot}$$

The first fragmentation represents C—C cleavage at the carbon next to oxygen (cf. fragmentation of alcohols in spectrum 3). The second fragmentation represents C—O cleavage with the charge remaining on the alkyl fragment. Thus long-chain ethers can become dominated by the hydrocarbon fragmentation pattern.

The ether $C_8H_{18}O$ appears to fragment by two different pathways, probably corresponding to (1) and (2). Loss of 43 mass units from the molecular ion gives a peak at m/e 87. This mass suggests an oxygenated ion. It is therefore unlikely that this ion loses a further 30 mass units to give the base peak at m/e 57 but, instead, this peak probably occurs by direct loss of 73 mass units from the molecular ion. Thus the peak at m/e 87 is derived by the loss of a propyl radical from the molecular ion according to route 1, whereas the loss of a butyl oxy radical by route 2 gives the peak at m/e 57. These alkyl units do not appear to be branched since there is a peak at m/e 101 corresponding to the loss of an ethyl radical from the molecular ion. Mass spectrum 4 is the spectrum of di-n-butyl ether.

ANSWER TO MASS SPECTRUM PROBLEM 5

The molecular formula $C_8H_{18}S$ gives the molecular ion at m/e 146. Note the associated isotope peak at m/e 148 due to the ^{34}S contribution.

This peak at $M+2$ (m/e 148) has an abundance of about 4·5% of the molecular ion peak (m/e 146) compared with the 0·4% relative abundance expected from carbon and hydrogen isotope contributions only.

For reasoning similar to that used in problem 4, $C_8H_{18}S$ is acyclic and saturated, and the first problem is to establish whether the compound is a mercaptan or a dialkyl sulphide. Since there is no peak at m/e 112 corresponding to the loss of hydrogen sulphide from the molecular ion, a mercaptan structure can be excluded.

The cleavage modes of dialkyl sulphides resemble those of dialkyl ethers as shown by a comparison of mass spectra 4 and 5. In this case, the molecular ion also loses 43 mass units which corresponds to the loss of a propyl radical.

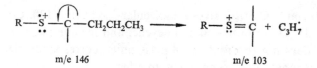

The peaks of even mass number at m/e 90 and 56 probably arise from rearrangement processes and from the intensities of peaks at two higher mass units it appears that in the fragment corresponding to m/e 90 sulphur is present whereas for m/e 56 it is certainly absent. A rearrangement (as illustrated in the scheme for n-dibutyl sulphide)

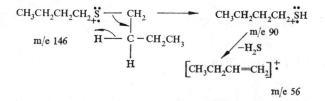

can account for the ion with m/e 90 by loss of butene. Subsequent elimination of hydrogen sulphide from the mercaptan then gives an olefin with m/e 56.

To account for the base peak at m/e 61 the following process has been postulated:

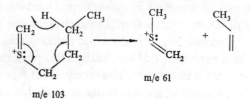

The peaks at m/e (due to $CH_2=CH-CH_2^+$) and at m/e 29 (due to $C_2H_5^+$) are a consequence of hydrocarbon fragmentations. Compound $C_8H_{18}S$ is, in fact, n-dibutyl sulphide.

ANSWER TO MASS SPECTRUM PROBLEM 6

The compound $C_6H_{12}O$ gives a molecular ion at m/e 100. The formula shows that it contains one double bond equivalent (see p. 31); therefore the compound is either monocyclic or acyclic with one double bond.

The base peak would be expected to represent a fragment containing oxygen, since with this atom, resonance stabilization of the positive charge is highly favoured. Thus the base peak at m/e 43 is largely associated with $CH_3-C=O^+$, while a propyl fragment with the same mass will only make a minor contribution. The peak at m/e 58 must arise from a rearrangement process since it is of even mass number. Such a rearrangement process can arise from an α-methylaldehyde or a methylketone.

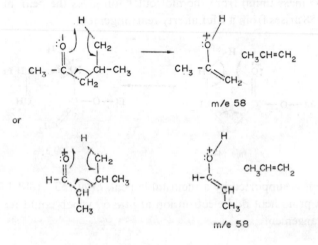

Since the base peak at m/e 43 is attributed to $CH_3C=O^+$, the product must be a methyl ketone. The loss of a methyl group from the molecular ion at m/e 100 is supported by the presence of a metastable peak at m^* 72·2 ($85^2/100$). In fact mass spectrum 6 represents methyl isobutylketone. The spectrum does not unambiguously distinguish between the isobutyl and n-butylketones.

ANSWER TO MASS SPECTRUM PROBLEM 7

The molecular ion for the compound $C_4H_8O_2$ gives a peak at m/e 88. The base peak at m/e 60 is attributed to the elements of acetic acid and must arise from the rearrangement

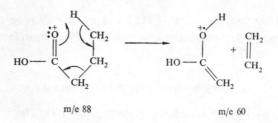

m/e 88　　　　　　　　　　　　　　　　m/e 60

For this rearrangement to occur all the four carbon atoms must be in a straight chain, and therefore the mass spectrum represents butyric acid. This McLafferty rearrangement in the mass spectrum of butyric acid is supported by the metastable peak at m^* 41 ($60^2/88$).

ANSWER TO MASS SPECTRUM PROBLEM 8

The molecular ion for the compound $C_8H_{16}O_2$ is at m/e 144. The peaks at m/e 99 and 88 can be attributed to the presence of an ethyl ester group in the molecule. Thus loss of an ethoxy radical (45 mass units) from the molecular ion gives the peak at m/e 99, while the base peak at m/e 88 arises from a McLafferty rearrangement.

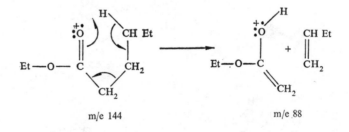

m/e 144　　　　　　　　　　　　　　m/e 88

The rearrangement is supported by a metastable peak at m^* 53·7 ($88^2/144$).

There is also a prominent odd-electron ion at m/e 60 which could result from a second McLafferty rearrangement.

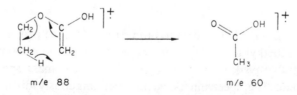

m/e 88　　　　　　　　　　　　　　m/e 60

From the ratio of the $M-15$ and $M-29$ peaks there is no suggestion of branching in the alkyl chain. Mass spectrum 8 represents ethyl n-hexanoate.

ANSWER TO MASS SPECTRUM PROBLEM 9

A small molecular ion is observed at m/e 158 with a more intense $(M-1)$ peak at m/e 157. The molecular formula is $C_9H_{18}O_2$ indicating a compound with one double bond or ring. The base peak at m/e 74 must be due to a rearrangement process, and since the intensity of this peak is so high relative to the other peaks it is highly likely that it contains at least one of the oxygens. The next more intense peak at m/e 87, i.e. $M-71$ is probably due to loss of C_5H_{11} from the molecular ion. Of special significance is the m/e 127 peak ($M-31$) which is characteristic of loss of OCH_3. Thus the indications so far are a saturated hydrocarbon unit of at least C_5H_{11}, an —OCH_3 group, and another oxygen and ring or double bond. One possibility for the other oxygen is, of course, a carbonyl group, and this could initiate a McLafferty type rearrangement to give m/e 74:

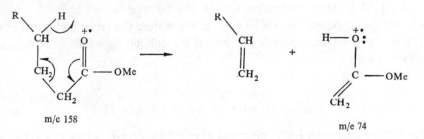

m/e 158   m/e 74

The metastable peak at $m*$ 34·7 ($74^2/158$) supports this process. Compound 9 is methyl octanoate.

ANSWER TO MASS SPECTRUM PROBLEM 10

The molecular ion is at m/e 126, and the formula C_9H_{18} suggests that the compound is either an olefin or a cycloalkane. In agreement with this, and in accordance with reference mass spectra, the molecular ion has a greater relative intensity than is found in n-alkanes.

A homologous series of peaks are prominent at m/e 27, 41, 55, 69, and 83. This series is thus similar to that found in the spectra of n-alkanes except that it is shifted two mass units down scale.

The metastable peak at $m*$ 36·5 corresponds to the ion at mass 83 fragmenting to give the ion at mass 55, and that at $m*$ 24·4 is derived from the ion at mass 69 which decomposes to give an ion at mass 41.

The fragmentation is thus consistent with that for an olefin. Cycloalkanes have a characteristic series of peaks of general formula C_nH_{2n-1} and C_nH_{2n-2}, and the latter series are not usually significant in acyclic olefins.

The olefin is probably branched with the possibility of a terminal isopropyl group as shown by peaks at m/e $M-15$ and $M-43$. In support of this assignment, the $M-29$ peak is smaller than the $M-15$ peak which contrasts to the fragmentation of n-hydrocarbons. Thus loss of an ethyl radical is unfavourable.

Due to migration of the double bond, either thermally during insertion of the sample or after ionization, it is notoriously difficult to locate the ethylenic linkage. The figure is the spectrum of 2,6-dimethylheptene-3.

$$CH_3—CH—CH=CH—CH_2—CH—CH_3$$
$$\quad\quad\quad|\quad\quad\quad\quad\quad\quad\quad\quad\quad\quad\;|$$
$$\quad\quad\quad CH_3\quad\quad\quad\quad\quad\quad\quad\quad CH_3$$

However, postulation of the olefin as 7-methyloctene-2 gives predicted strong peaks at m/e 55 and 83, and the ion at m/e 69 could quite conceivably lose 28 a.m.u.

Furthermore, a McLafferty rearrangement on the heptene would lead to peaks at m/e 42 and 84, and on the octene to m/e 70 and 56, suggesting the latter olefin rather than the former. Therefore from the spectrum it would be difficult to exclude 7-methyloctene-2 as a possible answer.

ANSWER TO MASS SPECTRUM PROBLEM 11

A rather intense molecular ion is observed at m/e 106 corresponding to C_8H_{10} indicating a fairly stable molecule. The base peak m/e 91 is aromatic and thus a stable ion. It is, in fact, the benzyl ion which deuteration studies have shown to rearrange to the tropylium ion, since all seven hydrogens are equivalent.

An intense m/e peak is characteristic of alkylbenzenes.

The very ready loss of methyl to give m/e 91 indicates an ethyl substituent on a benzene ring.

Compound 11 is ethyl benzene.

Note also the peaks at m/e $51\frac{1}{2}$ and $52\frac{1}{2}$ due to doubly charged ions. Such peaks are characteristic of aromatic or other highly stable compounds.

The isomeric xylenes will also give prominent peaks at m/e 106 and 91. However, the xylene spectra, compared with ethylbenzene, show peaks at m/e 105 with approximately five times greater relative intensity.

ANSWER TO MASS SPECTRUM PROBLEM 12

This spectrum shows a molecular ion at m/e 198 corresponding to C_8H_7OBr. Note the intensity ratio of the peaks in the molecular ion region, e.g. the peak m/e 200 is almost as intense as that at m/e 198. This is due to the two bromine isotopes ^{79}Br and ^{81}Br of nearly equal relative abundance. This isotope ratio at the molecular ion is diagnostic of the presence of one bromine atom in the molecular formula.

It is also noteworthy that none of the other intense peaks in the spectrum show the same pattern and, therefore, do not contain bromine. For example, the peak at m/e 105 must arise by loss of bromine plus 14 mass units from the molecular ion, i.e. loss of CH_2Br.

A very intense m/e 105 peak is frequently due to the benzoyl ion. The other possibility is $(C_6H_5CH_2CH_2)^+$, but this would almost certainly be accompanied by a more intense m/e 91 peak. The intense m/e 77 peak in this spectrum is due to loss of CO from m/e 105. This fragmentation is supported by the metastable peak at m^* 56·5 ($77^2/105$).

All the evidence, therefore, supports the structure $PhCOCH_2Br$. The m/e 91 peak is, however, too big to ignore and must be due to elimination of CO immediately before or after loss of bromine.

Compound 12 is phenacyl bromide.

ANSWER TO MASS SPECTRA PROBLEMS 13A AND 13B

The molecular formula for the methyl hydroxybenzoates is $C_8H_8O_3$ and the molecular ions in both cases appear at m/e 152. The base peak in spectrum 13a is at m/e 120 and the even mass number suggests that the fragment ion arises from rearrangement. In spectrum 13b the base peak is at m/e 121 which is due to loss of methoxyl radical (31 mass units) from the molecular ion.

Thus for 13a:

Whereas for 13b:

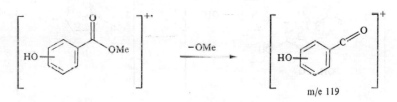

Both of these fragmentations are supported by metastable peaks: thus spectrum 13a has a metastable peak at m^* 94·7 ($120^2/152$), while for 13b m^* is at 96·3 ($121^2/152$).

Both fragment ions subsequently lose carbon monoxide (28 mass units) to give ions at m/e 92 and 93 respectively. The fragmentation from mass 121 → 93 is supported by a metastable peak at m^* 71·5. Alternatively, both molecular ions can lose the methoxycarbonyl radical in one step to give the fragment ion at m/e 93.

From the rearrangement process it is clear that spectrum 13a represents methyl 2-hydroxybenzoate. Spectrum 13b can therefore represent either methyl 3- or 4-hydroxybenzoate but the mass spectrum alone will not differentiate between these isomers. Spectrum 13b has been obtained from methyl 4-hydroxybenzoate.

ANSWER TO MASS SPECTRUM PROBLEM 14

The mass spectrum is that of aromadendrin: under mass spectral conditions the chalcone tautomer gives an identical result.

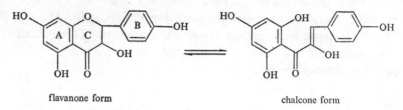

flavanone form   chalcone form

The molecular ion is at m/e 288 and in accordance with the fragmentation of phenols it loses 29 mass units (CHO) to give a fragment ion at m/e 259 (metastable peak at m/e 233). This fragment ion then loses the elements of phenol to give a peak at m/e 165 (metastable peak at m/e 105). These two fragmentations thus suggest the presence of two phenolic moieties in the molecule. The base peak at m/e 153 may arise from ketone fragmentation of the chalcone form

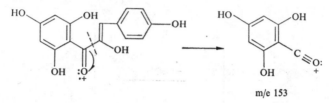

m/e 153

The peak of even mass no. at m/e 136 must arise from a rearrangement process, and this is consistent with a retro-Diels–Alder reaction of the flavanone form.

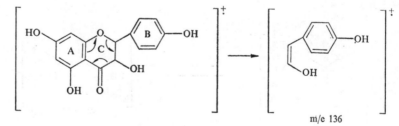

m/e 136

The peak at m/e 134 would correspond to the same fragmentation in the corresponding flavone. It is difficult to rationalize the origin of this peak from the flavanone, but similar fragmentations have been observed in other flavanones (H. Audier, *Bull. Soc. Chim. France*, 1966, 2892). The peak at m/e 107 is due to the *p*-hydroxybenzyl cation (or its tropylium equivalent) and has previously been observed in the mass spectra of related flavanones. The mass spectral data requires that two hydroxyl groups must be located in ring A, one ring B, and one in the heterocyclic ring, but the exact orientation of these groups is not given. However, biogenetic and statistical considerations of hydroxylation patterns of flavanones would favour a phloroglucinol nucleus for ring A with the rest of the molecule being derived from *p* hydroxycinnamic acid.

An Introduction to Ultraviolet Spectroscopy with Problems.

D. H. MAASS

Department of Chemistry, University of Salford

With the advent of newer and more structurally specific spectroscopic methods, less attention is focused on ultraviolet spectroscopy than of old, although the underlying theoretical principles continue to present a powerful challenge to the theoreticians.

For the organic chemist ultraviolet spectroscopy is mainly concerned with electronic transitions in conjugated systems; the positions and intensities of the absorption band maxima depend to a large extent on the particular system under consideration. The electronic transitions are very sensitive to structural changes and reflect the strains imposed on the system by steric and electronic interactions. Unlike infrared spectroscopy functional group absorptions cannot be assigned to fixed and specific regions of the ultraviolet visible wavelength scale, and considerable experience is required in the interpretation of the spectra.

While it is true that no comprehensive correlation tables of the type used in infrared spectroscopy can be constructed, there exists a large volume of data covering most structural types which absorb in the ultraviolet and visible region, that is between the wavelength of 200 and 800 nm, and within any family of compounds it is usually possible to correlate changes in the spectra with changes in structure with a fair degree of success. With some classes of compounds empirical rules have been worked out which allow one to calculate, with some degree of accuracy, the absorption maxima of particular structures, and these can be used to distinguish between and/or confirm structural features.

In addition to such empirical calculations, many classes of compounds have spectral features which are susceptible to analysis by fairly simple molecular orbital theory. The degree of accuracy with which these calculations describe the systems depends on the sophistication of the chosen method.

One of the great strengths of ultraviolet spectroscopy is the relative simplicity and inexpensiveness of the instrumentation coupled with great sensitivity; solutions of 10^{-3}–10^{-5} molar are commonly measured by this technique. The ease with which quantitative measurements can be made is also a great advantage. This combination of properties makes ultraviolet spectroscopy a favourite method whenever kinetic problems or quantitative analyses are being considered.

This chapter will deal in some detail with both instrumentation and the behaviour of absorbing systems in the 200–800 nm region in an elementary manner and makes no claim to be exhaustive in its coverage. Leading references will, however, be given to other reference books and texts which give wider and deeper examinations of the problems which can only be briefly introduced here.

The Absorption Process

The amount of incident radiation absorbed by a translucent medium can be related to the amount of the medium by two empirical relationships:

(a) *Beer's law:* the fraction of incident monochromatic light absorbed by a homogeneous medium is proportional to the quantity of the absorbing medium.

(b) *Lambert's law:* the proportion of monochromatic light absorbed by a homogeneous medium is independent of the intensity of the incident light and each successive layer absorbs an equal fraction of the light incident upon it.

When dealing with solutions it is apparent that the amount of absorption will depend on the concentration of the solution and the thickness of the layer in the light path. By combining the Beer and Lambert laws the fraction of the incident light absorbed is related to the concentration and the path length through the solution as follows:

$$\log_{10} I_0/I = \varepsilon c l = A, \tag{1}$$

where

I_0 is the incident monochromatic light intensity,
I the emergent monochromatic light intensity,
ε the molar absorptivity (molar extinction coefficient),
c the concentration in moles/litre of the absorbing solute in a non-absorbing solvent,
l the thickness of the solution in centimetres, and
A the absorbance (or optical density).

Rearranging the equation and putting in the dimensions moles and length (L) leads to:

$$\varepsilon = \frac{A}{c\,l} = \frac{1}{\text{moles } L^{-3} L} = \frac{L^2}{\text{moles}} = \frac{\text{cm}^2}{\text{mole}} = \text{cm}^2 \text{ mole}^{-1}.$$

Thus ε has the dimensions of area/mole and could be regarded as a photon capture cross-section for a mole of absorbing material at any particular wavelength.

Spectrophotometers may be constructed to read directly either absorbance (optical density) or transmittance (T). The following relations may be helpful:

$$A = \varepsilon c l = \log_{10} I_0/I = \log_{10} 1/T,$$

$$T = I/I_0 \qquad \%T = 100 I/I_0,$$

$$A = \infty \rightarrow 0 \quad \text{as} \quad T = 0 \rightarrow 1 \quad \text{or} \quad \%T = 0 \rightarrow 100.$$

Similarly,
$$A = \log_{10} [100/\% T].$$

When working in terms of absorption instead of transmittance, we have the following relations:

$$\% \text{ Absorption} = 100(1-T) = 100(1-I/I_0).$$

Hence

$$T = I/I_0 = 1 - \frac{\%\text{Absorption}}{100}$$

and

$$A = -\log_{10}\left(1 - \frac{\%\text{ Absorption}}{100}\right).$$

For the majority of materials, the combined Beer–Lambert relationship (Eqn. 1) holds over a wide concentration range. However, deviations do occur in plots of A against $(c\ l)$, which indicate that some change is occurring in the absorbing species such as association, dissociation, or a photochemical reaction.

The preceding discussion indicates that ε is a quantity which is characteristic of a particular molecule at a particular wavelength, and a plot of ε against wavelength will produce a characteristic curve. However, since A is proportional to ε, electronic methods can be used to plot A (or T) against wave number $\bar{\nu}$ (or wavelength λ) automatically as the spectrum is scanned, but it is more usual to quote only the positions of the absorption maxima together with their molar absorptivities for the purposes of diagnosis. The complete absorption curve can be of great value in the interpretation of spectra since the shapes of the band envelopes are often characteristic for the absorbing molecules (Fig. 1).

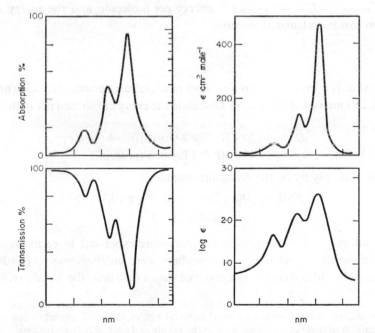

FIG. 1. Various methods for presenting the ultraviolet absorption curve of a substance mol. wt. $= 200$; $c = 8\cdot0$ g litre^{-1}; $l = 0\cdot05$ cm. (Reproduced by permission from *Physical Methods in Organic Chemistry*, Ed. I. C. P. Schwartz, Oliver and Boyd, 1964.)

The wavelength at which absorption occurs is a measure of the energy dE associated with the electronic transition involved, thus

$dE = h\nu = hc\,\bar{\nu} = hc/\lambda.$
$dE =$ energy (Joules J)/molecule.
$\quad c =$ speed of light $2\cdot998\times10^8$ ms^{-1}.
$\quad h =$ Planck's constant $6\cdot624\times10^{-34}$ J s.
$\quad \nu =$ frequency s^{-1}.
$\quad s =$ seconds.
$\quad \bar{\nu} = 1/m^{-1}.$

e.g. calculate the energy of a photon of wavelength 200 nm.

$$dE = \frac{6\cdot624\times10^{-34}\times2\cdot998\times10^8}{2\times10^{-7}}\ \frac{\text{Js. ms}^{-1}}{\text{m}} = 9\cdot933\times10^{-1}\ \text{J}.$$

It is clear from the above equation that energy is directly proportional to the frequency ν or the wave number $\bar{\nu}$. For theoretical work, it would be more convenient to have the read out of spectrophotometers in wave numbers instead of wavelengths.

Now in dealing with electronic transitions we are interested in the differences in energy between the ground state and the excited state.†Assuming that all the energy of the exciting radiation at a particular wavelength is used to raise an electron from the ground to the excited state, the energy equation may be rewritten.

$E_{\text{excited}}-E_{\text{ground}} = dE = hc\bar{\nu}$ absorbed $=$ energy per molecule, and the energy absorbed in the transition can be calculated directly.

Example

Biphenyl exhibits a broad absorption band in cyclohexane solutions with an absorption maximum at 250 nm ($4\cdot0\times10^6\,\mu\,\text{m}^{-1}$). Calculate the energy absorbed per mole for this transition.

$$dE = 6\cdot624\times10^{-34}\times2\cdot998\times10^3\times4\cdot0\times10^6$$
$$= 79\cdot47\times10^{-20}\ \text{J for one molecule.}$$

For one mole, multiply by N, the Avogadro number,

$$N\,dE = \Delta E = 79\cdot47\times10^{-20}\times6\cdot023\times10^{23}$$
$$= 478\cdot6\ \text{kJ.}$$

Before the adoption of SI units, wave numbers were expressed in reciprocal centimetres (cm^{-1}), and in the ultraviolet region the numbers were cumbersome to handle, thus $\nu = 250\,\text{nm} \equiv \bar{\nu} \equiv 1/\lambda = 40,000\,\text{cm}^{-1}$. As a result of this, a new unit, the Kayser (K), was defined

† An earlier description of absorbing systems made a distinction between chromophores—systems which absorb radiation in the visible and near ultraviolet regions—and auxochromes, which, though having no prominent absorptions of their own in the visible and near ultraviolet regions, when conjugated with a chromophore bring about a shift in the absorption maximum, usually to a longer wavelength. Typical auxochromes are heteroatoms directly attached to the chromophore, e.g. OCH_3, Cl; or multiply-bonded systems with no characteristic absorption in the 200–800 nm region, eg.—C≡N, ⟩C=O.

so that $1\,K = 1\,cm^{-1}$, thus $\lambda = 250\,nm \equiv \bar{\nu} = 40\,kK$. With SI units, this unit becomes unnecessary since if we consider the visible and ultraviolet range to extend from 1000 nm to 200 nm, then the wave-number range is, conveniently, 1–5 μm^{-1}; thus in the example given above, 250 nm = 0·250 μm = 4·0 μm^{-1}. A nomogram for converting μm^{-1} to nm is given below.

μm^{-1}	1·0	1·25	1·5	1·75	2·0
nm	1000	800	666·7	571·4	500

2·0	2·25	2·5	2·75	3·0	3·5	4·0	4·5	5·0
500	444·5	400	370·4	333·3	285·7	250	222·2	200

Note that $E = h\bar{\nu}c = 1·986 \times 10^{-19} \times \bar{\nu}\,J$, where $\bar{\nu}$ is in μm^{-1}.

The energy should be quoted in any appropriate SI units, and the following equivalences are useful in interconverting the most frequently used units:

$$1\,eV = 23·063\,kcal\,mole^{-1} = 96·53\,kJ\,mol^{-1}.$$

Classification of Electronic Absorption Transitions

Ultraviolet absorption spectra result from the interaction of light quanta with electrons. These interactions under favourable conditions raise the potential energy of the electron, and the system containing such an electron is referred to as an excited state. The energy thus acquired is dissipated largely by collisions with other molecules.

Since the electrons in a molecule are not identical in energy it follows that the energy absorbed in the excitation process may bring about one or more transitions according to the type of electrons involved. These transitions are classified and discussed below for singlet to singlet transitions.

1. N → R TRANSITIONS ($\pi \rightarrow \sigma^{*} \rightarrow$ IONIZATION)

These have been little studied by the organic chemist since they occur in the far ultra-violet, i.e. < 180 nm, and their study requires special instrumentation. This region is known as the Schuman or vacuum ultraviolet region. The absorption closely resembles that of atoms in that an electron is promoted to higher energy levels until finally ionization occurs thus giving rise to a series of absorption bands and finally a continuum. This may be represented by

$$Z = Z \rightarrow (Z \dot{-} Z)^{+}$$

Transitions of this nature can be described by a Rydberg equation.

2. N → V TRANSITIONS ($\pi \rightarrow \pi^{*}$)

This class makes up a large proportion of the most useful and characteristic absorptions of organic compounds and is generally associated with transitions to a polar excited state.

e.g. $\qquad Z = Z \rightarrow \overset{\pm}{Z}\ \overset{\mp}{Z},$ i.e. $\overset{+}{Z}-\overset{-}{Z}$ or $\overset{-}{Z}-\overset{+}{Z}$

In simple systems this transition occurs in the far ultraviolet, e.g. ethylene λ_{max}, ca. 160 nm, however, substitution by alkyl groups brings about a bathochromic (red) shift (Table 1) so that they are just accessible to a good conventional spectrometer with quartz optics, quartz cells with a short path length, very pure solvents, and a nitrogen flush to remove oxygen which would interfere with the observations in this region.

TABLE 1

	λ_{max} (nm)	ε_{max} (cm^2 mole^{-1})
$(CH_3)_2C{=}C(CH_3)_2$	ca. 185	ca. 10^3
$(CH_3)_2C{=}O$	188	900

Absorption of this type is due to the π-electrons absorbing energy and moving from a bonding molecular orbital to an anti-bonding molecular orbital as shown in Diagram I.

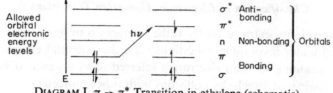

DIAGRAM I. $\pi \rightarrow \pi^*$ Transition in ethylene (schematic).

These $\pi \rightarrow \pi^*$ transitions show a red shift with substitution by either electron donating or withdrawing groups and with an increase in the dielectric constant of the solvent. In both cases this may be ascribed to stabilization of the polar excited state.

3. $n \rightarrow \pi^*$ TRANSITIONS

Transitions of this type involve the transition of the non-bonding electrons of heteroatoms to vacant anti-bonding π^*-molecular orbitals. These absorptions occur at long wavelengths and are of low intensity, e.g.

$$(CH_3)C_2 = O \quad \pi \rightarrow \pi,^* \quad \lambda\ 188\ nm, \quad \varepsilon\ 900\ cm^2\ mole^{-1},$$
$$n \rightarrow \pi,^* \quad \lambda\ 279\ nm, \quad \varepsilon\ 15\ cm^2\ mole^{-1}.$$

The $n \rightarrow \pi^*$ transition shows a hypsochromic (blue) shift in more polar solvents and with electron donating substituents.

4. $n \rightarrow \sigma^*$ TRANSITIONS

In addition to the foregoing systems in which transitions take place to anti-bonding molecular orbitals, saturated compounds containing heteroatoms show an absorption band which may be due to a transition of an electron from non-bonding orbitals of the heteroatom to anti-bonding σ^*-orbitals. Saturated compounds containing oxygen absorb below the 200 nm region; with increasing polarizability of the heteroatom the absorption maximum moves to longer wavelenths as shown in Table 2.

TABLE 2

Hetero-substituent	λ (nm)
OH	184
NH_2	215
SH	193
—S—	210

The absorption maximum in compounds containing more than one heteroatom is very sensitive to steric effects, and where overlap of the non-bonding orbitals can occur a marked red shift is seen. The iodo and sulphur compounds shown in Table 3 below illustrate this point.

TABLE 3

$>$C—I	λ (nm)	$>$C—S—S—C$<$	λ (nm)
CH_3I	258	acyclic	202, 252
CH_2I_2	292	six-membered ring	295
CHI_3	349	five-membered ring	334

The Effect of Conjugation

Up to now we have only considered simple absorbing systems in isolation or well insulated from one another by a saturated system. When this condition is not fulfilled, then the absorption which occurs may be considerably different from the parent system both in energy and intensity.

The conjugation of simple systems can take place in a number of ways, but only the three most commonly found ones will be dealt with here.

1. $\pi \rightarrow \pi^*$ CONJUGATION

This type of conjugation was recognized at an early stage in modern chemistry first from chemical and then from physical evidence, and is associated with systems containing alternating double and single bonds. More generally this type of transition occurs where π-orbital overlap can take place. When this occurs, the energy separation between the ground and excited states is reduced and the system absorbs at longer wavelengths and with a greatly increased intensity. This intense band is frequently referred to as the K-band (German *Konjugierte*). As a consequence of the lessening of the energy gap, the $n \rightarrow \pi^*$ transition due to the presence of a heteroatom also undergoes a red shift, but with little change in intensity and constitutes the *R*-band (German *Radikal*). Table 4 illustrates absorptions due to *K* and *R* transitions.

TABLE 4

Conjugated system	K-band		R-band	
	λ (nm)	ε (cm² mol⁻¹)	λ (nm)	ε (cm² mole⁻¹)
C=C—C=C	217	21,000		
C=C—C=O	217	16,000	321	20
C=C—C=N	220	23,000		

2. π–p CONJUGATION

The conjugation of a lone pair of electrons from a heteroatom with an unsaturated i.e. π-electronic system, constitutes π–p conjugation. This type of conjugation brings about a marked red shift of the absorption maximum, usually with only small changes in intensity as shown in Table 5.

TABLE 5

	λ_{max} (nm)	ε (cm² mole⁻¹)
C=C	*ca.* 160	*ca.* 10³
CH_3O—C=C	190	10³
$(CH_3)_2C$=C	230	10³
Benzene (*K*-band)	204	7400
Chlorobenzene (*K*-band)	210	7400
Bromobenzene (*K*-band)	210	7400

3. π–σ CONJUGATION (HYPERCONJUGATION)

Alkyl substituents on π-electron systems cause rather small red shifts of the order of 3–6 nm. This has been ascribed to hyperconjugation between the alkyl group and the π system involving the following structures:

$$CH_3—CH = CH_2 \longleftrightarrow CH_3—\overset{+}{C}H—\overset{-}{C}H_2 \longleftrightarrow H^+ \, CH_2 = CH—\overset{-}{C}H_2$$

An alternative view is that the small electron releasing tendency of the alkyl groups increases the polarizability of the π-electrons, thus reducing the energy of the π-π^* transition as illustrated in Table 6.

TABLE 6

Compound	λ_{max} (nm)	$\varepsilon \times 10^{-3}$ cm² mole⁻¹
$CH_2{=}CH{-}CH{=}CH_2$	217	21
$CH_3{-}CH{=}CH{-}CH{=}CH_2$	223	24
$RCH{=}CH{-}CH{=}CHR$	227	23
$CH_2{=}CR{-}CH{=}CH_2$	220	22
$CH_2{=}CR{-}CR{=}CH_2$	226	21
	237	7·7
	247	18

Empirical Rules for Computing λ_{max} for Non-aromatic Compounds Containing π-π^*, π-p, and π-σ Conjugated Electronic Systems

Many classes of compounds show regularities in their absorption spectra which are sufficiently well defined as to enable reasonably accurate calculations to be made of the position of absorption maxima of related compounds. Such calculations may be useful in structure determination by distinguishing between two or more isomers. The Woodward Fieser rules are a good example of this type of empirical calculation. Three types of system are dealt with:

(1) Acyclic dienes or dienes contained in non-fused ring systems (Table 7).

(2) Dienes or polyenes in fused ring systems (Table 8).

(3) Conjugated enones and dienones (Table 9).

TABLE 7
GROUP CONTRIBUTIONS FOR ACYCLIC AND NON-FUSED RING, CONJUGATED DIENES (Both double bonds not in the same ring)

	nm
Parent diene (not homoannular)	217
Alkyl group or ring residue (R)	5
Exocyclic double bond	5
Chlorine or bromine	17
Calculated λ_{max} EtOH	

Example 1

Calculate λ_{max} (EtOH) for exomethylene-4-isopropylcyclohex-2-ene.

	nm
Parent diene	217
$2 \times R-$	10
exocyclic double bond	5
Calculated	232
Observed	232

Problem 1

Calculate λ_{max}(EtOH) for the following compounds:

(a)   (b)

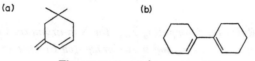

The answers are given on p. 141

TABLE 8

GROUP CONTRIBUTIONS FOR CONJUGATED
POLYENES CONTAINED WITHIN MONOCYCLIC
AND FUSED POLYCYCLIC RING SYSTEMS

	nm
Parent heteroannular diene	214
Parent homoannular diene	253
Double bond extending conjugation	30
Exocyclic double bond	5
Alkyl group (R) or ring residue	5
Polar groups O—COMe	0
OR	6
SR	30
Cl, Br	5
NR$_2$	60
Calculated λ_{max} (EtOH)	

N.B.—A homoannular diene is preferred as a parent to a heteroannular diene.
Solvent correction—see Table 9.
No correction is required for non-polar solutes.

Example 2

Calculate λ_{max} (EtOH) for 1,4-dimethylcyclohexane-1,3-diene.

	nm
Parent diene (homoannular)	253
$4 \times R-$	20
Calculated	273
Observed	265

Problem 2

Calculate λ_{max} (EtOH) for the following compounds:

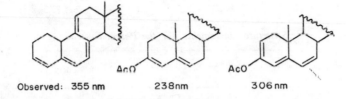

Observed: 355 nm   238nm   306 nm

Problem 3

Three isomeric trienes have the following absorption maxima in alcohol: (1) 284, (2) 315, (3) 323 nm. On the basis of these maxima, which of the following structures is plausible for each of them?

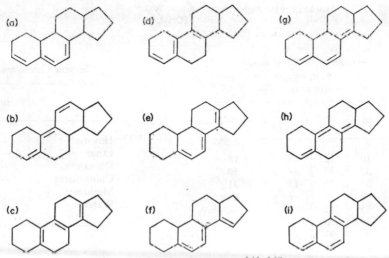

(a) (d) (g)

(b) (e) (h)

(c) (f) (i)

The answers are given on p. 141-143

TABLE 9
GROUP CONTRIBUTIONS FOR α, β-UNSATURA-
TED CARBONYL COMPOUNDS
(Acyclic, monocyclic, and polycyclic)

$$-\overset{\delta}{C}=\overset{}{\underset{\gamma}{C}}-\overset{\beta}{C}=\overset{}{\underset{\alpha}{C}}-\overset{X}{C}=O$$

Parent system	X=R alkyl or ring residue	H	HO or RO
	215	207	193
	215		
	202		

Structural Increments for the Parent System

	nm
Any exocyclic double bond $(CH_2)_n$—C=C	+5
Double bond endocyclic in 5- or 7-membered ring except cyclopent-2-enone.	+5
Double bond increasing conjugation	+30
	+39
Homoannular diene $(CH_2)_n$	

Increments for Substituents
α, β, etc.
(in nm)

	α	β	γ	$\delta(+1, +2, ...)$
R	10	12	18	18
OH	35	30	—	50
OR	35	30	17	31
OAc	6	6	6	6
Cl	15	12	—	—
Br	25	30	—	—
NR$_2$	—	93	—	—
SR	—	85	—	—

Solvent Corrections

	nm
Hexane	+11
Ether	+ 7
Dioxan	+ 5
Chloroform	+ 1
Methanol	0
Water	− 8

Example 3

Calculate λ_{max}(EtOH) for the structure shown.

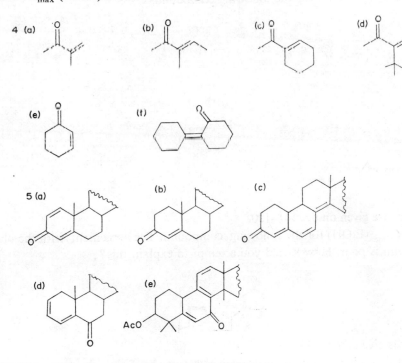

	nm
Parent cyclohexenone	215
2 Conjugated double bonds	60
1 Exocyclic double bond	5
Homoannular diene	39
β-alkyl	12
$(\delta+1)$ alkyl	18
2 $(\delta+2)$ alkyl	36
Calculated λ_{max}	385 (EtOH)
Observed	388

Problems 4–10

Calculate λ_{max} (EtOH) for the following compounds:

The answers are given on pp. 14 3147.

6. A natural product is known to have either structure A or B. The ultraviolet spectrum in alcohol has λ_{max} 252 nm. Which structure is the most likely one?

(A)   (B)

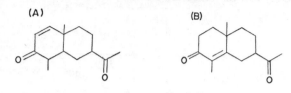

The answer is given on p. 147.

7. Calculate λ_{max}(EtOH) for compounds A and B

(A)   (B)

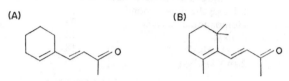

One of the spectra has ε_{max} 20,800 cm² mole⁻¹ and the other ε_{max} 10,700 cm² mole⁻¹. How may this difference be explained.

Comment on these extinction coefficients after reading section on steric inhibition of resonance (p. 121).

The answers are given on p. 148.

8. Calculate λ_{max} (EtOH) for the following compounds:

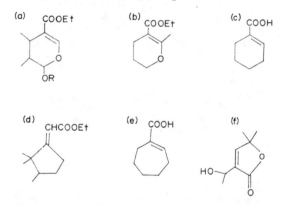

The answers are given on pp. 148–150.

9. Calculate λ_{max}(EtOH) for the following compound. The agreement with the observed value of 288 nm is poor. How would you attempt to explain this?

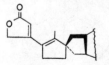

The answer is given on p. 150.

10. A natural product has been isolated which may have the structure shown below. λ_{max} (EtOH) = 257 nm (ε 10,700 cm² mole⁻¹).

On the addition of base to the solution, λ_{max} changes to 288 nm (ε 18,000 cm^2 mole^{-1}). Are these observations consistent with the structure shown? Explain the change in the spectrum when the pH is altered. The answer is given on p. 150.

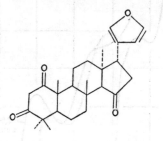

Aromatic Compounds

Three characteristic $\pi \to \pi^*$ electronic bands appear in the spectra of simple benzenoid molecules (Figs. 2A and 2B). In the vapour phase, the vibrational fine structure is marked, especially in the hydrocarbons, but tends to vanish in solution. The nomenclature of these bands is varied but the band maxima for benzene and the most common designations of them are shown below in Table 10.

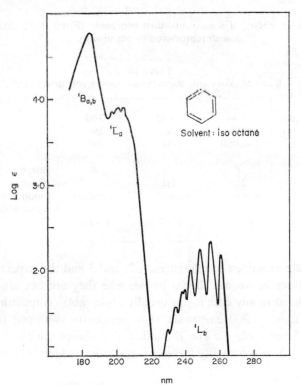

Fig. 2A. The ultraviolet absorption spectrum of benzene. (From Stevenson;[4a] reproduced by permission.)

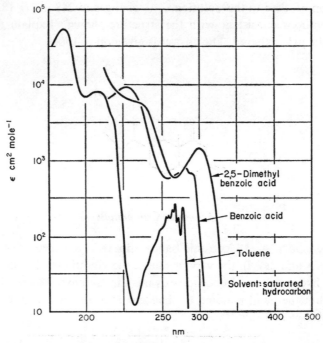

FIG. 2B. The ultraviolet spectra of some substituted benzenes. (From *U.V. Atlas of Organic Compounds* reproduced by permission.)

TABLE 10
BAND MAXIMA AND BAND ASSIGNMENTS FOR BENZENE

$\bar{v}$ μm^{-1}	5·44	4·95	3·89	
λ_{max} nm	184	202	254	
ε_{max}	46,000	7400	204	
	E	K	B	Burawoy[1]
	2nd	1st		
	primary	primary	secondary	Doub and Vandenbelt[2]
	$A \to B_b$	$A \to L_a$	$A \to L_b$	Platt[3]

These notations are explained in references 1, 2, and 3 and the papers cited therein.

Although these three bands are shown by benzene they are not always obvious in its derivatives. The E-band in any case is not usually observable on most instruments (< 200 nm) while the B-band ($\varepsilon < 300$ cm^2 mole^{-1}) is frequently swamped by the very strong K-band ($\varepsilon > 5 \times 10^3$ cm^2 mole^{-1}). The B-band, which shows the most fine structure and weak absorption (ε_{max}), is a formally forbidden transition which is allowed by vibrational distortion of the ring.

Clearly, in order to carry out any correlative work it is necessary to know which transitions one is dealing with in any particular case. This problem was investigated in detail by

Doub and Vandenbelt[2] who determined the spectra of a large number of substituted benzenes. From the data they obtained they were able to assign the various bands in these compounds and, in addition, formulated rules which can be used to aid in the assignment of absorption bands in other benzenoid molecules. These rules and examples of their application are shown below.

(1) The ratio between the $\bar{\nu}_{max}$ of the K- and B-bands $\bar{\nu}_K/\bar{\nu}_B$ is approximately 1·23, as shown in Table 11.

TABLE 11
$\bar{\nu}_K/\bar{\nu}_B$ FOR BENZENE AND SOME SUBSTITUTED BEN-ZENES

	$\bar{\nu}_K/\bar{\nu}_B$
Benzene	1·26
Chlorobenzene	1·26
Bromobenzene	1·24
sym. Trichlorobenzene	1·23
Hexachlorobenzene	1·25
Phenol	1·28
Styrene	1 14

(2) The ratio between the $\bar{\nu}_{max}$ of the E- and K-bands ($\bar{\nu}_E/\bar{\nu}_K$) is always less than 2 but increases with decreasing $\bar{\nu}_K$ as shown in Tables 12 and 13.

TABLE 12
$\bar{\nu}_E$, $\bar{\nu}_K$ AND ν_E/ν_K FOR SOME SUBSTITUTED BEN-ZENES IN ORDER OF DECREASING $\bar{\nu}_K$ (INCREAS-ING λ_K)

Substituent	$\bar{\nu}_E \, \mu m^{-1}$	$\bar{\nu}_K \, \mu m^{-1}$	$\bar{\nu}_E/\bar{\nu}_K$
C≡C	5·86	5·54	1·06
NH_2	6·10	5·46	1·12
COOH	5·78	5·41	1·07
CH=CH_2	5·96	5·32	1·12
C≡N	5·66	5·21	1·08
OH, CH_3 $(p-)$	5·65	5·12	1·10
$(m-)$	5·70	5·05	1·127
$(o-)$	5·66	5·00	1·130
OH	5·61	4·97	1·129
F, CH_3 $(p-)$	5·52	4·92	1·12
$(o-)$	5·55	4·86	1·14
I	5·74	5·29	1·09
1,3-di-Cl	5·78	5·12	1·13
Cl	5·61	5·01	1·12
Br	5·79	4·96	1·17

(3) The ratio between the absorptivities of the *E*- and *K*-bands decreases with decreasing $\bar{\nu}_K$ (Table 13).

The data used in Tables 11, 12, and 13 were obtained from solution spectra measured in aqueous or aqueous/alcoholic media.

TABLE 13

$\bar{\nu}_E/\bar{\nu}_K$ AND $\varepsilon_E/\varepsilon_K$ FOR SOME *p*-DISUBSTITUTED BENZENES

Substituents	$\bar{\nu}_E\ \mu m^{-1}$	$\bar{\nu}_K\ \mu m^{-1}$	$\bar{\nu}_E/\bar{\nu}_K$	$\varepsilon_E/\varepsilon_K$
HO, COOH	4·843	3·922	1·23	0·965
NH_2, CN	4·717	3·704	1·27	0·698
NH_2, $COCH_3$	4·329	3·220	1·35	0·604
NH_2, NO_2	4·425	2·625	1·69	0·497
NO_2, $O^{\ominus}$	4·425	2·491	1·78	0·338

With the improvements in commercially available instruments and use of the appropriate techniques, the 200 nm region, where the *E*-band is situated, should become more accessible, and thus the assignment of transitions should be made easier.

For monosubstituted benzenes these criteria must be used in conjunction with additional information derived from the effects of solvent, and of further substitution, on the bands under consideration. In the case of related groups of compounds, such as those in Table 13, however, the diagnostic success of the criteria is quite evident.

Having recognized the existence of the various absorption bands in benzenoid systems and devised criteria for their diagnosis, it was possible to attempt to correlate changes in the spectra with changes in the structure of the aromatic compounds.

Scott's Rules

Scott has devised a set of rules, analogous to the Woodward–Fieser rules, for derivatives of acyl benzenes by means of which one can calculate the position of the absorption maxima (*K*-band) in ethanol for a variety of monosubstituted benzaldehydes, benzoic acids (or their esters), and acetophenones (Table 14).

TABLE 14

SCOTT'S RULES FOR CALCULATING λ_{max}
(EtOH) OF THE *K*-BAND FOR PhCOR
DERIVATIVES

Parent chromophore PhCOR	nm
R = alkyl or alicyclic residue (R)	246
hydrogen (H)	250
hydroxyl or alkoxyl (OH or OR)	230

INCREMENTS IN nm FOR
FURTHER SUBSTITUENTS ON
THE BENZENE RING IN THE
ORTHO, META, AND PARA
POSITIONS

	o	m	p
R (alkyl)	3	3	10
OH, OR	7	7	25
$O^{\ominus}$	11	20	78
Cl	0	0	10
Br	2	2	15
NH_2	13	13	58
NHAc	20	20	45
NR_2	20	20	85

Example 4

Calculate the *K*-band absorption maximum in ethanol of p-$ClC_6H_4COCH_3$

	nm
Parent	246
p-Cl-	10
Calculated	256
Observed	254

Problems

Calculate the *K*-band absorption maxima in ethanol for:

Compound	λ_{max} Observed
p-HO C_6H_4CHO	283 nm (0·1 N HCl)
p-$NH_2C_6H_4COOH$	284 nm (pH 3·75)
m-Br C_6H_4COOH	232 nm
o-HO $C_6H_4COCH_3$	252 nm

Many workers have attempted to analyse benzenoid spectra by quantum mechanical methods with varying degrees of success. From these efforts relationships have emerged which allow the calculation of the positions of the absorption maxima of the *B*- and *K*-bands in mono- and polysubstituted benzenes.

TABLE 15

FÖRSTER'S FORMULAE FOR THE CALCULATION OF *B*-BAND
MAXIMA FOR MONO- AND DI-SUBSTITUTED BENZENES

Substitution type	Substituent positions[a]	Förster formulae
Mono	1	$\Delta\bar{\nu} = (l_1 + V_1^2)$
o-di	1,2–	$\Delta\bar{\nu}_{1,2} = (l_1 + V_1^2) + (l_2 + V_2^2) - V_1 V_2$
m-di	1,3–	$\Delta\bar{\nu}_{1,3} = (l_1 + V_1^2) + (l_3 + V_3^2) - V_1 V_3$
p-di	1,4–	$\Delta\bar{\nu}_{1,4} = (l_1 + V_1^2) + (l_4 + V_4^2) + 2 V_1 V_4$

[a]For clarity standard IUPAC numbering is used here and not that
used in the m.o. treatment by Förster and others.

TABLE 16

VALUES OF *l* AND *V* FOR USE WITH
FÖRSTER EQUATIONS

Group	$l\ \mu m^{-1} \times 10$	$V\ (\mu m^{-1} \times 10)^{\frac{1}{2}}$
CH_3	0·51	0·28
OCH_3	0·51	1·00
OH	0·64	1·00
Br	0·90	0·11
Cl	0·90	0·23
NH_2	1·11	1·44
COOH	1·17	−1·05
$COO^{\ominus}$	1·21	−0·63
CN	1·47	−0·73
$O^{\ominus}$	2·39	1·29
$COCH_3$	3·44	−0·52
CHO	3·69	−0·49
NO_2	5·16	−1·05

TABLE 17

Ortho CORRECTIONS TO $\Delta\bar{\nu}$ FOR THE FÖRSTER FORMULAE

Pair of substituents		Corrections (kK)	Pair of substituents		Corrections (kK)
CH_3	CH_3	−0·29	OCH_3	COOH	0·00
OH	OH	+0·26	OCH_3	$COO^{\ominus}$	−1·00
OH	$O^{\ominus}$	+0·59	OH	$COCH_3$	+1·61
NH_2	NH_2	0·00	$O^{\ominus}$	$COCH_3$	+0·81
Cl	Cl	−0·05	Cl	$COCH_3$	−0·85
CH_3	NH_2	−0·40	OH	CHO	+0·99
OCH_3	OCH_3	+0·37	$O^{\ominus}$	CHO	+1·48
OCH_3	OH	+0·32	Cl	CHO	+0·68
CH_3	CN	+0·07	Cl	NO_2	−0·29
Cl	COOH	−0·37	CH_3	NO_2	+0·98
Cl	$COO^{\ominus}$	−0·53	OH	NO_2	+1·54
OH	COOH	+0·73	$O^{\ominus}$	NO_2	+1·47
OH	$COO^{\ominus}$	−0·63	OCH_3	NO_2	+0·71
NH_2	COOH	+1·67	COOH	COOH	−0·77
NH_2	$COO^{\ominus}$	+1·07	$COO^{\ominus}$	$COO^{\ominus}$	−0·60

Förster has derived equations predicting λ_{max} of the *B*-bands for the mono- and di-substituted benzenes which work well for the *m*- and *p*-compounds but do not give good agreement with the experimental values for *ortho*-substituted benzenes. Petruska[4b] has attributed this *ortho* effect to stereoelectronic contributions which cannot be taken into account in a perturbation m.o. approach and has determined the corrections to be applied to the Förster prediction. It is of interest to observe that in most cases where hydrogen bonding between the substituents can bring about the formation of 5- or 6-membered ring, the *ortho* correction is positive, whereas in most cases where no hydrogen bonding is possible the correction is negative.

A striking example of this is found in the case of salicylic acid:

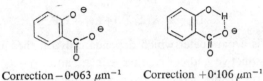

Correction $-0.063 \ \mu m^{-1}$ Correction $+0.106 \ \mu m^{-1}$

The Förster treatment also extends to trisubstituted benzenes and the interested reader is referred to reference 4. The formulae developed by Förster for mono- and di-substituted benzenes are shown in Table 15. $\Delta \bar{\nu}$ is calculated using two parameters, (1) *l* which includes the inductive and conjugative shift caused by a substituent, and (2) *V* which is characteristic of the substituent. This value is then substituted in eqn. (2).

$$\bar{\nu}_{calc} = 38.90 - \Delta \bar{\nu} \ kK \ (\mu m^{-1} \times 10), \qquad (2)$$

where

38·90 kK is the $\bar{\nu}_{max}$ of the benzene *B*-band, and $\Delta \bar{\nu}$ the red shift of $\bar{\nu}_{max}$ due to substitution.

Note that the calculations must be carried through in kK $= \mu m^{-1} \times 10$ because of the V^2 terms which appear in the formula. The values of *l* and *V* appear in Table 16 and *ortho* corrections to $\Delta \bar{\nu}$ are given in Table 17.

Example 5

Calculate the *B*-band maximum for *m*-hydroxybenzoic acid in aqueous solution at pH 9·0.

$$\Delta \bar{\nu}_{1,3} = (l_1 + V_1^2) + (l_3 + V_3^2) - V_1 V_3$$
$$= (0.64 + 1) + (1.21 + 0.397) - (-0.63)$$
$$= 3.88.$$

$$\bar{\nu}_{calc} = 38.90 - 3.88 = 35.02 \ kK \ (\text{observed } 34.84 \ kK)$$

$$\text{OR} \quad 3.502 \ \mu m^{-1} \equiv 285.9 \ nm.$$

(Observed $3.484 \ \mu m^{-1} \equiv 287.7 \ nm$).

Problem 11

Calculate the B-band absorption maxima for:
(a) Anisole (observed 37·17 kK).
(b) *p*-Toluidine (observed 34·97 kK at pH 11).
(c) *m*-Bromobenzoic acid (observed 35·09 kK at pH 3).
(d) *o*-Nitrophenol in acidic and in basic solution (observed at pH 3, 28·49 kK, in 0·1
    NaOH, 24·04 kK).
The answers are given on p. 151.

Stevenson has extended the work of Petruska to include the K-band in benzenoid systems:

$$\bar{v}_{calc} = 49\cdot1 - \Delta\bar{v}_{calc},$$

where

$$\Delta\bar{v}_{calc} = K(\Sigma_m l'_m) + (x_1 - x_4)^2.$$

In this equation, K is a parameter which depends only on the number of substituents, l'_m is the shift due to inductive and conjugative effects, and $(x_1 - x_4)^2$ is a charge transfer correction term which is only applied if there are substituents in the compound para to one another.

[(Stevenson[4a] uses $(x_0 - x_3)$, but this has been changed here to $(x_1 - x_4)$ to emphasize the fact that it is a correction for *p*-substituents only, i.e., 1,4- to one another].

TABLE 18
VALUES OF l' AND x FOR THE CALCULATION OF
K-BAND MAXIMA

Group	l' kK ($\mu m^{-1} \times 10$)	x (kK)$^{\frac{1}{2}}$ ($\mu m^{-1} \times 10$)$^{\frac{1}{2}}$
$O^{\ominus}$	6·58	1·62±0·08
NH_2	5·65	1·42±0·08
OCH_3	3·05	0·80±0·10
OH	1·63	0·68±0·10
Br	1·51	0·64±0·20
Cl	1·40	0·20±0·15
CH_3	0·71	0·12±0·20
$COO^{\ominus}$	4·49	−0·82±0·20
CN	4·49	−0·88±0·10
COOH	5·65	−1·02±0·15
CHO	9·05	−1·11±0·10
$COCH_3$	8·35	−1·13±0·10
NO_2	11·89	−1·50±0·06

Values of K

No. of Substituents	K
1	1·0
2	0·81
3	0·72

Example 6

Calculate the *K*-band absorption maximum for *p*-methoxyacetophenone.

$$\bar{\nu}_{calc} = 49 \cdot 10 - \Delta\bar{\nu}_{calc}.$$

$$\begin{aligned}
\Delta\bar{\nu}_{calc} &= K(\Sigma_m l'_m) + (x_1 - x_4)^2 \\
&= 0 \cdot 81(3 \cdot 05 + 8 \cdot 35) + (1 \cdot 93)^2 \\
&= 0 \cdot 81(11 \cdot 40) + (3 \cdot 73) \\
&= 9 \cdot 16 + 3 \cdot 73 = 12 \cdot 89 \text{ kK.}
\end{aligned}$$

$$\begin{aligned}
\bar{\nu}_{calc} &= 49 \cdot 10 - 12 \cdot 89 \\
&= 36 \cdot 21 \text{ kK} \quad \text{(observed in 0·1 M HCl 36·17 kK)} \\
&= 3 \cdot 621 \ \mu m^{-1} \quad \text{(observed in 0·1 M HCl 3·617 } \mu m^{-1}\text{)} \\
&= 277 \cdot 1 \text{ nm} \quad \text{(observed in 0·1 M HCl 278·3 nm)}
\end{aligned}$$

Problem 12

Calculate the *K*-band absorption maxima for:

(a) *m*-Bromobenzoic acid at pH 11 (observed 4·425 μm^{-1}).
(b) *m*-Nitroanisole (observed 3·656 μm^{-1}).
(c) *o*-Chlorobenzoic acid at pH3 (observed 4·367 μm^{-1}).
(d) *o*-Chloronitrobenzene (observed 3·846 μm^{-1}).
(e) *p*-Chloroaniline in 0·1 M NaOH (observed 4·184 μm^{-1}).
(f) *p*-Chlorophenol in 1 M NaOH (observed 4·098 μm^{-1}).

The answers are given on p. 152.

π-Aromatic Systems and Systems with Extended Conjugation

The wavelength at which a conjugated compound shows an absorption maximum depends on the energy difference between the ground and the excited states which may be affected by increasing conjugation in one of two ways as shown in Fig. 3.

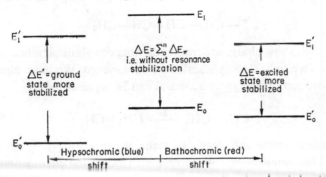

FIG. 3. The effect of resonance on the transition energies of conjugated systems.

The most usual result of extended conjugation is greater stabilization of the excited state as compared with the ground state and a concommitant red shift. In resonance terminology this is equivalent to saying that a larger number of canonical structures can be written in which charge separation occurs. This is well illustrated by the polyacenes and the *p*-poly-

phenylenes (Table 19) where it can be verified that the number of structures of the type

increases very rapidly with *n* the number of rings.

TABLE 19

THE INCREASE IN λ_{max}
WITH CHAIN LENGTH IN
THE POLYACENES AND
m-POLYPHENYLENES

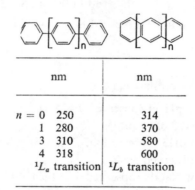

	nm	nm
$n = 0$	250	314
1	280	370
3	310	580
4	318	600
	^1L_a transition	^1L_b transition

In the linear polyenes and related compounds the number of resonance structures which may be written increases rapidly with the chain length. Thus, for example, for ethylene we can write

$$CH_2{=}CH_2 \leftrightarrow \overset{\pm}{C}H_2{-}\overset{\mp}{C}H_2$$

and for 1,3-butadiene

$$CH_2{=}CH{-}CH{=}CH_2 \leftrightarrow \overset{\pm}{C}H_2{-}\overset{\mp}{C}H{-}CH{=}CH_2 \leftrightarrow CH_2{=}CH{-}\overset{\pm}{C}H{-}\overset{\mp}{C}H_2 \leftrightarrow$$

$$\leftrightarrow \overset{\pm}{C}H_2{-}CH{=}CH{-}\overset{\mp}{C}H_2$$

The formulae which show charge separation do not contribute significantly to the ground state which is therefore only slightly stabilized by resonance. We have already seen, however, that the absorption of energy by a π-bond can be represented by

$$CH_2{=}CH_2 \xrightarrow{h\nu} \overset{\pm}{C}H_2{-}\overset{\mp}{C}H_2$$

and this type of excited state will clearly be stabilized much more by resonance than will the ground state. This stabilization will increase with increasing chain length since the number of possible ionic structures increase. As a result of this increased stabilization of the excited state the transition energy will be smaller and λ_{max} will appear at progressively longer wavelengths.

Compounds of this type show regular change in λ_{max} with changing chain length which fall into one of two classes.

CLASS I (e.g. LINEAR POLYENES)

Class I comprises those compounds in which the ground state resonance has virtually no contribution from charge separation or movement of charge. All other resonance structures involving charge separation contribute only to the stabilization of the electronically excited state. Thus in such compounds as the linear conjugated polyenes, while there is a slight increase in the stabilization of the ground state with an increase in the number of double bonds n, the number of dipolar resonance structures which can be written increases rapidly and the excited state is relatively much more stabilized. This situation is shown schematically in Fig. 4.

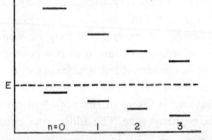

FIG. 4. Schematic representation of transition energies in Class I linearly conjugated molecules R—(CH=CH)$_n$R. (From A. Maccoll, *Quart. Rev.*, 1947, **1**, 16; reproduced by permission.)

Since $E = E_1 - E_0 = hc/\lambda$ observed, it is clear that λ_{max} moves to longer wavelengths with increasing n. Experimentally it is found for these compounds that $\lambda^2 = kn$.

CLASS II

In Class II are included polyenic compounds which are charged in their ground states as for example:

$$R_2N—(CH=CH)_n—CH=\overset{\oplus}{N}R_2 \leftrightarrow R_2\overset{\oplus}{N}=CH—(CH=CH)_n—NR_2$$

As n increases the number of structures of the type

$$R_2N—(CH=CH)_n—\overset{\oplus}{C}H—NR_2$$

increases rapidly and would be expected to stabilize the excited state, while at the same time the contributions of the ground states become relatively smaller. Experimentally it is found that for these compounds $\lambda = kn$, and it has been shown that the ground state is slightly destabilized with increasing n. This is shown in Fig. 5.

In this case the quantitative application of valence bond theory fails and leads to the

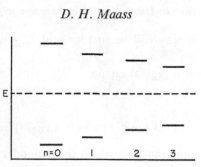

FIG. 5. Schematic representation of transition energies in Class II linearly conjugated molecules. (From A. Maccoll, *loc. cit.*; reproduced by permission.)

relation $\lambda^2 = kn$ for integral values of n greater than zero. Molecular orbital calculations on the other hand give good results and also give rise to the observed relationship $\lambda = kn$.

The polyacenes and polyenes have been the subjects of many theoretical investigations since they are particularly amenable to m.o. calculations. They have also been handled by a resonance method which though very simple in its approach yields excellent results and deserves closer attention from organic chemists.[5]

In many compounds the extent of conjugation may be limited by the geometry of the molecule. In some cases this will result in a family of compounds which will show a constant λ_{max} while, as more chromophores are added which do not extend the conjugated system, ε_{max} increases by increments characteristic of the added chromophore (Table 20). A good example of this type of behaviour is found in the *m*-polyphenylenes in which λ_{max} is determined by conjugation of the biphenyl type. The addition of benzenoid nuclei in *m*-positions cannot extend this conjugation.

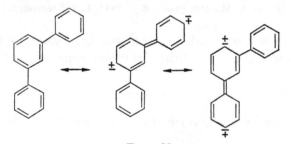

TABLE 20

THE CHANGE IN ABSORPTIVITY OF THE $^1L_a(\pi \rightarrow \pi^*)$ TRANSITION (250 nm) WITH INCREASING CHAIN LENGTH IN THE *m*-POLYPHENYLENES

Compound	No. of rings	$\varepsilon_{max} \times 10^{-3}$ cm^2 mole^{-1}	$\Delta\varepsilon$
Biphenyl	2	1·8	
m-Terphenyl	3	4·0	2·2
m-Quaterphenyl	4	5·9	1·9
m-Quinquephenyl	5	7·9	2·0
m-Scxiphenyl	6	10·2	2·3

In the case of molecules which consist of unsymmetrically substituted chromophores, it is possible to observe bands corresponding to the two different cross-conjugated systems as, for example, in the 4-substituted benzophenones shown in Table 21.

TABLE 21

THE ABSORPTION MAXIMA AND ABSORPTIVITIES OF SOME SUBSTITUTED BENZOPHENONES
(The systems giving rise to the x- and y-bands are shown in bold face)

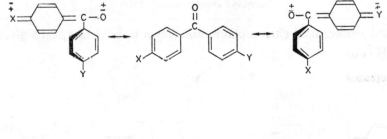

X	Y	X-band λ_x nm $\epsilon_x \times 10^{-3}$ cm^2 mole^{-1}	$\lambda_{x=y}$ nm $\epsilon_{x=y} \times 10^{-3}$ cm^2 mole^{-1}	Y-band λ_y nm $\epsilon_y \times 10^{-3}$ cm^2 mole^{-1}
H	H		248 20·0	
H	OCH₃	247 10·5		274 17·0
OCH₃	OCH₃		278 27·0	
Cl	OCH₃	257 15·0		280 17·5
Cl	Cl		261 27·5	

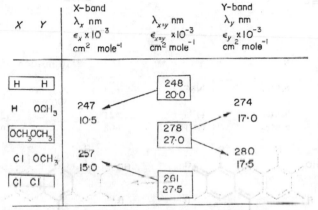

Such cross-conjugation also occurs in certain dienones when the application of the Woodward–Fieser rules leads to two predicted maxima. Usually the observed maximum is closest to the longer of the predicted wavelength values [e.g. (I)], although some anomalies do occur, for example with (II), where the shortwave band agrees with the calculated λ_{max}. Sometimes the agreement is not good, as in the case of the $\Delta^{4,6}$ dienone (III).

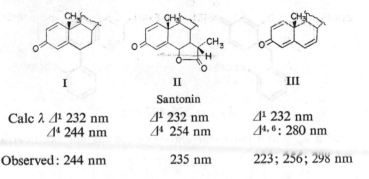

	I	II	III
		Santonin	
Calc λ	Δ^1 232 nm	Δ^1 232 nm	Δ^1 232 nm
	Δ^4 244 nm	Δ^4 254 nm	$\Delta^{4,6}$: 280 nm
Observed:	244 nm	235 nm	223; 256; 298 nm

A somewhat similar situation arises in many dyestuffs in which two (or more) chromophores of differing polarizability occur in the same molecule but insulated from one another. Once again two absorption bands may be observed. Examples of this behaviour are found in the triphenylmethane dyes and the phthaleins (Table 22).

TABLE 22

THE EFFECT OF DIFFERING POLARIZABILITIES
OF THE CHROMOPHORES IN THE TRIPHENYL-
METHANE AND PHTHALEIN DYES

The x- and y-substituted chromophores are shown in bold face; each gives rise to its own separate λ_{max}.

Triphenylmethanes

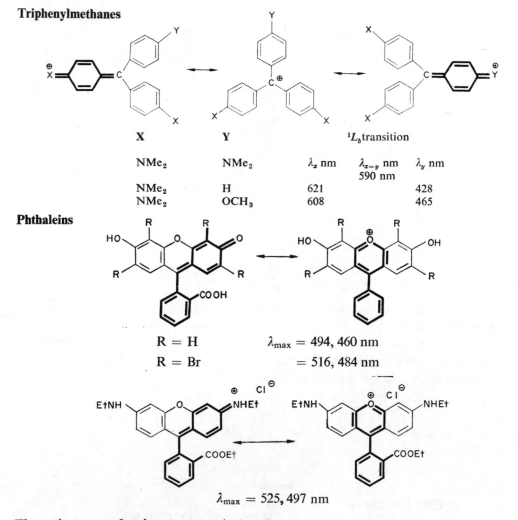

X	Y	^1L_b transition		
		λ_x nm	$\lambda_{x=y}$ nm	λ_y nm
NMe$_2$	NMe$_2$		590 nm	
NMe$_2$	H	621		428
NMe$_2$	OCH$_3$	608		465

Phthaleins

R = H λ_{max} = 494, 460 nm

R = Br   = 516, 484 nm

λ_{max} = 525, 497 nm

The assignment of a λ_{max} to a particular chromophore has not been carried out for the phthaleins though the different absorbing systems are shown in bold face.

Steric Inhibition of Resonance

It is convenient for the purposes of discussion to divide steric inhibition effects into three classes based on $\Delta\lambda$ the shift of the position of the absorption maximum in the hindered system relative to the unhindered system,

i.e. $\Delta\lambda = \lambda_{hindered} - \lambda_{unhindered}.$

Class I $\Delta\lambda < 0$ Hypsochromic (blue) shift.

Class II $\Delta\lambda \simeq 0$,

Class III $\Delta\lambda > 0$—Bathochromic (red) shift.

CLASS I

This effect is often the result of destabilization of the excited state brought about by the decreased ability of the conjugated system to attain a coplanar configuration. Although complete coplanarity of the chromophores leads to the greatest amount of conjugation,

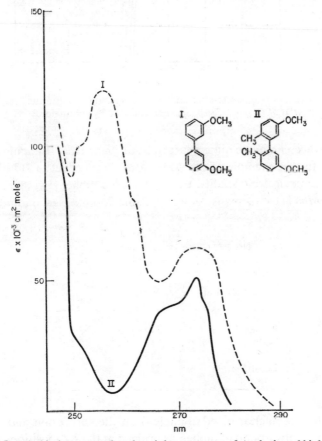

FIG. 6A. The effect of steric hindrance on the ultraviolet spectrum of a substituted biphenyl. (From Gillam and Stern;[1] reproduced by permission.)

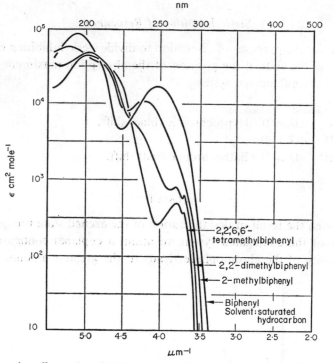

Fig 6b. The progressive effects of steric hindrance on the spectra of substituted biphenyls. (From *U.V. Atlas of Organic Compounds;* reproduced by permission.)

these systems can accommodate quite a large out-of-plane twist. In many cases the spectro-scopic effect is a further example of the well-known *ortho* effect of substituents.

One of the most extensively studied examples is the biphenyl system. The effect of the introduction of *ortho* substituents is shown in Table 23 and Figs. 6a and 6b.

TABLE 23

THE EFFECT OF *ortho*-SUBSTITUTION ON
THE λ_{max} AND ε_{max} OF BIPHENYL

Compound	λ_{nm} (^1L_a)	$\varepsilon \times 10^{-3}$ cm^2 mole^{-1}
Biphenyl	249	14·5
2-Me	237	10·5
2-Et	233	9·0
2,2′-diMe	220	—

The effect of the internuclear dihedral angle ϕ on the absorption maximum is shown in Table 24 in which the absorption of biphenyls bridged at the 2,2′-positions by polymethylene chains are given. ϕ was determined from molecular models of these compounds and the

TABLE 24
THE EFFECT OF THE INTERNUCLEAR
DIHEDRAL ANGLE ON λ_{max} IN THE 2,2'
BRIDGED BIPHENYLS

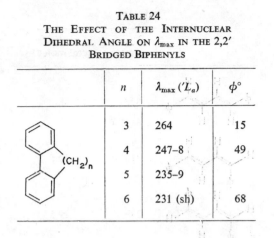

n	λ_{max} ($'L_a$)	$\phi°$
3	264	15
4	247–8	49
5	235–9	
6	231 (sh)	68

values are supported by gas phase electron diffraction measurements on other 2- and 2,2'-substituted biphenyls.

It is interesting to note that the compound

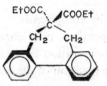

has been resolved into its enantiomers, thus demonstrating the non-coplanarity of the rings; the absorption characteristics, however, are still of the biphenyl type with λ_{max} 249 nm and ε_{max} 16,980 cm² mole⁻¹.

Olefinic compounds may also show Class I steric effects depending on the disposition of substituents about the double bonds; thus *cis*-stilbene absorbs at a shorter wavelength than *trans*-stilbene and with a lower intensity.

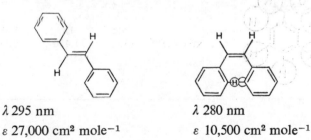

λ 295 nm λ 280 nm
ε 27,000 cm² mole⁻¹ ε 10,500 cm² mole⁻¹

1,4-Diphenylbutadiene (DPBD) shows the effect of changes in geometric isomerism and hence steric hindrance on the spectra of conjugated olefins. In this case, Table 25 shows the decrease in wavelength and the extinction coefficient with increase of steric hindrance in the dienes.

TABLE 25

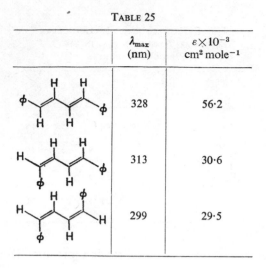

	λ_{max} (nm)	$\varepsilon \times 10^{-3}$ cm² mole⁻¹
	328	56·2
	313	30·6
	299	29·5

CLASS II

This type of hindrance is shown by compounds in which the absorption band is due to an electron transfer between two portions of a chromophoric system.

Distortion from the coplanar configuration in such cases leads to a drop in the molar absorptivity ε with no change in the position of the absorption band as illustrated in Table 26.

TABLE 26
STERIC INHIBITION OF RESONANCE IN ALKYLPHENONES WITH
CHANGES IN R

	λ_{max} ca. 238 nm		
R= ε_{max} cm² mole⁻¹	Me 13,000	Et 11,450	Buᵗ 8100

CLASS III

In contrast to Class I, where relative destabilization of the excited state led to a blue shift in the spectrum, in this class relative destabilization of the ground state leads to a rise in the ground state energy and thus a decrease in $\Delta E = E_{ex} - E_g$, which is observed as a red shift.

This behaviour is observed with compounds of type A where introduction of an *N*-alkyl group produces a red shift. On the other hand, in compounds of type B, which do not belong to Class III, *N*-alkylation brings about only a small blue shift (Table 27).

TABLE 27

THE EFFECT OF *N*-METHYLATION ON THE ABSORPTION SPECTRA OF A AND B

Substituents	λ_{max} (nm) A	$\varepsilon \times 10^{-4}$ cm^2 mol^{-1} A	λ_{max} (nm) B	$\varepsilon \times 10^{-4}$ cm^2 mol^{-1} B
$R_1=R_2=R_3=R_4=H$	470			
$R_1=CH_3$; $R_2=R_3=R_4=H$	510			
$R_1=H$; $R_2=R_3=R_4=CH_3$	446	3·5	536	6·3
$R_1=CH_3$; $R_2=R_3=R_4=CH_3$	479	1·25	534	7·6
$R_1=H=R_2=R_3$; $R_4=CH_3$	473	13·5	595	17·0
$R_1=CH_3$; $R_2=R_3=H$; $R_4=CH_3$	510	5·7	581	17·4

External Factors Affecting Spectra

The major external factors which modify the appearance of electronic spectra are temperature and solvent, both of which may have profound effects. Let us examine these two factors separately.

1. TEMPERATURE

In general, ultraviolet spectral bands are broad compared with those in infrared spectra, partly because of interactions with the solvent and partly because the energy of the absorbed proton not only raises an electron from the electronic ground state E_0 to the electronic excited state E_1, but brings about transitions between rotational r and vibrational V energy levels in these two states. Such closely spaced energy difference cannot be resolved in the spectrometer, and the groups of closely allied transitions appear to form a broad absorption peak (band envelope). For example, a group of transitions such as

$$E_0 V_0 r_0 \rightarrow E_1 V_1 r_1,$$
$$E_0 V_0 r_1 \rightarrow E_1 V_1 r_2,$$
$$E_0 V_1 r_1 \rightarrow E_1 V_2 r_2,$$

would form a single broad absorption peak.

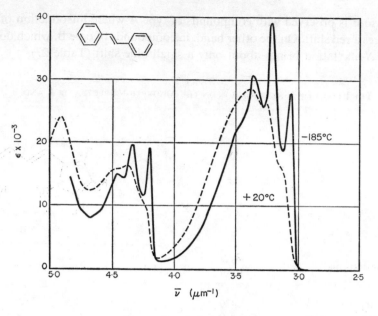

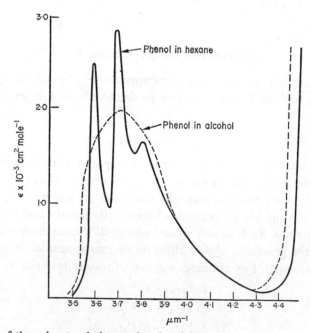

FIG. 7. Ultraviolet absorption spectrum of *trans*-stilbene in isopentane-methylcyclohexane (5 : I v/v). (Abscissa: wavenumber $\bar{\nu}$ (cm^{-1}); ordinate: molecular extinction coefficient (ε).) (From Beaven *et al.*;[8] reproduced by permission.)

FIG. 8. The effect of the solvent polarity on the ultraviolet spectrum of phenol. (From Gillam and Stern;[7] reproduced by permission.)

Lowering the temperature reduces the vibrational and rotational contributions to the bands and reveals, to some extent, the individual electronic transitions. This is clearly shown in Fig. 7 in the two spectra of *trans*-stilbene at 20°C and −185°C.

2. SOLVENT

The effect of the solvent on the spectrum is complex. The most obvious effect in changing from non-polar to polar solvents is the smoothing out of the spectrum, the fine structure frequently vanishing completely. The spectrum of phenol in hexane and in alcohol shown in Fig. 8 demonstrates this very effectively.

The fine structure of the ultraviolet spectrum is best observed in the vapour phase where intermolecular interactions are minimized. Even in a non-polar solvent, the fine structure is largely eliminated by collision interaction with solvent molecules. In polar solvents the fine structure is further concealed by solute–solvent interactions. The combined effects of temperature and solvent are exceptionally well illustrated by the spectra of symmetrical tetrazine shown in Fig. 9.

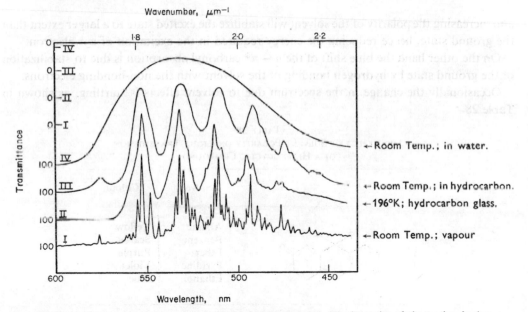

FIG. 9. Absorption spectrum of *sym*-tetrazine in the visible region. The intensity of absorption is shown in terms of transmittance (0–100%, reading downwards) with separate zero levels for each spectrum to avoid overlapping. (The wavelength scale refers to spectrum I; spectra II and III have been displaced by 150 and 250 cm^{-1} respectively to higher frequencies, and spectrum IV by 750 cm^{-1} to lower frequencies.) (From Beaven *et al.*;[8] reproduced by permission.)

In addition to the changes in the structure of the absorption bands with changes in the solvent, changes may occur in the positions of the absorption maxima. Generally $\pi \rightarrow \pi^*$ transitions undergo bathochromic (red) shifts in more polar solvents, whereas $n \rightarrow \pi^*$ transitions undergo hypsochromic (blue) shifts.

The blue shift of the $n \to \pi^*$ carbonyl transitions in acetone shows the magnitude of this effect.

Solvent	Hexane	CHCl$_3$	Dioxan	C$_2$H$_5$OH	H$_2$O
λ_{max}	280	278	277	270	265 ($\pm$2 nm)

These shifts in the positions of the absorption maxima with changing solvent polarity are not always easily accounted for, but in many cases are due to greater stabilization of the ground state or the excited state through dipole–dipole interaction or by solvation. Thus in the merocyanines the two extreme resonance structures are

$$—N—CH{=}CH—(CH{=}CH)_{n-1}—CH{=}O \text{ (ground state)}$$

and

$$—\overset{\oplus}{N}{=}CH(—CH{=}CH)_{n-1}—CH{=}CH—\overset{\ominus}{O}$$

and increasing the polarity of the solvent will stabilize the excited state to a larger extent than the ground state, hence reducing the energy required in the excitation of a π-electron.

On the other hand the blue shift of the $n \to \pi^*$ carbonyl absorption is due to stabilization of the ground state by hydrogen bonding of the solvent with the non-bonding electrons.

Occasionally the change in the spectrum due to solvent effects is startling, as shown in Table 28.

TABLE 28

THE EFFECT OF SOLVENT POLARITY ON LIGHT TRANSMISSION
OF A HETEROCYCLIC COMPOUND

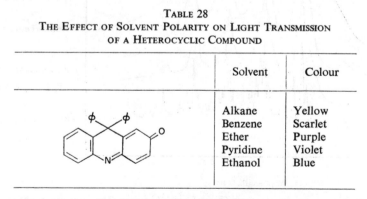

	Solvent	Colour
	Alkane	Yellow
	Benzene	Scarlet
	Ether	Purple
	Pyridine	Violet
	Ethanol	Blue

Transannular Effects

We have seen that the carbonyl group shows two absorptions $\pi \to \pi^*$ (*ca.* 185 nm, ε *ca.* 900) and $n \to \pi^*$ (*ca.* 280 nm, ε *ca.* 15). Conjugation with an α, β-double bond brings about a bathochromic shift to 220–260 nm and 300–350 nm respectively with little change in intensity of the $n \to \pi^*$ transition but a large increase in the intensity of the $\pi \to \pi^*$ transition. With systems in the molecule which are not conjugated with the carbonyl group in the classical sense we should not expect the absorption to be markedly different from that obtained by summation of the isolated chromophores. In some cases, however, and particularly in cyclic systems, the geometry of the molecules enables orbital overlap to occur, and this causes changes in the ultraviolet spectrum.

The orbital overlap may occur in two different ways. In π-type overlap the molecular orbitals (m.o.'s) and atomic orbitals (a.o.'s) overlap so as to retain the π-bond nodal plane thus:

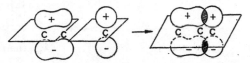

This type of overlap is shown by the two molecules illustrated below whose ultraviolet spectra closely resemble those of the α, β-unsaturated ketones.

	$\pi \to \pi^*$	$n \to \pi^*$
λ_{max} (nm)	214	284
ε (cm^2 mole^{-1})	1500	30

3-methylenecyclobutanone

	$\pi \to \pi^*$	$n \to \pi^*$
λ_{max} (nm)	209·5	298·4
ε (cm^2 mole^{-1})	3110	32

In σ-type overlap of π-orbitals only one lobe of each orbital is involved and the nodal planes no longer coincide as shown in the given examples:

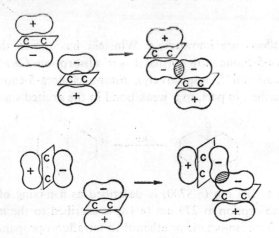

Thus in the saturated bicyclic ketone shown below, the carbonyl absorption is quite normal, whereas in its unsaturated analogue the olefinic π-orbital is able to take part in σ-type overlap with the carbonyl π-orbital giving rise to an absorption pattern very similar to that of the coplanar α,β-unsaturated ketones.

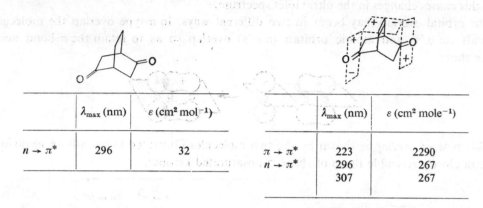

	λ_{max} (nm)	ε (cm² mol⁻¹)		λ_{max} (nm)	ε (cm² mole⁻¹)
$n \to \pi^*$	296	32	$\pi \to \pi^*$ $n \to \pi^*$	223 296 307	2290 267 267

Similarly, one of the phenyl groups in 2,2-diphenylcyclohexanone undergoes σ-overlap with the carbonyl orbital, a situation which clearly does not occur in 2-phenylcyclohexanone.

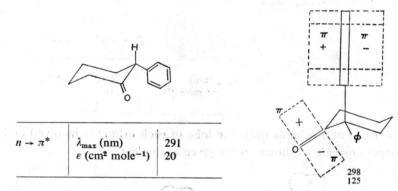

$n \to \pi^*$	λ_{max} (nm)	291
	ε (cm² mole⁻¹)	20

298
125

Even long-range effects are known; e.g. Winstein has shown that while cyclodecanone and *cis*-cyclodeca-5-enone show normal $n \to \pi^*$ absorptions (289·9 nm, ε 15 cm² mol⁻¹ and 288·3 nm, ε 15 cm² mol⁻¹, respectively), *trans*-cyclodeca-5-enone has additional ab, sorption which he ascribes, in part, to a weak bond in the excited state as shown:

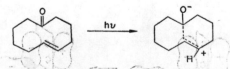

A strong absorption at 188 nm (ε 8700) is described as a mixing of $n \to \sigma^*$ and $\pi \to \pi^*$ transitions, and the absorption at 279 nm (ε 18) is ascribed to the $n \to \pi^*$ transition. On changing the solvent from iso-octane or ethanol to tetrafluoropropanol, a third absorption was revealed at 214·5 nm (ε 2300) which has been attributed by Winstein[6] to the excited state already illustrated and has been called a photodesmotic band.

Quantitative Analysis

Ultraviolet spectroscopy has several advantages in analytical work over some of the other instrumental methods. The instruments, which are relatively cheap and accurate, are very sensitive, and aqueous solutions may be examined without difficulty. The concentrations of single substances may be determined with little trouble provided that they have reasonnable absorptivities. With mixtures the procedure is a little more complicated. The species to be determined must have reasonably well-separated absorptions and must obey Beer's law for simplicity. If the components do not satisfy these conditions, the analysis is more complicated but still feasible.

One important application of this technique is the determination of the ionization constants of acids and bases, and, by similar procedures, the stability constants of complexes. A further application is the investigation of kinetic and mechanistic problems where the sensitivity of the spectrum to stereoelectronic effects is of great value.

When two compounds in equilibrium with one another are responsible for all the absorption in a given region, there will be at least one point in the spectrum where the absorbance will be independent of the ratio of the two concentrations. When the absorption bands overlap, these points will be at the frequencies at which the absorptivities of the two species is equal. These are called isosbestic points. The existence of an isosbestic point usually indicates the existence of only two absorbing species as illustrated in Fig. 10.

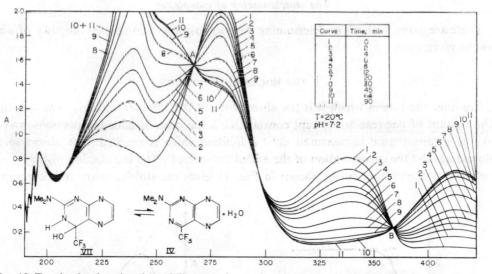

FIG. 10. Two isosbestic points (*A* and *B*) appear in a series of spectra recorded at different times for the reaction VII→IV. Reproduced by permission of Dr. J. Clark, University of Salford.)

When 2-*N*, *N*-dimethylamino-4-trifluoromethylpteridine (IV) is dissolved in 0·1 M HCl, a rapid protonation occurs to give (V) which then undergoes a slower hydration reaction to form the hydrate (VI). Addition of a solution of (VI) to a buffer solution (pH9) results in instantaneous deprotonation to give (VII) which subsequently eliminates water in a slow reaction to regenerate (IV), the anhydrous neutral molecule.

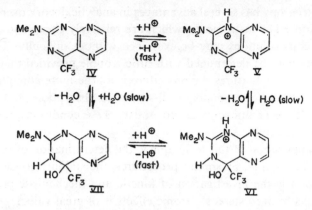

In Fig. 10 a series of curves is shown which have been plotted at different time intervals after the injection of a solution of (VI) into a buffer solution. This family of curves demonstrates how the absorption changes as the equilibrium between the hydrate (VII) and the anhydrous neutral molecule (IV) is established. This particular system exhibits two isosbestic points *A* and *B*.

The stoichiometry of complexes

There are many methods of determining the composition of complexes only two of which will be given here.

1. THE MOLAR RATIO METHOD

Consider the case in which it is the absorbance of the complex which is to be measured. The amount of one reactant is kept constant while the other is added in portions, and the change in absorbance is measured until a limiting value is reached. The absorbance is plotted against the concentration of the added component. The intersection of the two absorbance–concentration curves shown in Fig. 11 gives the stoichiometry of the complex.

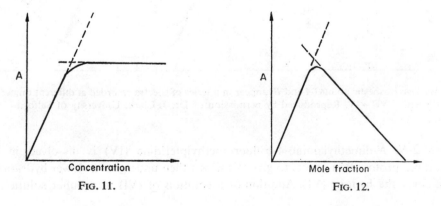

FIG. 11.   FIG. 12.

2. JOB'S METHOD

In this method the sum of the molar concentrations is kept constant while their ratio is varied and the absorbance is plotted against the mole fraction. The intersection point of the two branches of the curve shown in Fig. 12 indicates the stoichiometry of the complex and its position gives an indication of the size of the equilibrium constant.

Acid–Base Equilibria

The equilibrium constants of acids and bases which absorb in the ultra violet region may be determined spectrophotometrically easily and accurately. Many procedures are are available but only two are given here as illustrative of the methods adopted.

Consider the weak acid HA, in aqueous solution, undergoing dissociation,

$$HA \rightleftharpoons H^+ + A^-$$

The true equilibrium constant may be written

$$K_a = a_{H^+} \cdot a_A / a_{HA}$$

$$= \frac{[H^+][A^-]}{[HA]} \frac{f_{H^+} + f_{A^-}}{f_{HA}}$$

$$= K_a' \frac{f_{H^+} + f_{A^-}}{f_{HA}},$$

where [ ] is the concentration in moles litre $^{-1}$, a the activity, and f the activity coefficient.

In dilute solutions the activity coefficients approach unity and for many purposes it is sufficiently accurate to use concentrations; hence we can write

$$K_a' = \frac{[H^+][A^-]}{[HA]} = \frac{\alpha^2}{(1-\alpha)}, \quad \text{where } \alpha \text{ is the degree of dissociation,}$$

and after taking logarithms so that $p(F) = -\log(F) = 1/\log(F)$, where (F) is any function,

$$pK_a' = pH + \log \frac{[HA]}{[A^-]}$$

$$= pH + \log \frac{(1-\alpha)}{\alpha}.$$

To obtain α it is first necessary to measure the spectra of the undissociated weak acid (HA) and the anion (A$^-$) separately in buffer solutions of the appropriate pH values and obtain ε_{HA} and ε_{A^-} at suitable wavelengths. The acid is then dissolved in a buffer solution at a pH close to the pK_a' and ε is calculated from the absorbance at the same wavelength. Assuming that the absorbance is additive,

$$\varepsilon = \alpha \varepsilon_A + (\alpha - 1)\varepsilon_{HA},$$

hence

$$\alpha = \varepsilon - \varepsilon_{HA} / \varepsilon_{A^-} - \varepsilon_{HA},$$

and this value can be substituted into

$$pK'_a = pH_{buffer} + \log \frac{(1-\alpha)}{\alpha}.$$

A method for determining pK'_a for the salt of a weak base is similar in requiring the spectra of the free base and the almost undissociated salt in suitable buffers and, finally, the determination of the absorbance of mixtures of the acid and base of known composition $v_1 M_A$ and $v_1 M_B$ (v = volume, M = molarity).

The equilibrium is

$$B + H^+ \rightleftharpoons BH^+$$

and

$$K'_a = \frac{[B][H^+]}{[BH^+]} = \frac{(1-\alpha)(1-\alpha)}{\alpha}.$$

The concentrations of the added acid and base are:

$$C_A = [H^+] + [BH^+] = v_1 M_A / v_1 + v_2,$$
$$C_B = [B] + [BH^+] = v_2 M_B / v_1 + v_2,$$

and

$$\alpha = [BH^+] / C_B = \frac{\varepsilon - \varepsilon_{HB^+}}{\varepsilon_B - \varepsilon_{HB^+}}.$$

K'_a can be expressed as follows:

$$K'_a = \frac{[B][H^+]}{[BH^+]} = \frac{[B]\{C_A - [BH^+]\}}{[BH^+]}$$

from which

$$pK'_a = \log \frac{[BH^+]}{[B]} - \log \{C_A - [BH^+]\}$$

$$= \log \left(\frac{\alpha}{1-\alpha} \right) - \log \{C_A - [BA^+]\}.$$

Molecular Weight Determination

Molecular weights can be determined if a chromophore which has a known absorption and which is isolated from any other absorptions can be introduced into the molecule. This can be achieved, for example, by preparing the picrates of amines, the osazones of sugars, and the 2,4-dinitrophenylhydrazones and aldehydes and ketones. The molecular weight M can be calculated from the absorbance as follows:

$$A = \varepsilon c l = \frac{\varepsilon m l}{LM}$$

where m is the weight of the derivative in grams and L the volume of the solution in litres,

whence
$$M = \frac{\varepsilon ml}{AL}.$$

Instrumentation

A typical spectrophotometer can be represented as follows:

| Source | → | Dispersion and collimation | → | Sample and reference | → | Detector | → | Recorder |

See also Fig. 13.

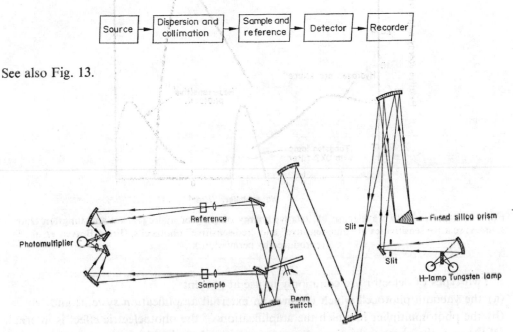

FIG. 13. Optical layout of a double-beam spectrophotometer.

The Source

In order to cover all the ultraviolet and visible regions it is necessary to have two radiation sources. The hydrogen or deuterium lamp is excellent up to 360 nm, but thereafter the intensity falls off rapidly; however, the tungsten filament lamp is particularly rich in red (low energy) radiation, and is used to span the long-wave ultraviolet and visible regions.

The Detector

Just as no single source is satisfactory over the complete range of wavelengths in which we are interested, there is a similar problem with the detectors. Fortunately the detector sensitivities can be matched to the energy output of the sources so that only a "red" and "blue" detector are required. This is clearly shown in Fig. 14.

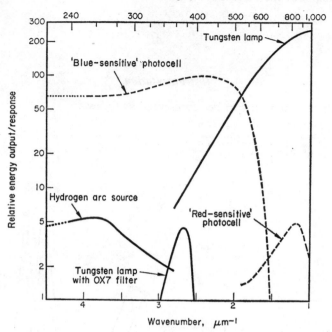

FIG. 14. Variation with wavelength of: relative energy outputs of hydrogen arc and tungsten lamp sources; relative sensitivities of "blue-sensitive" and "red-sensitive" photocells. (From Beaven *et al.*;[8] reproduced by permission.)

Two types of detector are commonly in use at present:
(a) the vacuum photocell which requires an external amplification system; and
(b) the photomultiplier in which the amplification of the photoelectric effect is internal. Of the two, the photomultiplier is the more sensitive but also the more expensive.

The Monochromator

The monochromator, probably more than any other single component, determines the utility of the spectrometer. It is here that the incident radiation is dispersed and collimated to provide a beam of, as nearly as possible, monochromatic light. Dispersion systems depend largely on the wavelength range to be studied. In the visible region a glass prism is adequate and inexpensive. In the ultraviolet region, however, it becomes necessary to use quartz prisms since ordinary glass absorbs ultraviolet radiation strongly; and since the ultraviolet source itself is not powerful, no radiation would pass through the prism. (Note that this means that in instruments working down into the ultraviolet, any part of the optical system which must transmit radiation must be made of quartz and any reflecting surfaces must be silvered.)

An alternative dispersion system employs a diffraction grating. These have become much more attractive in recent years with the introduction of replicate gratings.

Whichever system is used, the aim is to obtain a monochromatic beam of radiation. Since the dispersion of a single prism or grating is quite small, it is not possible to isolate and collimate very narrow bandwidths, and much ingenuity has been expended in designing and constructing multiple-pass monochromators. In general, the light from the first dispersion passes through a slit and thence to a second disperser. After the second disper sion, the light passes to the exit slit. In this way two advantages accrue: (a) the second dispersion increases the bandwidth of the emergent light so that the beam which finally passes through the exit slit is more nearly monochromatic, (b) most of the stray light from the source is suppressed.

With the rapid progress which is taking place in laser technology, it seems quite probable that it will not be long before truly monochromatic light sources which can be tuned over a wide wavelength range will be available to the spectroscopist.

The Slit

The exit slit of the optical system can be varied in width either manually or by a mechanical link with the recorder. Since the detector and recorder systems require a minimum input of energy to enable them to operate, the slit mechanism can be programmed to maintain a constant signal strength to the recorder. This method has certain disadvantages, particularly at high transmittances, when the slit may close entirely and cut off the radiation completely. A better system is to allow the gain control in the amplifier to control the signal to the recorder while the slit opening is adjusted continuously to maintain a constant bandwidth.

The Recorder

In order to make use of the data obtained from the optical part of the machine it is necessary that the detector response is presented in an easily appreciated form such as a plot of absorbance against wavelength. This can be done by means of a recorder. Two types of recorder are available: (a) human, and (b) mechanical. In terms of speed, reliability, ease of maintenance, and cheapness of operation, mechanical operators are superior, although for some purposes human recorders may be more suitable.

In both cases the essential operations are the same; the electrical signal generated in the detector is fed to a potentiometer which has been calibrated to read directly in absorbance (optical density); after balancing the potentiometer, the absorbance is recorded at that particular wavelength. This is repeated over the whole of the spectral range under investigation, and the results are recorded in tabular or, more usually, in graphical form. It is obvious that a repetitive chore of this sort is much more efficiently handled by an automatic recording device.

The Light Path

The light from the source eventually passes through the sample compartment where the sample solution and the pure solvent are contained in cells with optically flat windows. It is essential that each absorbance measurement on the solution is accompanied by a simul taneous measurement on the pure solvent. There are a number of ways of accomplishing this. In the case of a manual instrument, there is a single light path and the cells are placed in the beam by means of a sliding cell carriage. The majority of instruments, however, are self-recording and are constructed on the double-beam principle. The most commonly used system is the single monochromator/double beam arrangement in which the light, after passing through the monochromator, is directed in alternating pulses through the two cells by means of a set of oscillating mirrors. The light which subsequently reaches the detectors generates signals which are separated by a pair of phase modulated amplifiers and fed to the opposite branches of a Wheatstone bridge circuit and thence to the recorder. The alternative system, which duplicates the whole of the optical set-up, is found only in a few instruments, the outstanding example being those manufactured by Cary Incorporated.

Care of the Instrument

Spectrometers are expensive though, fortunately, fairly robust instruments, and some care taken over their use and maintenance will ensure that they will function without trouble for long periods. The most important requirement is that the machine be kept clean and as free as possible from dust, which will play havoc with the optical system and the mechanical linkages. Ideally, the instrument should be kept in a room which is continuously purged with dry, dust-free air. This counsel of perfection can, of course, seldom be achieved in practice. Nevertheless, even under the most unfavourable conditions a dust-cover can be kept over the machine when it is not in use, and the location can be as far away from the corrosive atmosphere of the laboratory as possible. The wavelength calibration should be checked at regular intervals.

The same sort of consideration should extend to the cells, which should be handled with the greatest of care; even slight scratching can turn optically worked cells into expensive pieces of junk. They should never be handled by the optical surfaces and only the very gentlest wiping with a solvent wet tissue should be used to clean the windows. The internal surfaces should be washed clean with solvent or, if necessary, with chromic acid.

References

1. A. BURAWOY, *J. Chem. Soc.*, 1939, 1177.
2. L. DOUB and J. M. VANDENBELT, *J. Amer. Chem. Soc.*, 1947, **69**, 2714; 1949, **71**, 2414; 1955, **77**, 4535.
3. J. R. PLATT, *J. Chem. Phys.*, 1949, **17**, 484.
4. TH. FÖRSTER, *Z. Naturforsch.*, 1947, **2a**, 149.
4a. P. E. STEVENSON, in *Systematics of the Electronic Spectra of Conjugated Molecules*, Wiley, 1964.
4b. J. PETRUSKA, *ibid.*
5. L. S. BARTELL, *Tetrahedron*, 1964, **20**, 139.
6. S. WINSTEIN, L. DE VRIES, and R. ORLOSKI, *J. Amer. Chem. Soc.*, 1961, **83**, 2020.
7. A. E. GILLAM and E. S. STERN, *Electronic Absorption Spectroscopy*, 3rd edn (ed. E. S. STERN and C. J. TIMMONS), Arnold, 1970.
8. G. H. BEAVEN *et al.*, *Molecular Spectroscopy*, Heywood, 1962.

Further Selected Reading

H. H. Jaffe and M. Orchin *Theory and Applications of Ultraviolet Spectroscopy*, Wiley, 1962.

A. I. Scott, *Interpretation of the Ultraviolet Spectra of Natural Products*, Pergamon, 1964.

C. N. R. Rao, *Ultraviolet and Visible Spectroscopy*, Butterworths, 1967.

R. P. Bauman, *Absorption Spectroscopy*, Wiley, 1962.

R. A. Friedel and M. Orchin, *Ultraviolet Spectra of Organic Compounds*, Wiley, 1950.

J. H. Murrell, *The Theory of the Electronic Spectra of Organic Molecules*, Methuen, 1963.

E. A. Braude and F. C. Nachod, *Determination of Organic Structures by Physical Methods*, Academic Press, 1955.

A. Weissberger, *Techniques of Organic Chemistry*, Vol. 9, Interscience, 1960.

U. V. Atlas of Organic Compounds, Butterworths/Verlag Chemie, 1966–continuing.

L. Lang (ed.) *Absorption Spectra in the Ultraviolet and Visible*, Akadémiai Kiadó, Budapest, 1959–continuing.

H. M. Hershenson, *Ultraviolet and Visible Absorption Spectra*, Academic Press, 1956 continuing.

Organic Electronic Spectral Data, Interscience, 1960.

K. Hirayama, *Handbook of Ultraviolet and Visible Absorption Spectra of Organic Compounds*, Plenum Press, 1967.

Sadtler Ultraviolet Spectra. Published periodically.

Catalogue of Ultraviolet Spectral Data, American Petroleum Institute, 1968–continuing.

K. Yamaguchi, *Spectral Data of Natural Products*, Elsevier, 1970.

J. A. Eidus *et al.*, *Atlas of Spectra of 5-Nitrofuran Compounds*, 1970.

R. A. Friedel and M. Orchin, *Ultraviolet Spectra of Aromatic Compounds*. Wiley, 1951.

J. G. Grasselli, *Atlas of Spectral Data and Physical Constants for Organic Compounds*, Chemical Rubber Company Press, 1973.

Answers to problems in ultraviolet spectroscopy

D. H. MAASS

PROBLEM 1

(a)

	nm
Parent diene	217
Exocyclic double bond ×1	5
Ring residue ×2	10
	232
Observed	232

(b)

	nm
Parent diene	217
Ring residue ×4	20
	237
Observed	236

PROBLEM 2

(a)

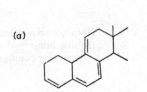

	nm
Parent homoannular diene	253
Double bond extending conjugation ×2	60
Exocyclic double bond×3	15
Ring residue ×5	25
	353
Observed	355

141

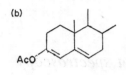

(b)

	nm
Parent heteroannular diene	214
Exocyclic double bond $\times 1$	5
Ring residue $\times 2$	15
Polar group $\times 1$	0
	234
Observed	238

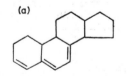

(c)

	nm
Parent homoannular diene	253
Double bond extending conjugation	30
Exocyclic double bond $\times 1$	5
Ring residue $\times 3$	15
Polar group	0
	303
Observed	306

PROBLEM 3

(a)

	nm
Homoannular diene	253
Double bond extending conjugation	30
Exocyclic double bond $\times 2$	10
Ring residue $\times 4$	20
	313

(b)

	nm
Heteroannular diene	214
Double bond extending conjugation	30
Exocyclic double bond $\times 3$	15
Ring residue $\times 5$	25
	284

(c)

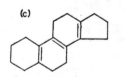

	nm
Homoannular diene	253
Double bond extending conjugation	30
Ring residue $\times 8$	40
	323

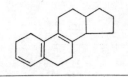

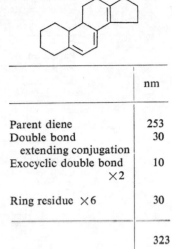

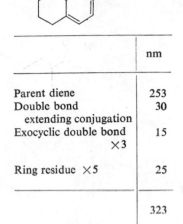

	nm
Parent diene	253
Double bond	30
extending conjugation	
Exocyclic double bond	0
Ring residue ×6	30
	313

	nm
Parent diene	253
Double bond	30
extending conjugation	
Exocyclic double bond ×2	10
Ring residue ×6	30
	323

	nm
Parent diene	253
Double bond	30
extending conjugation	
Exocyclic double bond ×3	15
Ring residue ×5	25
	323

	nm
Parent diene	214
Double bond	30
extending conjugation	
Exocyclic double bond ×3	15
Ring residue ×5	25
	284

	nm
Parent diene	214
Double bond	30
extending conjugation	
Exocyclic double bond ×5	25
Ring residue ×7	35
	304

	nm
Parent diene	253
Double bond	30
extending conjugation	
Exocyclic double bond ×1	5
Ring residue ×5	25
	313

PROBLEM 4

(a)

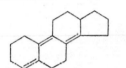

	nm
Parent enone X=R	215
α − R	10
	225
Observed	220

(b)

		nm
Parent enone	X=R	215
α − R		10
β − R		12
		237
Observed		236

(c)

		nm
Parent enone	X=R	215
α − R		10
β − R		12
		237
Observed		233

(d)

		nm
Parent enone	X=R	215
α − R		10
β − R×2		24
		249
Observed		243

(e)

		nm
Parent enone	X=R	215
β − R		12
		227
Observed		225

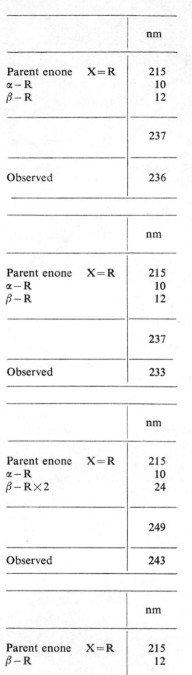

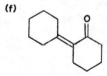

(f)

		nm
Parent enone	X=R	215
α−R		10
β−R×2		24
Exocyclic double bond		10
	×2	
		259
Observed		256

PROBLEM 5

	nm
Parent enone	215
β R	12
	227
Observed	244

(a)

alternatively

	nm
Parent enone	215
β−R×2	24
Exocyclic double bond	5
	244
Observed	244

	nm
Parent enone	215
Exocyclic double bond	5
Double bond extending conjugation	30
$\beta - R$	12
$\delta - R$	18
	280
Observed	284

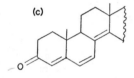

	nm
Parent enone	215
Exocyclic double bond $\times 3$	15
Double bond extending conjugation $\times 2$	60
$\beta - R$	12
$(\delta + 1)$	18
$(\delta + 2) \times 2$	36
	356
Observed	348

(d)

	nm
Parent enone	215
Exocyclic double bond	5
Double bond extending conjugation	30
Homoannular diene	39
$\alpha - R$	10
$\delta - R$	18
	315
Observed	314

	nm
Parent enone	215
Exocyclic double bond	5
$\beta - R \times 2$	24
	244

alternatively

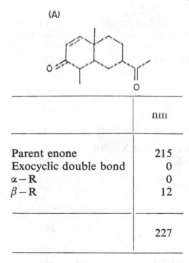

(e)

	nm
Parent enone	215
Double bond extending conjugation	30
Homoannular diene	39
$\alpha - R$	10
$\beta - R$	12
$\gamma - R$	18
	324
Observed	256
	327

PROBLEM 6

(A)

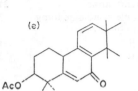

(B)

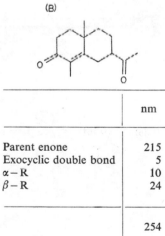

	nm
Parent enone	215
Exocyclic double bond	0
$\alpha - R$	0
$\beta - R$	12
	227

	nm
Parent enone	215
Exocyclic double bond	5
$\alpha - R$	10
$\beta - R$	24
	254
Observed	252

PROBLEM 7

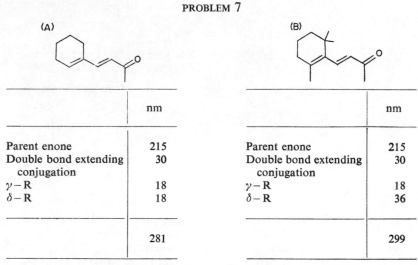

(A)	nm
Parent enone	215
Double bond extending conjugation	30
$\gamma - R$	18
$\delta - R$	18
	281

(B)	nm
Parent enone	215
Double bond extending conjugation	30
$\gamma - R$	18
$\delta - R$	36
	299

Observed 281 nm; ε 20,800 cm² mole⁻¹   Observed 296 nm; ε 10,700 cm² mole⁻¹

B is reduced compared with A by steric inhibition arising from interference between β—H and δ—CH₃.

PROBLEM 8

(a)

	nm
Parent enone X=OR	193
$\alpha - R$	10
$\beta - OR$	30
	233
Observed	236

(b)

	nm
Parent enone X=OR	193
$\alpha - R$	10
$\beta - R$	12
$\beta - OR$	30
	245
Observed	248

(c) COOH

	nm
Parent enone X=OH	193
α − R	10
β − R	12
	215
Observed	212

(d) CHOOEt

	nm
Parent enone X=OR	193
Exocyclic double bond	5
β − R × 2	24
	222
Observed	222

(e) COOH

	nm
Parent enone X=R	193
α − R	10
β − R	12
Exocyclic double bond in C₇ ring	5
	220
Observed	222

	nm
Parent enone X=OR	193
Double bond endocyc- lic five-membered ring	5
α−R	10
β−R	12
	220
Observed	218

(f)

PROBLEM 9

	nm
Parent enone X=R	193
Endocyclic double bond in five-membered ring	10
Double bond increasing conjugation	30
β−R	12
γ−R	18
δ−R×2	36
	299
Observed	288

The hypsochromic (blue) shift is due to strain in the five-membered rings and steric hindrance between the α-H and the (δ+1)-H of the methyl group.

PROBLEM 10

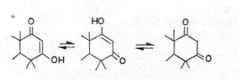

	nm
Parent enone X=R	215
β−R	12
β−OH	30
	257

In alkaline solution

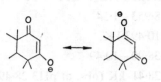

In which the π-system is resonance stabilized so that the transition will occur at lower energy, i.e. longer wavelength.

PROBLEM 11

(a) Anisole

From Table 15: $\Delta\bar{v} = (l_1+v_1^2)$

From Table 16: $l_1 = 0{\cdot}51, \quad v_1 = 1{\cdot}00.$

Hence $\Delta\bar{v} = (0{\cdot}51+1{\cdot}00)$

$= 1{\cdot}51 \text{ kK},$

$\bar{v} = 38{\cdot}90-1{\cdot}51$

$= 37{\cdot}39 \text{ kK} = 3{\cdot}739 \ \mu\text{m}^{-1} \text{(obs. } 3{\cdot}717 \ \mu\text{m}^{-1}\text{)}.$

(b) *p*-Toluidine

From Table 15: $\Delta v_{1,4} = (l_1+v_1^2)+(l_4+v_4^2)+2v_1v_4.$

From Table 16: $CH_3 \ l_1 = 0{\cdot}51, \quad v_1 = 0{\cdot}28,$

$NH_2 \ l_4 = 1{\cdot}11, \quad v_4 = 1{\cdot}44,$

$\Delta\bar{v} = (0{\cdot}51+0{\cdot}26)+(1{\cdot}11+2{\cdot}074)+0{\cdot}806$

$= 4{\cdot}76 \text{ kK},$

$\bar{v} = 38{\cdot}90-4{\cdot}76 = 34{\cdot}14 \text{ kK (obs. } 34{\cdot}97 \text{ kK).}$

(c) *m*-Bromobenzoic acid

From Table 15: $\Delta\bar{v}_{1,3} = (l_1+v_1^2)+(l_3+v_3^2)-v_1v_3.$

From Table 16: $Br \ l_1 = 0{\cdot}90, \quad v_1 = 0{\cdot}11,$

$COOH \ l_3 = 1{\cdot}17 \quad v_3 = -1{\cdot}05,$

$\Delta\bar{v}_{1,3} = (0{\cdot}90+0{\cdot}0121)+(1{\cdot}17+1{\cdot}103)-(-0{\cdot}1155)$

$= 3{\cdot}30 \text{ kK}.$

$v = 38{\cdot}90-3{\cdot}30 = 35{\cdot}60 \text{ kK (obs. } 35{\cdot}09 \text{ kK).}$

(d) *o*-Nitrophenol

From Table 15: $\Delta\bar{v}_{1,2} = (l_1+v_2^1)+(l_2+v_2^2)-v_1v_2.$

From Table 16: $NO_2 \ l_1 = 5{\cdot}16, \quad v_1 = -1{\cdot}05,$

$OH \ l_2 \ \ = 0{\cdot}64, \quad v_2 = 1{\cdot}00,$

$O^{\ominus} \ l_2 = 2{\cdot}39, \quad v_2 = 1{\cdot}29,$

$\Delta\bar{v}_{1,2} = (5{\cdot}16-1{\cdot}103)+(0{\cdot}64+1{\cdot}00)-(-1{\cdot}05)$

$= 5{\cdot}933 \text{ kK}.$

From Table 17:   *Ortho* correction NO_2—OH $= 1.54$ kK.

Hence $\Delta\bar{v}_{1,2} = 8.953 + 1.54$

$= 10.493$ kK,

$\bar{v} = 38.90 - 10.49$

$= 28.41$ kK (obs. at pH3 28.49 kK).

$\Delta\bar{v}_{1,2} = (5.16 + 1.103) + (2.39 + 1.664) - (-1.355)$

$= 11.672$ kK.

From Table 17: *Ortho* correction NO_2—$O^{\ominus} = 1.47$ kK.

Hence $\Delta\bar{v}_{12} = 11.672 + 1.47$

$= 13.142$ kK,

$\bar{v} = 38.90 - 13.14$

$= 25.76$ kK (obs. in 0.1 м NaOH 24.04 kK).

PROBLEM 12

(a) *m*-Bromobenzoate anion

$$\Delta\bar{v} = K(\Sigma_m l'_m).$$

From Table 18: $K = 0.81, \quad l'_{Br} = 1.51, \quad l'_{COO^{\ominus}} = 4.49,$

$\Delta\bar{v} = 0.81 (6.00) = 4.86$ kK.

$\bar{v} = 49.10 - 4.86 = 44.24$ kK $= 4.425 \ \mu m^{-1}$ (obs. 4.425 μm^{-1}).

(b) *m*-Nitroanisole

$$\Delta\bar{v} = 0.81(11.89 + 3.05) = 12.10 \text{ kK},$$

$\bar{v} = 49.10 - 12.10 = 37.00$ kK $= 3.700 \ \mu m^{-1}$ (obs. 3.656 μm^{-1}).

(c) *o*-Chlorobenzoic acid

$$\Delta\bar{v} = 0.81 (1.40 + 5.65) = 0.81(7.05) = 5.71 \text{ kK},$$

$\bar{v} = 49.10 - 5.71 = 43.39$ kK $= 4.339 \ \mu m^{-1}$ (obs. 4.367 μm^{-1}).

(d) *o*-Chloronitrobenzene

$$\Delta\bar{v} = 0.81 (1.40 + 11.89) = 0.81(13.29) = 10.765 \text{ kK},$$

$\bar{v} = 49.10 - 10.77 = 38.33$ kK $= 3.833 \ \mu m^{-1}$ (obs. 3.846 μm^{-1}).

(e) *p*-Chloroaniline

$$\Delta\bar{v} = K(\Sigma_m l'_m) + (x_1 - x_4)^2$$

$$= 0.81(1.40 + 5.65) + (1.42 - 0.20)^2$$

$$= 0.81 (7.05) + (1.96) = 7.67 \text{ kK}.$$

$\bar{v} = 49.10 - 7.67 = 41.43$ kK $= 4.143 \ \mu m^{-1}$ (obs. 4.184 μm^{-1}).

(f) *p*-Chlorophenoxide anion

$$\Delta\bar{v} = 0.81(1.40 + 6.58) + (1.62 - 0.2)^2$$

$$= 6.464 + 2.016 = 8.480 \text{ kK},$$

$\bar{v} = 49.10 - 8.48 = 40.62$ kK $= 4.062 \ \mu m^{-1}$ (obs. 4.098 μm^{-1}).

An Introduction to Electron Spin Resonance Spectroscopy

H. W. WARDALE

Department of Chemistry and Applied Chemistry, University of Salford

Principles

1. INTRODUCTION: SCOPE AND LIMITATIONS

Many of the applied research projects undertaken during the Second World War produced results which have subsequently benefited research workers in pure science. One of the best examples of this unusual line of development is provided by the research carried out on radar. This resulted in such great improvements in microwave technology that by the end of the war all the necessary components for the investigation of a previously unexplored region of the electromagnetic spectrum had become available. Two new branches of spectroscopy have emerged and both chemists and physicists have been provided with a wealth of subtle information. The extension of orthodox absorption spectroscopy into the microwave region has resulted in the precise determination of the structure of many molecules, but information is obtainable only from the rotational spectra of gaseous polar molecules at low pressures. The development of electron spin resonance[†] (e.s.r.) spectroscopy using microwave components has greatly improved our understanding of the interactions of electrons in molecules, particularly in liquids and solids.

Successful e.s.r. experiments were first performed by Zavoisky. His results were published[1] in 1945 just a few months before the first reports of successful n.m.r. experiments. Physicists at the Clarendon Laboratory in Oxford were responsible for much of the systematic development of e.s.r. spectroscopy during the immediate post-war period, and for a number of years the technique remained almost entirely a province of the physicists. Consequently there was scant appreciation of the potential applications of e.s.r. spectroscopy to the solution of problems in chemistry. The necessary stimulus required by chemists to persuade them that here was a new technique requiring exploitation was provided by the introduction of commercial high resolution spectrometers in the middle of the fifties. Since that time e.s.r. spectroscopy has not only been applied with considerable success

---

[†] Electron paramagnetic resonance, electron magnetic resonance and paramagnetic resonance are alternative names that may be encountered

to many branches of chemistry but has also become a valuable physical aid in biological and medical research.

Just as n.m.r. is restricted to systems containing nuclei having a resultant nuclear angular momentum so e.s.r. can only be observed in systems having a resultant electronic angular momentum. Whereas a resultant nuclear angular momentum arises solely from the intrinsic spin of nuclei, a resultant electronic angular momentum is made up of contributions from both the intrinsic spin and the orbital motion of electrons. In many substances all the electrons are paired and there is no resultant electronic angular momentum because the spin angular momentum of one electron is cancelled by that of the electron with which it is paired and the orbital angular momentum is often zero. Consequently e.s.r. is normally only observed in systems containing unpaired electrons. This may seem to be a severe restriction on the technique as paramagnetic substances are usually considered to be rather unusual, but the many successful applications of e.s.r. spectroscopy demonstrate that paramagnetism is more frequently encountered in chemical systems than is still generally realized. Most of the early workers concentrated their attention on single crystals of compounds of the transition and inner-transition metals containing ions having unpaired d- or f-electrons. More recently studies have been made of conduction electrons in metals and semiconductors, of solvated electrons in solutions of the alkali metals in liquid ammonia, amines, and ethers, and of trapped electrons in colour centres and other damage sites produced by irradiation. Organic chemists have discovered that e.s.r. spectroscopy provides valuable information about free radicals and triplet state species in both liquids and solids. A well resolved spectrum can contain a wealth of analytical, structural, and kinetic data. The analytical data may serve to identify the radical and allow its concentration to be estimated. Concentrations as low as 10^{-14}M can be detected by the best spectrometers. The structural data indicate the stereochemistry in the neighbourhood of the unpaired electron, and in certain circumstances bond angles can be calculated. The kinetic data measure the lifetime of the radical and enable the rates of very fast radical reactions to be determined.

2. THE RESONANCE CONDITION

In the absence of a magnetic field the $(2J+1)$ energy levels associated with a chemical entity having a total electronic angular momentum J are degenerate. When a magnetic field H is applied to this system, the degeneracy is lifted giving $(2J+1)$ distinct levels having energies

$$E = E_0 + m_J g\beta H, \tag{1}$$

where E_0 is the energy of the degenerate levels, m_J the total electronic magnetic quantum number taking the values $(J), (J-1), \ldots, -(J-1), -(J), \beta$ the Bohr magneton having the value $9 \cdot 2732 \times 10^{-28}$ J gauss^{-1}, and g the dimensionless spectroscopic splitting factor that is constant for each radical and is a measure of the rate of divergence of the energy levels in the applied magnetic field. The interaction of the radical with its environment in liquid or solid matrix almost completely quenches the orbital contribution to the total electronic angular momentum. In this situation it is a good approximation to consider the total electronic angular momentum to be made up of the spin contribution only. The energy levels in the applied field are then given by

$$E = E_0 + m_S g\beta H,\tag{2}$$

where m_S is the electronic spin quantum number taking the values $(S), (S-1), \ldots, -(S-1), -(S)$. Most free radicals are doublet states having only one unpaired electron. In such cases $S = \frac{1}{2}$, $m_S = \pm\frac{1}{2}$, and there are just two diverging energy levels (Fig. 1) with an energy difference of $g\beta H$. Transitions between the levels can be induced by irradiating the

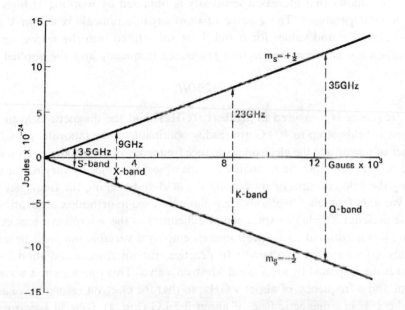

FIG. 1. Diverging electron spin energy levels in a magnetic field. Transitions correspond to $g = 2$ at various commonly employed frequencies.

sample with electromagnetic radiation of a frequency such that the energy of this radiation is precisely the same as the energy difference between the levels. The fundamental equation of e.s.r. spectroscopy is

$$h\nu = g\beta H.\tag{3}$$

When this relationship between the frequency ν and the applied magnetic field H is realized, resonance occurs. Electrons in the lower level absorb energy from the incident radiation and are excited into the upper level, while electrons in the upper level emit energy and fall back to the lower level. Transitions in either direction are equally probable, but if the system is in thermal equilibrium with its surroundings then the lower level has a slightly greater population than the upper one and absorption of radiation occurs slightly more frequently than stimulated emission. The overall result is that there is a net absorption of energy from the incident radiation, and it is this loss of incident radiation that is detected, amplified, and displayed by the spectrometer.

The sensitivity of e.s.r. spectroscopy is governed by the difference in population between the two levels. This is given by the Maxwell–Boltzmann equation

$$\frac{N(-\frac{1}{2})}{N(+\frac{1}{2})} = e^{g\beta H/kT} \approx 1 + \frac{g\beta H}{kT}, \tag{4}$$

where $N(-\frac{1}{2})$ and $N(+\frac{1}{2})$ are the populations of the $m_S = -\frac{1}{2}$ and the $m_S = +\frac{1}{2}$ levels respectively. It follows that increased sensitivity is obtained by working at high magnetic fields and low temperatures. The g value of most organic radicals is within 1% of 2·00. When this value of g and values for h and β are substituted into the resonance equation (3), the relationship between the electron resonance frequency and the applied magnetic field becomes

$$v = 2·80H, \tag{5}$$

where the frequency is measured in gigahertz (GHz)[†] and the magnetic field in kilogauss (kG). Magnetic fields of up to 40 kG are readily obtainable in the laboratory, and for fields of this order of magnitude the electron resonance frequency occurs in the microwave region of the spectrum. In principle the resonance may be observed either by varying the frequency and keeping the field constant or by varying the field and keeping the frequency constant. Although variable frequency methods are commonly used in orthodox absorption spectroscopy, these present formidable experimental difficulties in the microwave region, and for this reason all conventional e.s.r. spectrometers employ a variable magnetic field and operate at a constant microwave frequency. In practice, the microwave radiation most commonly used is that supplied by an X-band klystron valve. This operates at a wavelength of about 3 cm and a frequency of about 9 GHz, so that the electron resonance of an organic free radical occurs in a magnetic field of about 3·2 kG (Fig. 1). Q-band spectrometers are gaining popularity as a greater sensitivity is achieved by operating at a higher frequency (35 GHz), but much of this gain is lost because of the need to use much smaller samples. Instruments operating at 23 GHz (K-band) and 3·5 GHz (S-band) are also available commercially. No further discussion of instrumentation will be given here, but detailed accounts of modern spectrometers and spectrometer circuits can be found in books by Poole[2] and Wilmshurst,[3] in reviews by Anderson[4] and Fraenkel,[5] and in the technical literature available from the various manufacturers.

The e.s.r. transition probability is proportional to the population of the level. Consequently the thermal equilibrium is upset during resonance, and the populations of the two levels tend to equalize. This results in the eventual disappearance of the signal as there is no net absorption of radiation. The loss of resonance in this way is called saturation and is promoted by a high incident microwave power. Saturation is not usually observed if the

[†] 1 GHz = 10^9 Hz $\equiv 10^9$ c/s.

microwave power is kept low as the thermal equilibrium between the two levels is maintained by non-induced transitions from the upper to the lower level. The excess energy of the upper state is dissipated by processes other than the stimulated emission of a microwave quantum. Any process that contributes to the maintenance of the thermal equilibrium between the levels is called a relaxation process. In the most important of these—the spin lattice relaxation process—the excess energy is transferred to the surroundings, or lattice, where it causes rotational or vibrational excitations.

The position of the e.s.r. line is measured by the ratio of the microwave frequency to the external field at resonance, but as β and h are constant this ratio is effectively labelled by the g value. The g value of an atom having a total angular momentum of J is given by the Landé formula

$$g = \frac{3J(J+1)+S(S+1)-L(L+1)}{2J(J+1)} . \tag{6}$$

This gives a g value of 2 for an atom having the orbital contribution to the electronic angular momentum completely quenched. The actual free spin g value for radicals is 2·00232 as there is a small quantum electrodynamic correction. There is only one possible e.s.r. transition in the simple two-level system (Fig. 1). This transition has $\Delta m_S = 1$ and the spectrum consists of a single line having the free spin g value. Most spectrometers are constructed so that the signal is normally presented as either the first or the second derivative of the absorption line. The illustrations in this chapter are all of first derivative spectra.

The fact that all e.s.r. lines have a finite width implies that the resonance occurs over a range of values of the external field rather than at the precise value suggested by the resonance equation (3). The e.s.r. energy levels are not well defined in transient radicals. The shorter the lifetime of the radical, the shorter is the time available for measuring the energy of the levels and the greater is the uncertainty in the determination. This uncertainty in the energy levels (Fig. 2) is reflected in an uncertainty in the magnetic field required to observe the resonance so that the observed linewidth is closely related to the lifetime of the radical. Relaxation processes can make an important contribution to the linewidth of stable radicals. By shortening the lifetime of the upper level, efficient relaxation makes the energy of this level less certain and hence produces line broadening. The interaction of the unpaired electron with its environment can also contribute to the linewidth of stable radicals. These environmental interactions produce random local fields in the sample such that the resultant field experienced by the unpaired electron is not necessarily equal to the external field and the resonance occurs over a range of values of the external field. Line broadening by this process is of particular importance for radicals in the solid state. In liquids the random local fields are usually averaged to zero by rapid molecular tumbling. The above discussion seems to indicate that for the two-level spin only system (Fig. 1) there are only two parameters,—the line intensity and the linewidth—from which information about the radical and its environment can be extracted. Although both parameters can in suitable circumstances provide worthwhile information, it is probable that if they were the only quantities obtained from e.s.r. spectra then this branch of spectroscopy would have remained a mere curiosity.

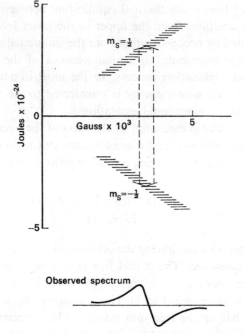

Fig. 2. Broadening of the resonance resulting from uncertainty in the electron spin energy levels.

Electron spin resonance spectra provide much more information when the local magnetic fields produced within the sample not only result in the broadening of the resonance but also bring about a shift of the resonance from the expected position or even lead to a splitting of the resonance into a number of components. These internal fields may be conveniently divided into two categories: fields that arise from interactions between the unpaired electron and neighbouring magnetic nuclei, and fields that result from the incomplete quenching of the orbital angular momentum. The former are, to a first approximation, independent of the magnitude of the external field while the latter are proportional to it.

3. THE *g* SHIFT

A shift of the resonance from the free-spin *g* value indicates that the spin only approximation is invalid. The application of the external field can induce orbital motion in the plane perpendicular to the direction of the external field (Fig. 3). In terms of molecular orbital theory this orbital momentum may be visualized as the mixing of excited state configurations with the ground state of the radical. If the induced excitation is that of an unpaired electron from the half-filled level into a higher empty orbital (Fig. 3), then the orbital momentum produces a magnetic field that is opposed to the external field and the resultant field at the unpaired electron is smaller than the external field. Resonance occurs at a larger value of the external field than that required when there is no orbital angular momentum and the measured *g* value is less than the free-spin value. Positive *g* shifts are observed when the induced excitation is that of an electron from a filled level into the half-filled level.

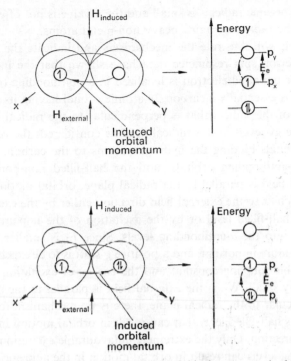

FIG. 3. Induced orbital momentum produces local fields that apparently shift the *g*-value.

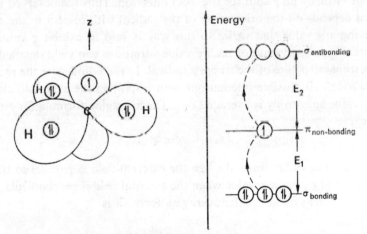

FIG. 4. Electronic structure and energy levels of the methyl radical.

In this situation the orbital motion produces a local field that supplements the external field. The size of the g shift is inversely proportional to the energy of excitation and directly proportional to the spin orbit coupling constant of the atom on which the unpaired electron is localized. This constant is a measure of the effectiveness of the interaction between the orbital motion and the electron spin; its magnitude is very strongly dependent on the atomic number of the atom and on the number of electrons in the valence shell. Consequently the *g* shift in most organic radicals is small and the *g* value is not often a very informative parameter unless the radical contains heavy non-metal atoms.

A simple example will illustrate the mechanisms and indicate the significance of quite small *g* shifts. Electron spin resonance data have shown that the methyl radical CH_3 is planar and that the unpaired electron is localized in a non-bonding orbital on the carbon atom. This orbital is essentially a carbon $2p$ atomic orbital having its axis along the three-fold rotation axis of the radical that is perpendicular to the radical plane (Fig. 4). Only three groups of energy levels in this radical need be considered: the very stable fully occupied *σ*-bonding orbitals binding the hydrogen atoms to the carbon, the correspondingly unstable empty *σ*-antibonding orbitals, and the half-filled non-bonding level (Fig. 4). When the external field is parallel to the radical plane, orbital motion can be induced in the plane perpendicular to the external field direction either by the excitation of *σ*-bonding electrons into the half-filled level or by the excitation of the unpaired electron from the non-bonding level into the *σ*-antibonding levels. Since E_1 is smaller than E_2, the former process will be the more important, and a positive *g* shift is to be expected. As carbon has a very small spin orbit coupling constant and the energy of excitation is very large, this *g* shift should be very small. When the external field is parallel to the axis of the half-filled level and perpendicular to the radical plane, there is no mechanism for promotion into or excitation from this half-filled level that can result in orbital motion in the plane perpendicular to the field direction. Only the extremely unfavourable excitation from the *σ*-bonding to the *σ*-antibonding levels can result in orbital motion in the appropriate plane, and hence there should be virtually no *g* shift for this field direction. Thus the observed *g* value of the methyl radical depends on the orientation of the radical with respect to the external field. A radical having a *g* value that varies in this way is said to exhibit *g* value anisotropy. In the solid state the observation of such *g* value variations can yield detailed information about the electronic structure of the trapped radical. In the liquid state the radical is tumbling rapidly through all possible orientations with respect to the external field, and consequently the *g* value anisotropy is averaged out. The averaging formula for an axially symmetrical radical is

$$3g_{av} = g_{\parallel} + 2g_{\perp}, \tag{7}$$

where $g_{\parallel}$ is the *g* value of the radical when the external field is parallel to the axis of the half filled orbital and $g_{\perp}$ is the *g* value when the external field is perpendicular to that axis. In the absence of axial symmetry, the averaging formula is

$$3g_{av} = g_x + g_y + g_z, \tag{8}$$

where x, y, and z are the mutually perpendicular principal axes of the radical. The average *g* values observed for hydrocarbon radicals in solution are generally within the range 2·0025 to 2·0029.

The similarity between the *g* shift in e.s.r. and the chemical shift in n.m.r. can now be appreciated. Both effects are the result of electronic orbital motion induced by the external field, and in both cases the induced local field is proportional to the external field. Some confusion in the interpretation of spectra can arise unless it is always remembered that a chemical shift is associated with magnetic nuclei and the *g* shift with unpaired electrons. A high resolution proton magnetic resonance spectrum has a chemical shift associated with each magnetically different proton in the molecule, but an e.s.r. spectrum can have only one *g* shift because there is only one unpaired electron in the radical.

4. ELECTRON-NUCLEAR HYPERFINE INTERACTIONS

The description of the resonance phenomenon given so far is applicable only to radicals that do not contain any magnetic nuclei. Most organic radicals possess magnetic nuclei, in particular protons and nitrogen nuclei, and by far the most chemically interesting modifications to e.s.r. absorptions are those that result from the interaction of the spin of the unpaired electron in the radical with the spins of neighbouring magnetic nuclei. This interaction produces a splitting of the energy levels characterized by the electronic spin quantum number, m_S, and results in the splitting of the resonance absorption into a number of hyperfine components. The hyperfine coupling is, to a first approximation, independent of the size of the external field and is the e.s.r. equivalent of the spin–spin coupling in n.m.r. spectroscopy.

The application of an external magnetic field H to a system having an electronic angular momentum S and containing a single nuclear magnetic spin I, lifts the degeneracy of the system and gives a total of $(2S+1)(2I+1)$, different energy levels having energies given by

$$E = E_0 + m_S g \beta (H + m_I A). \tag{9}$$

The nuclear spin quantum number m_I can have the $(2I+1)$ values (I), $(I-1)$, $\ldots$, $-(I-1)$, (I), for each of the $(2S+1)$ allowed values of m_S. The parameter A is called the hyperfine coupling constant. It is measured in gauss when defined by eqn. (9), and this is the unit most frequently encountered. Some spectroscopists prefer to quote hyperfine couplings in energy units, and both megahertz (MHz) and reciprocal centimetres (cm^{-1}) are used. The relationships between these units are

$$1\,G = g\beta/10^6 h = g/0.71449\,\text{MHz} \tag{10}$$

and

$$1\,G = g\beta/hc = g/2.1420 \times 10^{-4}\,\text{cm}^{-1}. \tag{11}$$

These reduce to

$$1\,G = 2.802\,\text{MHz} = 9.348 \times 10^{-5}\,\text{cm}^{-1} \tag{12}$$

when the radical has a free spin *g* value. There are $2(2I+1)$ different energy levels for radicals containing a single unpaired electron. The energy levels of radicals having a single proton $(I = \frac{1}{2})$ and a single $^{14}_{7}$N nucleus $(I = 1)$ are shown in Figs. 5 and 6 respectively. Transitions between these levels can be induced in the usual way, but the only allowed e.s.r. transitions are those in which $\Delta m_S = \pm 1$ and $\Delta m_I = 0$. Transitions in which $\Delta m_S = 0$ and $\Delta m_I = \pm 1$ are n.m.r. transitions, while those involving changes in both m_S and m_I are forbidden. It is apparent from eqn. (9) that the actual field experienced by the unpaired electron at resonance H^* is no longer equal to the external field H but is given by

$$H^* = H + m_I A. \tag{13}$$

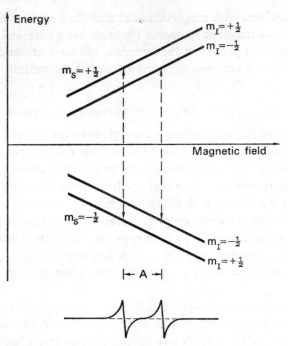

FIG. 5. Energy levels and transitions for the hyperfine interaction between an unpaired electron and a nucleus of spin $^1/_2$.

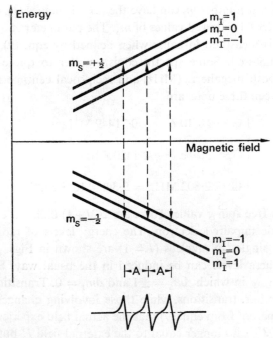

FIG. 6. Energy levels and transitions for the hyperfine interaction between an unpaired electron and a nucleus of spin 1.

Since m_I can take $(2I+1)$ different equally spaced values, then there are $(2I+1)$ different values of H that can produce a resultant field of H^* at the unpaired electron. Thus the spectrum consists of $(2I+1)$ equally spaced lines with a splitting of A gauss between adjacent lines. The splitting of the electron resonance levels by the magnetic nucleus is usually very much smaller than the overall splitting of the electron levels by the external field. Consequently the populations of energy levels having the same value of m_S, but different values of m_I are almost equal, and the hyperfine lines arising from an interaction with just one magnetic nucleus are of equal intensity. The g value of a series of hyperfine lines is found from the equation

$$hv = g\beta H^*. \tag{14}$$

The field value H^* is found by taking the mean of the observed values of the external field H for each of the $(2I+1)$ lines.

Interaction with more than one magnetic nucleus results in the further splitting of the energy levels and leads to further complications in the observed spectrum. In general the interaction of one unpaired electron with a series of magnetic nuclei having spins $I_1, I_2, \ldots, I_n$ leads to the observation of $(2I_1+1)(2I_2+1) \ldots (2I_n+1)$ allowed transitions. A six-line spectrum should be observed from a radical containing a single $^{14}_{7}N$ nucleus ($I = 1$) and a single proton ($I = \frac{1}{2}$), while a radical containing a $^{17}_{8}O$ nucleus ($I = \frac{5}{2}$), a $^{19}_{9}F$ nucleus ($I = \frac{1}{2}$) and a proton should have a twenty-four-line spectrum. If the unpaired electron interacts with a number of similar nuclei which by virtue of their geometrical conformation within the radical are magnetically equivalent, that is they produce exactly the same magnetic effect at the unpaired electron, then an overlapping of transitions occurs and different transitions are observed at the same value of the external field. This overlapping of transitions results in a decrease in the number of lines observed and a characteristic modification of the intensities of the lines. The interaction of an unpaired electron with n equivalent nuclei of spin I gives rise to $(2nI+1)$ equally spaced lines whose intensities are given by the coefficients of the expansion $(x+x^2+x^3 \ldots x^m)^n$, where $m = (2I+1)$. The three equivalent protons in the methyl radical produce a four line spectrum with relative intensities 1:3:3:1 (Fig. 7). A radical containing six equivalent protons, such as the radical anion of benzene, gives rise to a spectrum having seven equally spaced lines with the relative intensities 1:6:15:20:15:6:1, while the perdeutero form of this radical should have thirteen equally spaced lines with the relative intensities 1:6:21:50:90:126:141:126:90:50:21:6:1. A comparatively simple analysis of the number and relative intensities of the hyperfine lines in an e.s.r. spectrum can frequently result in the unambiguous identification of the radical. A more detailed analysis of the magnitude of these splittings may provide information about the structure of the radical.

There are two separate interactions that give rise to hyperfine coupling between an unpaired electron and a neighbouring magnetic nucleus: an orientation independent or isotropic interaction, and an orientation dependent or anisotropic interaction. The observed hyperfine coupling of radicals in solution is normally due only to the isotropic term as the rapid random tumbling of the radical through all possible orientations with respect to the direction of the external field results in the anisotropic term being averaged to zero. In the

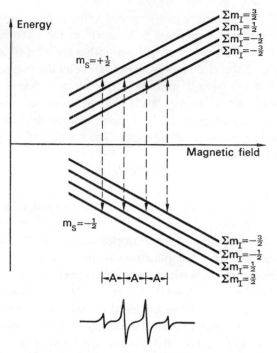

Fɪɢ. 7. Energy levels and transitions for the hyperfine interaction between an unpaired electron and three equivalent nuclei of spin $^1/_2$ as encountered in the methyl radical.

solid state or in highly viscous solutions, this tumbling either does not occur or occurs only slowly, and the observed hyperfine coupling is due to the combination of isotropic and anisotropic terms.

The isotropic coupling is a manifestation of the Fermi contact interaction. This interaction is directly proportional to the probability of finding the unpaired electron at the interacting magnetic nucleus. The probability, and hence the contact term, can only be non-zero when the orbital of the unpaired electron has some s character on that nucleus for all other orbitals have nodes through the nucleus. The orientation independence of this contribution to the hyperfine coupling is a result of the spherical symmetry of s orbitals. The magnitude of the isotropic interaction is given by

$$a_{\text{iso}} = \frac{8\pi}{3} g_N \beta_N |\psi(0)|^2, \tag{15}$$

where g_N is the g value of the magnetic nucleus β_N is the nuclear magneton, and $|\psi(0)|^2$ is the probability of finding the electron at that nucleus. By comparing the experimentally determined values of the isotropic coupling a_{iso} with the parameter a_0 corresponding to the expected isotropic coupling for an unpaired electron entirely localized in that s orbital, it is possible to estimate the s character of the orbital of the unpaired electron on that magnetic nucleus. Accurate values for this parameter (a_0) have been obtained experimentally for hydrogen and alkali metal nuclei from molecular beam measurements. Values of a_0 for other magnetic nuclei are calculated from computed values of $|\psi(0)|^2$ obtained from self-consistent-field wave functions. Comparatively large values of the a_{iso}/a_0 ratio, greater than 0·1, arise from radicals in which the unpaired electron is mainly localized in a σ molecular orbital. These radicals are called σ radicals. The value of the a_{iso}/a_0 ratio indicates the extent of hybridization in the unpaired electron orbital. Frequently somewhat smaller values of this ratio, less than 0·05, are observed in radicals in which no isotropic coupling is expected as the unpaired electron should be located in a π orbital having a nodal plane through the magnetic nucleus. Radicals of this type are called π radicals. Molecular orbital theory provides a satisfactory explanation of the origin of these splittings in terms of the configurational mixing of excited states having the unpaired electron in an orbital possessing some s character with the ground state of the radical. A further consequence of this mixing is that a small isotropic coupling to a second magnetic nucleus in the nodal plane that is bonded to the first nucleus is also observed. The spectrum of the $^{13}_{6}CH_3$ radical illustrates both these effects. Although the unpaired electron is located in an atomic p orbital on the carbon atom with all four nuclei lying in the nodal plane of that orbital (Fig. 4), an eight-line isotropic spectrum is observed. The interaction of the unpaired electron with the $^{13}_{6}C$ nucleus ($I = \frac{1}{2}$) produces a doublet splitting of 38 G, representing an a_{iso}/a_0 ratio of 0·034. Each of these compoents is further split into a characteristic quartet by the interaction with the three equivalent protons, and the observed 23 G splitting represents an a_{iso}/a_0 ratio of 0·045. An alternative description of the mechanism of this interaction can be more easily visualized than the molecular orbital description. The unpaired electron in the p orbital brings about polarization of the spins of the bonding electron pairs. A consideration of Hund's rule indicates that the situation shown in Fig. 8 is more favourable energetically. The isotropic coupling at the carbon nucleus is the result

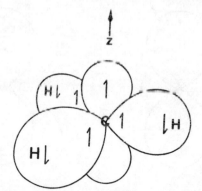

FIG. 8. Spin polarization of the electrons in C–H bonds in the methyl radical.

of a Fermi contact interaction between that nucleus, and the spin polarized electrons having their spins parallel to the spin of the electron in the p orbital, whereas the isotropic coupling at the protons is a result of the Fermi contact interaction between the protons and the spin polarized electrons having their spins antiparallel to that of the electron in the p orbital. The former situation is said to result in a positive isotropic coupling and the latter in a negative isotropic coupling Normally it is impossible to distinguish between positive and negative electron-nuclear coupling constants using e.s.r. methods, but the relative signs of such interactions may be determined using nuclear magnetic resonance (n.m.r.) and electron nuclear double resonance (e.n.d.o.r.) spectroscopy.

The anisotropic hyperfine coupling is the result of a classical dipole–dipole interaction between the electronic and nuclear magnetic moments. The magnitude of this interaction is inversely proportional to the cube of the distance between the electron and the nucleus but is anisotropic as it also depends on the relative orientation of the two spins. This orientation-dependent term averages to zero when the radical is rapidly tumbling in solution. The anisotropic coupling measures the non-*s* character of the orbital of the unpaired electron. The contributions of *d* and *f* orbitals to the unpaired electron orbital of organic radicals can be neglected, so that the anisotropic coupling may be considered to measure the *p* orbital character of the unpaired electron on that magnetic nucleus. The magnitude of the field produced at a point *r* from a magnetic nucleus is

$$H_{\text{local}} = g_N \beta_N r^{-3} (3 \cos^2 \theta + 1)^{\frac{1}{2}}, \tag{16}$$

where θ is the angle between the line joining that point to the magnetic nucleus and the direction of the applied field. When $H \gg H_{\text{local}}$, the effective anisotropic coupling to an electron at that point produced by a magnetic nucleus is the component of the local field parallel to the applied field;

$$a_{\text{aniso}} = g_N \beta_N r^{-3} (3 \cos^2 \theta - 1), \tag{17}$$

and has principal values of $2g_N \beta_N r^{-3}$ and $-g_N \beta_N r^{-3}$, when $\theta = 0°$ and $90°$ respectively (Fig. 9). The *p* orbital character of the unpaired electron can be estimated by comparing the observed principal values of the anisotropic coupling with the principal values that can

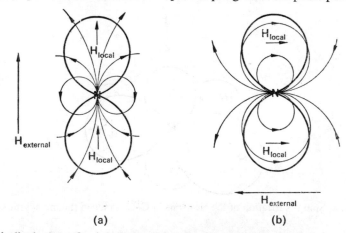

(a)   (b)

FIG. 9. Anisotropic dipolar hyperfine interaction between an electron in a *p* orbital and a magnetic nucleus: (a) external field parallel to the axis of the orbital; (b) external field perpendicular to the axis of the orbital.

be calculated for an unpaired electron entirely localized in that orbital. For any orbital these principal values can be obtained from the expression

$$b_0 - \left(\frac{2l}{2l+3} \right) g_N \beta_N \langle r_{nl}^{-3} \rangle, \tag{18}$$

where $\langle r_{nl}^{-3} \rangle$ is the average value of the cube of the reciprocal of the distance between the electron in the atomic orbital having an orbital quantum number of l and a principal quantum number of n and the magnetic nucleus. This expression becomes

$$b_0 = \tfrac{2}{5} g_N \beta_N \langle r_{np}^{-3} \rangle \tag{19}$$

for a p orbital and has a principal value of $2b_0$ when the external field is parallel to the axis of the orbital, and $-b_0$ when the external field is perpendicular to that axis. Values of $\langle r_{np}^{-3} \rangle$ have been calculated from the self-consistent-field wave functions for the appropriate orbital and, like the calculated values of $|\psi(0)|^2$ for s orbitals, it is not easy to estimate the accuracy of these terms, and the reliability of values of b_0 is difficult to assess. Measurements of both the isotropic and the anisotropic hyperfine interactions for organic radicals can yield estimates of both the s and p orbital character of the unpaired electron on that magnetic nucleus. It is simple to calculate the bond angles about that magnetic nucleus from this hybridization ratio.

5. HYPERFINE SPLITTING OR g VALUE EFFECTS?

A frequent dilemma encountered in the interpretation of e.s.r. spectra is that of deciding whether certain features of a spectrum are produced by different radicals with different g values or by a single radical with hyperfine splittings. The nature of the dilemma and the method of resolving it may be demonstrated by the following example. An X-band spectrometer produces a spectrum from a liquid sample that consists of two lines of equal intensity having g values of 2·000 and 2·010 respectively. At this microwave frequency the splitting between these lines is about 16 G. There are two possible interpretations of this spectrum. The spectrum may be the result of the interaction of the unpaired electron with a magnetic nucleus having $I = \tfrac{1}{2}$ in which case a_{iso} is 16 G and g is 2·005. Alternatively the spectrum may be due to two radicals with different g values and no hyperfine splitting. In certain cases it may be convenient to remove this ambiguity by the isotropic substitution of a magnetic nucleus having a different nuclear spin. If in this example the spectrum results from the interaction of the unpaired electron with a proton, then substitution of this proton by a deuteron results in the observation of an isotropic triplet with a splitting of 2·5 G and a g value of 2·005. The much smaller value of the isotropic splitting is observed because the magnetic moment of the deuteron is considerably smaller than that of the proton. These substitutions are often very difficult to accomplish, and in many cases suitable alternative magnetic nuclei do not exist. Since hyperfine couplings are essentially independent of the magnitude of the external field while g values are related to the ratio of the radiation frequency to the external field, it is possible to distinguish unambiguously between hyperfine coupling and g value effects by obtaining spectra at two substantially different radiation frequencies. If the features in the chosen example are due to two radicals, then the spectrum obtained from a Q-band spectrometer still consists of two lines with g values of 2·000 and 2·010, but since the operating frequency is nearly four times greater, the splitting between the lines is about 60 G. If the two features are the result of a hyperfine interaction, then the Q-band spectrum still consists of two lines separated by about 16 G but no longer having the same g values although the mean g value is still 2·005. Measurements at different frequencies resolve ambiguities in g value and hyperfine coupling information in much the same way as frequency changes in n.m.r. spectroscopy resolve ambiguities in spin–spin coupling and chemical shift information.

6. PARAMETERS ASSOCIATED WITH TRIPLET STATE MOLECULES

The considerable current interest in organic molecules having either triplet ground states or low-lying triplet excited states means that the e.s.r. spectra of these species are of interest to organic chemists. There are two unpaired electrons in these molecules so that $S = 1$ and the allowed values of M_S are $+1$, 0, and -1. The divergence of the three energy levels in the applied field is shown in Fig. 10; the two transitions having $\Delta m_S = 1$ both occur at the same value of the external field and the $\Delta m_S = 2$ transition is strictly forbidden.

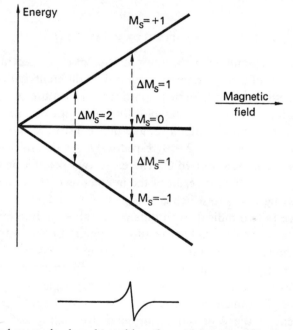

FIG. 10. Simplified energy levels and transitions for a triplet state with no zero-field splitting.

The two unpaired electrons in triplet state species are sufficiently close that the energy levels are considerably modified by the dipolar spin–spin interaction between the two unpaired electrons. This interaction lifts the degeneracy of the levels even in the absence of the external field so that the two $\Delta m_S = 1$ transitions occur at different values of the external field (Fig. 11). Since the zero field splitting originates from a dipolar interaction it is strongly anisotropic and varies with the orientation of the triplet molecule in the external field. Consequently the two $\Delta m_S = 1$ transitions are also anisotropic and the resonance may be

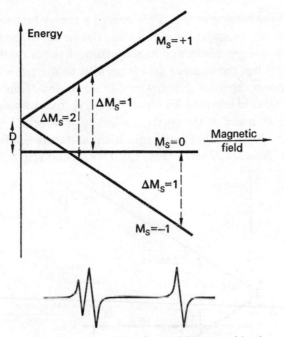

FIG. 11. Simplified energy levels and transitions for a triplet state with a large zero-field splitting.

spread over such a wide range of magnetic field, frequently several thousand gauss, that it is not easily detected. A further consequence of the zero field splitting is that the half-field transition corresponding to $\Delta m_s = 2$ is no longer strictly forbidden. The only anisotropy associated with this transition is the normal g value anisotropy, so that although this transition is weaker than those associated with $\Delta m_s = 1$ it is far less anisotropic and is often the most easily observed feature in the spectrum of a randomly oriented triplet state molecule (Fig. 12).

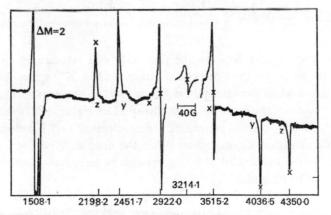

FIG. 12. Electron spin resonance spectrum of a randomly oriented triplet state molecule (naphthalene-d_8). (Reproduced with permission, from the *Journal of Chemical Physics*.)

The zero field splitting parameter D is the interaction energy between the two spins when the external field is along the major symmetry axis, the z-axis, of the triplet. This energy is normally measured in wave numbers, and its magnitude depends on the average separation of the two electrons. When the external field is parallel to the z-axis the splitting between the two $\Delta M_S = 1$ resonances is $2D'$ gauss, where D' gauss and D cm^{-1} are related by eqn. (11). When the external field is perpendicular to the z-axis the splitting between these resonances is reduced to D' gauss. If the z-axis has less than threefold symmetry, then the x and y directions are no longer magnetically equivalent and there is a further small zero field splitting of the energy levels of $2E$ cm^{-1} (Fig. 13). This second zero field-splitting parameter

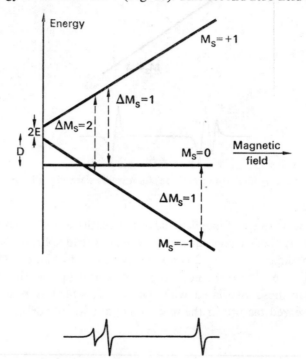

Fig. 13. Simplified energy levels and transitions for a triplet state having two zero-field splittings.

further modifies the splitting between the two $\Delta M_S = 1$ resonances. When the external field is parallel to the y-axis the observed splitting is $(D'+3E')$ gauss, but the splitting is only $(D'-3E')$ gauss when the external field is parallel to the x-axis. The magnitude of E measures the deviation of the molecule from threefold or higher axial symmetry, and hence the value of E is of considerable structural utility. Both D and E are determined by the electron distribution within the molecule. When the nuclear hyperfine interactions of the two unpaired electrons is also resolvable, a great deal of information about the structure of the triplet state can be obtained.

Seven e.s.r. parameters, the line intensity, the linewidth, the g value, the isotropic and anisotropic hyperfine couplings, and the two zero field-splitting parameters of the triplet state, have now been introduced. In the remainder of this chapter a number of examples of the successful application of e.s.r. methods to structural problems in organic chemistry are outlined. Applications involving intensity and linewidth measurements will not be discussed as these parameters normally provide quantitative analytical and kinetic information. Reviews by Caldin[6] and Carrington[7] describe the application of e.s.r. spectroscopy to the study of reaction rates and relaxation processes respectively.

Applications to Structural Organic Chemistry

1. TRANSIENT FREE RADICALS IN SOLUTION

(a) *The geometry of substituted methyl radicals*

A very simple example of the structural utility of the isotropic hyperfine coupling constant has been given by Fessenden and Schuler.[8] They have studied many transient species produced in small stationary concentrations by the continuous radiolysis at low temperatures of liquid aliphatic compounds.[9] Radiolysis of liquid methane and its fluoro derivatives with high energy electrons produces the radicals $\cdot CH_3$, $\cdot CH_2F$, $\cdot CHF_2$, and $\cdot CF_3$. These were identified from the number and relative intensities of the hyperfine lines due to the interaction of the unpaired electron with the $_1^1H$ and $_9^{19}F$ nuclei. Further experiments were conducted in which the compounds were labelled with an appreciable concentration of $_6^{13}C$ nuclei, and this resulted in the resolution of additional hyperfine lines from the interaction of the unpaired electron with this magnetic nucleus. The origin of the 38 G isotropic splitting due to the $_6^{13}C$ nucleus in $_6^{13}CH_3$ has already been discussed. Since the a_0 value for a carbon $2s$ orbital is about 1100 G, the unsubstituted methyl radical is clearly a π radical having a planar structure with the unpaired electron occupying a pure p atomic orbital on carbon. The progressive substitution of the hydrogen atoms by fluorine atoms results in the progressive increase of the isotropic coupling to the $_6^{13}C$ nucleus. This coupling reaches 272 G in $_6^{13}CF_3$, representing an a_{iso}/a_0 ratio of about 0·25, and shows that trifluoromethyl is a σ radical. The orbital of the unpaired electron has about 25% $2s$ character on carbon, and if there is little delocalization of the unpaired electron on to the fluorine atoms, then the unpaired electron must have about 75% $2p$ character on carbon. This suggests that the unpaired electron occupies an sp^3 hybrid orbital on carbon in $\cdot CF_3$. Consequently this radical must have a pyramidal structure with an FCF bond angle close to the tetrahedral value.

(b) *Radical intermediates in redox reactions*

The study of the transient radical intermediates generated during the oxidation or reduction of a large variety of organic compounds is proving to be one of the most fascinating and fruitful applications of e.s.r. spectroscopy. The most widely used method of observing the spectra of these radicals is that devised by Dixon and Norman.[10] This is a rapid flow method in which the reactants, driven by hydrostatic pressure, are mixed immediately before entering the cavity of the spectrometer. The spectrum of this mixture is obtained within 20 ms of the mixing. By far the most popular oxidizing agent used in flow systems is a mixture of hydrogen peroxide with acidified titanium(III) ions, although the exact nature of the reactive oxidizing species is not yet fully understood. This mixture appears to act as a source of hydroxyl ($\cdot$OH) and hydrogen superoxide ($HO_2 \cdot$) radicals. Oxidations in flow systems have also been accomplished by cerium(IV) ions in acid solution, by ferricyanide in basic solution, and by lead tetra-acetate in organic solvents. One of the most satisfactory reducing agents for reactions in flow cells is sodium dithionite. The anion dissociates into SO_2^- radical anions in aqueous solution. The unpaired electron can then be transferred to the organic molecule. Electrolytic and chemical reduction by electropositive metals are not easily adapted to flow system reactions.

The mixing of methanol with the titanium/peroxide reagent results in the replacement of the lines attributed to complexes of the hydroxyl and hydrogen superoxide radicals with aquated titanium ion by a six-line spectrum (Fig. 14). This consists of a $1:2:1$ triplet

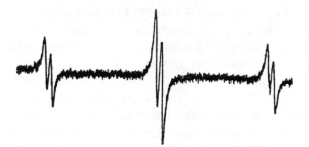

Fig. 14. Electron spin resonance spectrum of the $\dot{C}H_2OH$ radical. (Reproduced with permission from *Laboratory Practice*, 1964, **13**, 1086.)

with a splitting of 17 G from two equivalent protons subdivided into doublets having a 1 G splitting from a third proton. This suggests that the reactive species from the titanium/peroxide reagent, almost certainly the hydroxyl radical rather than hydrogen superoxide, abstracts a hydrogen atom from the methyl group of methanol and that the observed spectrum is due to the hydroxymethyl radical $\cdot CH_2OH$. The size of the α-proton coupling suggests that $\cdot CH_2OH$ is a π-radical with a planar structure about the carbon atom, but this coupling is numerically smaller than that observed in the unsubstituted methyl radical. This indicates that there is some delocalization of the unpaired electron on to the electronegative hydroxyl group. This delocalization on to oxygen, an atom with a larger spin orbit coupling constant than carbon, is also reflected in the small shift of the g value from 2·0026 in $\cdot CH_3$ to 2·0033 in $\cdot CH_2OH$.

An eight-line spectrum (Fig. 15) is obtained when ethanol is mixed with the titanium/peroxide reagent. Hyperfine splitting from four protons is present. Three of these are equivalent and produce a $1:3:3:1$ quartet with a splitting of 22 G. The splitting from the fourth proton is 15 G. Abstraction of a hydrogen atom from the methylene group produces the radical $CH_3\dot{C}HOH$ and the four protons bonded to carbon atoms can give rise to the eight-line spectrum observed. The expected splitting from the proton on the hydroxyl group

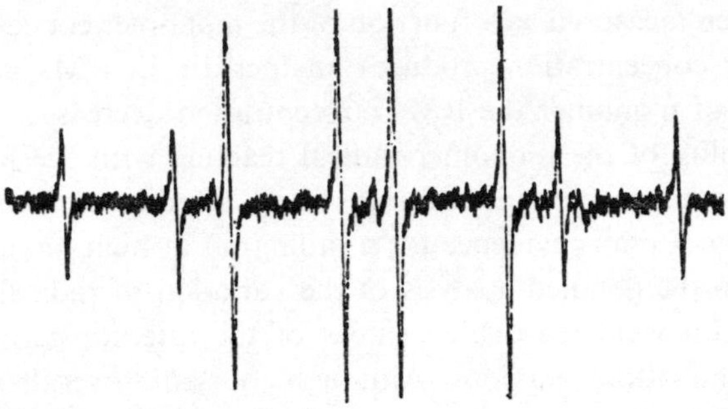

FIG. 15. Electron spin resonance spectrum of the $CH_3\dot{C}HOH$ radical. (Reproduced with permission from *Laboratory Practice*, 1964, **13**, 1086.)

is too small to resolve at room temperature but has been resolved at $-30°$ C by Livingston and Zeldes.[11] In this case the radical is generated in a slow-flow cell by the intense ultraviolet irradiation of the alcohol containing about 1% hydrogen peroxide. When concentrated solutions of reactants are used the spectrum of a second radical can be detected. This is a nine-line spectrum, a triplet of triplets, and arises from a small concentration of the $\cdot CH_2CH_2OH$ radical formed by hydrogen abstraction from the methyl group of ethanol. This mode of attack seems to be quite general for the reaction of aliphatic alcohols and ethers with the titanium/peroxide reagent. Thus the main signal from isopropanol is attributable to the $(CH_3)_2\dot{C}OH$ radical and the weaker signal to $\cdot CH_2CH(OH)CH_3$ whilst attack on diethyl ether produces mainly $CH_3\dot{C}HOCH_2CH_3$ and very little $\cdot CH_2CH_2OCH_2CH_3$.

(c) Radical-induced polymerization of olefins

The reactions of olefins with various radicals in flow systems have been studied very extensively particularly by Fischer.[12a] Abstraction of hydrogen atoms is most unusual in these systems, the normal reaction being radical addition across the double bond. When ethylene reacts with the titanium/peroxide reagent, the radical $\cdot CH_2CH_2OH$ is obtained, and if the hydrogen peroxide is replaced by hydroxylamine, the radical $\cdot CH_2CH_2NH_2$ is formed. Fischer *et al.*[12b] have examined the radical produced during the liquid phase polymerization of various vinyl monomers induced by the radicals $\cdot OH$, $\cdot NH_2$, $\cdot CH_3$ and $\cdot CH_2OH$. The methyl radicals were produced by using t-butylhydroperoxide instead of hydrogen peroxide in the titanium reagent. Well-resolved spectra are observed and normally three types of radical can be distinguished: the initiating radical $R\cdot$, the secondary radical $RM\cdot$, formed by addition of $R\cdot$ to the vinyl monomer, and the polymerization radical $P\cdot$, a growing chain radical of the type RM_n, where $n \geqslant 2$. The results obtained from these studies have confirmed the usual views held by polymer chemists about the general mechanism of such radical polymerizations. Radical attack on vinyl monomers of general formula CH_2CXY occurs predominantly at the unsubstituted carbon atom producing the monomer radical $RCH_2\dot{C}XY$ rather than $RCXY\dot{C}H_2$. Furthermore, the polymerization radicals are exclusively of the type $R(CH_2CXY)_n\dot{C}H_2CXY$. The relative concentrations of

RM· and P· have been measured as a function of the monomer concentration. Initially an increase in monomer concentration produces an increase in RM· concentration but at higher concentration of monomer the RM· concentration decreases. This corresponds to the increased probability of the monomer radical reacting with the monomer. The polymerization radical increases in concentration fairly steadily with increasing monomer concentration although there is evidence for a tailing off at high monomer concentration. Calculations based on the detailed analysis of the variation of radical concentration with monomer concentration yield reasonable values of the rate constants for the initiation, propagation, and termination reactions. Although the same overall pattern of variation of radical concentration with monomer concentration is observed with all four initiating radicals, some differences in detail occur. For example, the concentration of RM· builds up more rapidly when methyl attacks the monomer than when hydroxyl attacks, but this is accompanied by a slower build-up of the concentration of P·. Methyl radical initiation is also more specific than that of hydroxyl. Lines associated with the spectrum of the monomer radical HOCXYĊH₂ formed by hydroxyl attack at the substituted end of the vinyl monomer can be detected in low concentration, but the spectra of radicals of the type CH₃CXYĊH₂ have not been observed. The spectra also provide information about the conformations of the radicals. The inequivalence of the β-CH₂ protons indicates that there is restricted rotation about the carbon-carbon bond adjacent to the radical site. As the lifetime of the radical conformers, about 10^{-6} s, is rather longer than the time required for a bond to form between the radical and another monomer, detailed study of the conformation of the radicals can provide insight into the rates and probability of formation of isotactic or syndiotactic linkages.

(d) *Identification of radicals from the g value*

Organic radicals often do not contain atoms having large spin–orbit coupling constants. Consequently the g value is only rarely an informative parameter. Nevertheless, sufficient data have now been obtained, particularly from studies of transient substituted methyl radicals in flow systems, to enable some correlation to be made between the size of small g shifts and the nature of the substituents. When a hydroxyl or alkoxyl group is bonded to the tervalent carbon atom g values about 0·0010 greater than the free-spin value are observed while attachment of an aldehydic or ketonic carbonyl group to the tervalent carbon results in g values of about 2·0040. On the basis of these observations, Norman and Pritchett[13] offer an interpretation of the ambiguous spectrum obtained when the titanium/peroxide reagent reacts with butan-2,3-diol. The thirty-two-line spectrum consists of a large quartet ($a_H = 22·5$ G) due to three equivalent protons, a small quartet ($a_H = 2·05$ G) due to a further three equivalent protons, and a doublet ($a_H = 18·6$ G) due to a single proton. If it is assumed that hydroxyl proton interactions are too small to resolve, then this spectrum could arise from the radical CH₃Ċ(OH)CH(OH)CH₃ formed by abstraction of a hydrogen atom from the diol, but a hydroxyl substituted radical of this type should have a g value of about 2·0033, whereas the observed g value of 2·0039 is more in accord with that expected for a carbonyl substituted radical. The carbonyl radical $CH_3COCHCH_3$, produced by the elimination of the constituents of a water molecule from the diol radical, should have a more appropriate g value and produce similar proton hyperfine couplings. Confirmation that this carbonyl radical is responsible for the spectrum obtained when butan-2,3-diol is oxidized was obtained by showing that an identical spectrum is observed when methyl ethyl ketone is oxidized by the titanium/peroxide reagent. A review by Norman and Gilbert[14] provides a very comprehensive coverage of the application of e.s.r. to the study of transient organic radicals.

2. THE STRUCTURE OF SUBSTITUTED AROMATIC RADICAL IONS

Studies of π-radical systems soon revealed that the magnitude of isotropic hyperfine splitting is intimately related to the unpaired electron spin density. In particular it was shown that the magnitude of the isotropic hyperfine coupling to ring protons in aromatic radicals and radical ions is directly related to the spin density in the $2p_\pi$ orbital on the adjacent carbon atom. This relationship is expressed by the McConnell equation[15]

$$a_{iso}(H_\alpha) = Q_\alpha \varrho(C),\tag{20}$$

where $\varrho(C)$ is the unpaired electron density in the $2p_\pi$ orbital on carbon adjacent to the H_α atom and Q_α is a constant having a value of about -23 G. A detailed examination of the experimental evidence suggests that Q_α is not strictly a constant for all carbon π radicals. The spin density on the carbon atom in the methyl radical is unity, and the Q_α value must equal the proton isotropic hyperfine coupling of -23 G, but the isotropic proton coupling of -3.75 G observed for the benzene anion C_6H_6 suggests a Q_α value of 22.5 G as the spin density on each carbon must be $1/6$. A number of explanations of the slight variations of Q_α have been offered. The main modifying factors appear to be the partial charge on each carbon atom and variations in carbon to hydrogen bond length. Nevertheless, a Q_α value of -23 G is satisfactory for most aromatic radicals. The great utility of the McConnell equation is that it enables experimental values of spin densities obtained from e.s.r. measurements to be compared with values calculated by using various quantum chemical approximations. The numerous e.s.r. studies of substituted benzene radical ions have not only provided subtle tests of the predictions of molecular orbital theory but also yielded detailed information about the electronic characteristics of the substituent groups.

Combination of the $2p_\pi$ atomic orbitals on each carbon atom of benzene produces three bonding and three antibonding molecular orbitals (Fig. 16). The highest bonding and the lowest antibonding levels occur as degenerate pairs. In the benzene anion seven electrons occupy these π levels and the unpaired electron is equally shared between the two degenerate antibonding orbitals. The calculated spin density at each carbon atom in these two orbitals is shown in Fig. 17 and the orbitals are labelled symmetric (ψ_S) and antisymmetric (ψ_A) according to their behaviour on reflexion in the plane perpendicular to the ring passing through the atoms at positions 1 and 4. The observed seven line spectrum of the benzene anion being characteristic of six equivalent protons confirms that the unpaired electron is

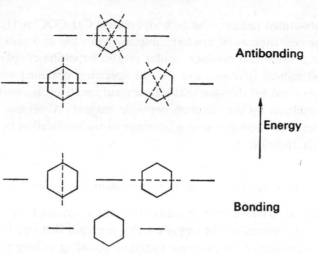

FIG. 16. Energy levels for π molecular orbitals of the benzene anion. Dashed lines indicate nodal planes.

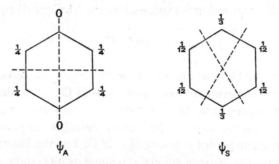

FIG. 17. Calculated spin densities for the degenerate π antibonding molecular orbitals of the benzene anion. Dashed lines indicate nodal planes.

shared between ψ_S and ψ_A and that the spin densities for these two orbitals are averaged 1/6 at each position.

The introduction of a substituent group on to the benzene ring lifts the degeneracy of ψ_S and ψ_A. If the substituent group is electron repelling, then the unpaired electron occupies ψ_A as this orbital keeps the electron away from the substituent group. The spin density on the ring carbon atom adjacent to the substituent group should be zero. The spectrum of such a radical anion will be dominated by a quintet (1 : 4 : 6 : 4 : 1) with a splitting of about 5·7 G ($Q_\alpha/4$) due to the interaction of the unpaired electron with the four-ring protons adjacent to carbon atoms each having a spin density of $\frac{1}{4}$. If the substituent group is electron attracting, then ψ_S lies lower than ψ_A as ψ_S has high spin density on the carbon atom adjacent to the substituent. In this situation the spectrum will be dominated by a doublet (1 : 1) with a splitting of about 7·6 G ($Q_\alpha/3$) due to the proton *para* to the substituent, but these components will be further split into quintets by the remaining ring protons. The quintet splitting will be about 1·9 G ($Q_\alpha/12$).

The spectrum of the t-butyl benzene radical anion (Fig. 18) is dominated by a large quintet splitting, about 4·7 G, indicating that ψ_A is the singly occupied orbital and that the t-butyl

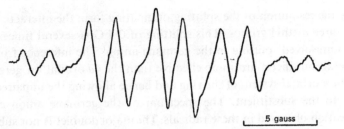

FIG. 18. Electron spin resonance spectrum of the t-butylbenzene anion. (Reproduced with permission from *Orbital Degeneracy and Spin Resonance in Free-radical Ions* by A. Carrington, RIC London, 1964.)

FIG. 19. Electron spin resonance spectrum of the trimethylphenylsilane anion. (Reproduced with permission from *Orbital Degeneracy and Spin Resonance in Free-radical Ions* by A. Carrington, RIC London, 1964.)

FIG. 20. Electron spin resonance spectrum of the trimethylphenylgermane anion. (Reproduced with permission from *Orbital Degeneracy and Spin Resonance in Free-radical Ions* by A. Carrington, RIC London, 1964.)

group is electron repelling. The spectrum is a little more complex than that predicted for an electron-repelling substituent as there is a small doublet splitting of 1·7 G due to the *para* proton. This suggests that ψ_S and ψ_A are very nearly degenerate and that there is considerable mixing of ψ_S into the unpaired electron orbital. The interaction with the nine equivalent protons in the t-butyl group is too small for resolution and must be less than 0·1 G. The spectra of the trimethylphenylsilane anion (Fig. 19) and the trimethylphenyl germane anion (Fig. 20) are strikingly different from that of the t-butylbenzene anion. Both spectra are dominated by a large doublet interaction, 8·1 G for the silicon radical and 7·6 G for the germanium radical, that is in very good agreement with the estimated value for the *para* proton when the unpaired electron occupies the symmetrical molecular orbital. This suggests that these two substituents, unlike the t-butyl group, are essentially electron attracting with respect to a benzene ring. The spectrum of the silane anion is further complicated by the resolution of the splitting originating form the interaction with the nine protons in the three methyl groups. This splitting of 0·4 G is several times larger than the corresponding unresolved splitting in the germane anion. The inference from this must be that the 3d orbitals on silicon are more effective than the 4d orbital on germanium at conjugating with the π orbital system of the ring and hence allowing the unpaired electron to be delocalized on to the substituent. The spectrum of the germane anion clearly shows a further complication observed in these radicals. The major doublet is not subdivided simply into quintets by the *ortho* and *meta* protons. The observed pattern is produced because the *ortho* and *meta* protons are no longer equivalent and the coupling to *ortho* protons is rather larger than that to the *meta* protons. Further examples of the effect of substituent groups on benzene anion spectra may be found in reviews by Carrington,[16, 17] and Coulson[18] has discussed these substituent effects in terms of the relative roles of the inductive and hyperconjugative effects.

Studies of aromatic radical ions containing unsaturated substituents have provided interesting information about internal motion and rotational isomerization in radicals. The spectrum of the *p*-nitrobenzaldehyde anion in dimethyl formamide shows couplings to five magnetically different protons. This observation is interpreted as resulting from the constraint of the formyl group to the plane of the ring. Such a constraint lowers the symmetry of the radical and results in the magnetic inequivalence of all four ring protons. Similarly, the complex spectrum of the terephthalaldehyde radical anion is interpreted as resulting from the superposition of the spectra of the *cis* and *trans* forms of the radical (I, II). Each

I   II

conformer contains three sets of two equivalent protons, but the two conformers are not present in equal concentrations. The observed concentration ratio is about 7 to 5, but an absolute assignment can only be made by selective deuteration. The *trans* configuration of the terephthalaldehyde-2,5-d_2 radical anion has two equivalent deuterons. A total forty-five lines should be obtained from this configuration. These deuterons are not equivalent in the *cis* configuration and 108 lines should result. Analysis of the observed spectrum from this dideutero derivative shows that it is the *trans* isomer that is present in greater concentration. Some indication of the rate of isomerization can be obtained from these spectra. A simple quintet splitting from the ring protons should be observed if the rate is fast compared with the magnitude of the splitting when measured in frequency units. The experimental observation of two distinct but overlapping spectra demonstrates that the rate is considerably slower than the splitting frequency. If the rate of isomerization is similar to the splitting frequency (about 6×10^6 s^{-1} in this example) then a complex pattern of lines of varying widths is observed. A situation of this type has been encountered in the spectrum of the naphthazarin semiquinone cation. Measurement of the temperature dependence of the linewidth has resulted in an estimate for the rotational activation energy of

4 kcal mole^{-1}. Studies of a large variety of formyl and acetyl substituted aromatic radical anions have now been made. In all cases the inequivalence of the ring protons suggests that the carbonyl group is locked in the plane of the ring. The lifetime of the planar configuration has been estimated from linewidth measurements to be rather greater than 10^{-6} s.

Evidence for steric hindrance has also been obtained from the e.s.r. spectra of substituted nitrobenzene radical anions. In such species the unpaired electron is mainly localized in the π orbital system of the nitro group, but can delocalize into the π orbital system of the ring so long as the ring and the nitro group remain coplanar. The isotropic nitrogen coupling of 10·3 G in the unsubstituted nitrobenzene anion is typical for this situation. However, nitrogen coupling constants of 17·8 G and 20·4 G are observed for the anions of 2,6-dimethyl nitrobenzene and 2,3,5,6-tetramethylnitrobenzene respectively, but the ring proton coupling constants in these radicals are considerably smaller than the corresponding splittings in the unmethylated anion. The greatly increased nitrogen splitting suggests that the methyl groups prevent the delocalization of the unpaired electron on to the ring. The comparable nitrogen coupling constants in nitroaliphatic anions, where there is virtually no delocalization, are about 25 G. The most likely explanation of this effect is that the methyl groups overcrowd the nitro group and force it to twist out of the plane of the ring and thus prevent it from effectively conjugating with the ring. The many studies of substituted nitrobenzene anions have been discussed in some detail by Geske[19] in a review that covers most aspects of the application of e.s.r. spectroscopy to conformation problems.

3. RADICALS RELATED TO NATURAL PRODUCTS

(a) *Sea urchins*

The above discussion of substituted benzene radical anions demonstrates the extreme sensitivity of the e.s.r. spectra of organic radicals to functional group substitutions. Some recent studies[20] of a variety of naphthazarins and naphthoquinones indicate that in favourable circumstances e.s.r. can rival n.m.r in the elucidation of the constitution and structure of complex organic molecules. Investigation of the structure of spinochromes isolated from various sea urchins has been hampered by the difficulties encountered in recovering sufficient quantities of these pigments for structural analysis by conventional methods. In such a situation e.s.r. can be an extremely useful technique as it is about 1000 times more sensitive than n.m.r. and only requires about 10^{-11} moles of material. Although spectra can only be obtained from molecules containing unpaired electrons, the univalent polarographic oxidation or reduction of the diamagnetic molecule can produce closely related paramagnetic species suitable for e.s.r. studies.[21] The spinochromes are generally polyhydroxyl derivatives of naphthazarin, and in order to investigate the feasibility of identifying unknown spinochromes by e.s.r. methods, Piette *et al.*[21] systematically studied the spectra obtained by the *in situ* polarographic reduction of a large number of synthetic and naturally occurring naphthoquinones and naphthazarins.

There are fifteen lines in the spectrum of univalently reduced naphthazarin; a 2·43 G quintet due to the four equivalent ring protons subdivided into a 0·51 G triplet due to the equivalence of the two hydroxyl protons. This spectrum confirms the equivalence of the four tautomers (III–VI) in the unsubstituted reduced naphthazarin. This equivalence is removed when a ring proton is replaced by another group. An electron releasing group confers quinoidal properties on the ring to which it is directly bonded, while an electron withdrawing group confers benzenoidal properties. The former situation favours tautomer III, while the latter favours tautomer VI. Support for these conclusions is obtained from

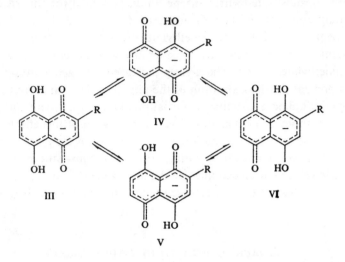

both the directions of the chemical shifts of the remaining protons in the n.m.r. of the diamagnetic species and the magnitude of the e.s.r. proton coupling constants in the radical form. It has already been observed that an electron-repelling substituent in a benzene radical anion produces an increase in the hyperfine coupling to the proton *ortho* to it. Furthermore, the localization of the unpaired spin in the ring containing the substituent lowers the spin density in the unsubstituted ring, and this is reflected in a decrease of the proton hyperfine couplings at positions 6 and 7. These expectations are generally realized by a variety of electron-releasing substituents. The spectra from various di-, tri-, and tetra-substituted naphthazarins and naphthoquinones were also examined. Some of these compounds are naturally occurring spinochromes, and in most cases satisfactory interpretations of the spectra of the reduced form were made. Deuteration of the acidic protons by direct exchange with deuterium oxide considerably simplifies and greatly assists interpretation of the more complicated spectra (Fig. 21).

In a second paper[22] the factors governing the observation of resolvable proton splittings from various methoxyl substituted naphthazarin, naphthoquinone, and benzoquinone radical anions were examined. Couplings to such protons are observed only when the methoxyl group is not sterically hindered by other substituents. In this situation the filled *p* orbital on the methoxyl oxygen atom can conjugate with the π orbitals of the ring enabling unpaired electron spin density to leak on to this atom and produce a measurable coupling

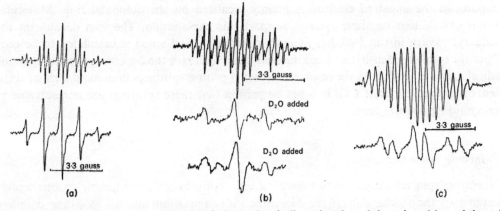

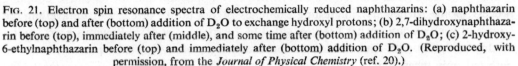

FIG. 21. Electron spin resonance spectra of electrochemically reduced naphthazarins: (a) naphthazarin before (top) and after (bottom) addition of D_2O to exchange hydroxyl protons; (b) 2,7-dihydroxynaphthazarin before (top), immediately after (middle), and some time after (bottom) addition of D_2O; (c) 2-hydroxy-6-ethylnaphthazarin before (top) and immediately after (bottom) addition of D_2O. (Reproduced, with permission, from the *Journal of Physical Chemistry* (ref. 20).)

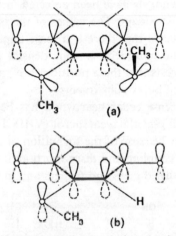

FIG. 22. Representation of the conformers in methoxy-substituted quinones, indicating π orbital overlap: (a) disubstituted, and (b) monosubstituted. (Reproduced with permission from the *Journal of Chemical Physics* (ref. 22).)

with the methoxyl protons. The presence of a substituent adjacent to the methoxyl group forces the methoxyl methyl group to twist away from the plane of the ring (Fig. 22). This destroys the conjugation of the oxygen atom with the ring, prevents unpaired spin density reaching the oxygen atom, and should result in the loss of any coupling to the methoxyl protons. This explanation is in accord with the experimental results only if the methoxyl group is bonded to a quinoidal ring. No methoxyl proton splittings are observed in naphthoquinones having the substituent in the benzenoid ring even when this group is not sterically hindered. Although the methoxyl group can conjugate with the benzenoid ring, the spin density on the oxygen atom is insufficient to produce a resolvable methoxyl proton coupling as the unpaired electron is mainly localized on the quinoidal ring. Molecular orbital calculations on these systems support this explanation. The spin density on the methoxyl oxygen in the 5-methoxynaphthoquinone radical anion is calculated to be only 1% of the total spin density, whereas the equivalent figure for the 2-methoxynaphthoquinone radical anion is 13·4%. As the resolved methoxyl proton splittings in quinoidal ring substituents is rather less than 1 G, it is not surprising that these splittings are unresolvable in benzenoid ring substituents.

(b) *Incense cedar heartwood*

Electron spin resonance studies have recently contributed considerably to our understanding of the mechanism of the biosynthetically important phenol oxidative coupling process. Some particularly interesting examples of the utility of e.s.r. in unravelling the course of such oxidations has been provided by Fitzpatrick *et al.*[23] Previous e.s.r. studies of this oxidation with lead dioxide demonstrated that relatively stable phenoxyls can be detected only when bulky substituents occupy at least one of the positions *ortho* to the hydroxyl group. Transient phenoxyls have been detected in flow systems using cerium(IV) ions in dilute sulphuric acid as oxidant. The values of the isotropic ring proton couplings in these transient radicals indicate that there is a large unpaired spin density on the *para* carbon atom ($a_H = 10$ G) and considerable spin density ($a_H = 6\cdot6$ G) on each of the *ortho* carbon atoms. This suggests that these positions are particularly susceptible to coupling reactions in the absence of bulky substituents.

The decay resistance of incense cedar heartwood has been attributed to the presence of *p*-methoxythymol (VII) and *p*-methoxycarvacrol (VIII). It has been suggested that oxidation of these phenols is slow because of the formation of stable phenoxyl intermediates. The radicals undergo various coupling and disproportionation reactions which result both in the production of phenol ether dimers and trimers and in the regeneration of the parent

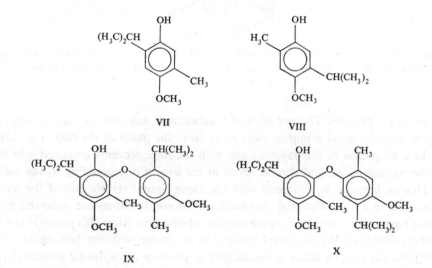

phenols. The decrease in the decay resistance of this wood on ageing can be correlated with an increase in the concentration of the phenol ether dimers, libocedrol (IX) and heyderiol (X), and a decrease in the concentration of the parent phenols. Fitzpatrick *et al.* have used e.s.r. to investigate this proposed mechanism of oxidative detoxification of this wood by preparing the phenoxyl intermediates in benzene solution using lead dioxide as an oxidant. Although they contain unsubstituted *ortho* groups, both the monomer phenoxyls are stable for many hours in deoxygenated solutions under nitrogen. The reduction in the size of the isotropic proton coupling at the unsubstituted *ortho* position to about 1 G reflects the stability of these radicals. The phenoxyls produced by a similar oxidation of the phenyl ether dimers are even more stable. Solutions in benzene are only slightly decomposed after 24 hours' exposure to air. In non-hydroxylic solvents all these phenoxyls exist in equilibrium with their quinol ether dimers, an essentially pure dimer was prepared by evaporation of the solvent. These dimers were characterized by their infrared spectra. They show very weak e.s.r. absorption from the monomer, but redissolution in benzene results in the regeneration of a large concentration of monomer radicals. The high electron density in the *ortho* and *para* positions of transient phenoxyls accounts for the rapid polymerization by way of coupling reactions. It was suggested that the relative stability of the cedar heartwood monomer radicals is due to the steric shielding of the unsubstituted *ortho* position by the adjacent *meta* substituent. Although these steric factors may make an important contribution to the stability of the radicals, the size of the proton coupling at this *ortho* position clearly indicates that the nature of the substituents in the cedar heartwood phenols deactivates the unsubstituted *ortho* positions in the corresponding phenoxyls. This deactivation prevents the coupling reactions encountered in transient phenoxyls occurring with any great facility. The monomer radicals do appear to react slowly in this way and produce quinol ether dimers. Oxidation of these dimers produces phenoxyls having unsubstituted ring positions that are even more deactivated than the monomer radicals. The stability of these dimer radicals is such that there is a considerable tendency for them to disproportionate into quinones and phenols rather than form higher coupling products. Although the studies of phenoxyls were carried out *in vitro* they appear to provide very satisfactory confirmation of the suggestion that the stability of the radical intermediates produced during the detoxification reactions of incense cedar heartwood accounts for its decay-resistant properties.

4. RADICALS TRAPPED IN ORGANIC SOLIDS

Exposure of solids to high energy radiation normally results in the trapping of radicals in the crystalline matrix. Almost invariably these radicals are oriented in specific directions within the crystal, and consequently there has been considerable investigation of organic radicals trapped in irradiated single crystals by e.s.r. spectroscopy. The crystal may be as large as a 4 mm cube when studies are made at X-band frequencies, but satisfactory results can be obtained from much smaller crystals particularly when higher frequencies are used. After irradiation the crystal is mounted on a teflon rod and placed at the point of maximum microwave magnetic field within the cavity. This rod is made part of a goniometer so that the crystal can be accurately rotated about an axis perpendicular to the external magnetic field. Spectra are recorded at intervals of 10°, or less if the orientation dependence is particularly striking, for 180° from a known origin. Measurements are made about three mutually perpendicular axes or about the crystallographic axes. These sets of spectra illustrate the anisotropy of both the g value and the hyperfine coupling, and from them the values of g and of the hyperfine coupling for the principal symmetry axes of the radical can be calculated. If the crystal structure of the matrix is known it is possible to relate the directions of the principal symmetry axes of the radical to the orientations of the host molecules within the crystal. Most of the early studies of radicals trapped in organic single crystals were carried out on samples irradiated at room temperature. Irradiation at much lower temperatures usually results in the trapping of different radicals. Subsequent warming of such crystals does not always produce the radicals observed after room temperature irradiation. Irradiated single crystals of amino and carboxylic acids and their salts have proved to be particularly fruitful systems for e.s.r. studies. The comprehensive review by Morton[24] describes progress made in the study of both organic and inorganic radicals trapped in solids.

The significance of both isotropic and anisotropic interactions with the magnetic nucleus at the radical centre, and isotropic couplings with α nuclei in π radicals has been examined already. Spectra from radicals in single crystals are normally sufficiently well resolved that hyperfine couplings to more remote nuclei may be measured. The ethyl radical will be used to exemplify these interactions. It is structurally similar to the methyl radical with a planar atomic framework about the radical centre but having one of the α protons replaced by a pyramidal methyl group (Fig. 23). The isotropic coupling to the β protons in this group

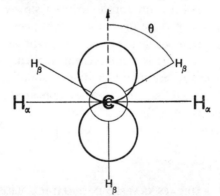

FIG. 23. The dihedral angle θ between the axis of the half-filled p orbital and β protons.

may be related to the spin density at the radical centre by an equation similar to the McConnell equation (20) for α protons

$$a_{\text{iso}}(H_\beta) = Q_\beta \varrho(C), \qquad (21)$$

where $\varrho(C)$ is the spin density on the tervalent carbon, $a_{\text{iso}}(H_\beta)$ is the isotropic β proton splitting, and Q_β is the Q value for β protons. Unlike Q_α, Q_β is not approximately constant but a function of the dihedral angle θ between the plane containing the axis of the half-filled p orbital and the C—C bond, and the plane containing the C—C bond and the β proton (Fig. 23). If the methyl group is freely rotating, an averaged value of the isotropic β proton splitting is observed, and hence an averaged Q_β value is obtained. It has been found that for radicals of the general type $CH_3\dot{C}X_1X_2$ the *average* value of Q_β is a unique constant having a value of 29·3 G where the spin density on the tervalent carbon is given by the Fischer equation

$$\varrho(C) = \prod_{i=1}^{3} (1 - \Delta X_i). \tag{22}$$

The parameter ΔX is a measure of the spin withdrawing power of the α substituent X. Typical values of ΔX are as follows:

X	H	CH_3	CH_2OH	CH_2NH_2	CN	OH	OCHO
ΔX	O	0·081	0·079	0·034	0·148	0·160	0·136

The agreement between observed values of $a_{iso}(H_\beta)$, both in solution and in the solid state, and those calculated from the Fischer equation and Q_β, is so good that the equation may be used for predicting spectra and as an aid in the analysis of complex spectra. When the substituent group is not freely rotating, the angular variation of Q_β is given by

$$Q_\beta = Q_n + Q_e \cos^2 \theta, \tag{23}$$

where Q_n is the Q value associated with the isotropic hyperfine coupling to a β proton when $\theta = \pi/2$ and the proton is in the nodal plane of the half-filled p orbital, and $(Q_n + Q_e)$ is the Q value associated with the isotropic hyperfine coupling to a β proton when $\theta = 0$ and the β proton is eclipsed by the half-filled p orbital. Observed values of isotropic β proton couplings suggest that Q_n is very small, certainly less than 5 G, while Q_e is about 50 G. Thus the isotropic coupling to β protons can vary from more than 50 G to less than 5 G as the angle θ varies when there is approximately unit spin density on the tervalent carbon. In certain circumstances the isotropic coupling to more remote magnetic nuclei may be resolved, but such couplings are small and generally have little structural significance.

The size of the anisotropic coupling between an unpaired electron in a p orbital and a magnetic nucleus in the α position is determined by the magnitude of the magnetic moment of that nucleus, the distance between that nucleus and the electron, and the angle between the direction of the external field and a line joining the electron and the nucleus. Both the angle and the distance vary as the electron moves in its orbit, but some estimate of the size of the anisotropic coupling can be made by using the most probable values of these quantities. Figure 24 illustrates the situation for a C—H fragment. When the external field is parallel to the C—H bond (Fig. 24a) most of the p orbital is concentrated in a region where the component of the field produced by the proton at the electron along the external field is parallel to this field. A large positive anisotropic coupling should result. When the external field is parallel to the axis of the p orbital (Fig. 24b) the regions of maximum electron density in the p orbital are such that the field due to the proton of the electron has virtually no component along the direction of the external field. The anisotropic coupling in this direction

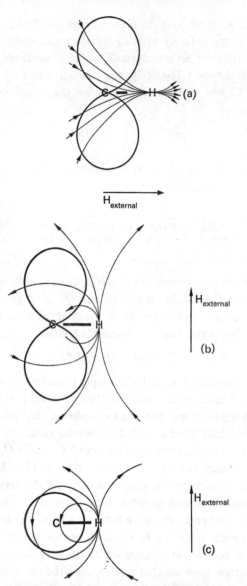

FIG. 24. Anisotropic dipolar hyperfine interaction between an electron in a *p* orbital and an α proton: (a) external field parallel to the C—H bond; (b) external field parallel to the *p* orbital axis; (c) external field perpendicular to both (a) and (b).

should thus be close to zero. When the external field is along the direction mutually perpendicular to the orbital axis and the bond (Fig. 24c), the *p* orbital is almost entirely concentrated in a region where the field due to the proton at the electron is anti-parallel to the external field. This should result in a large negative anisotropic coupling in this direction. A fairly sophisticated calculation of the anisotropic coupling along these three main axes suggests that the values should be $+15.5$ G, -1.5 G, and -14.0 G. The anisotropic coupling to α protons has been measured in a number of substituted methyl radicals where the spin population in the p orbital is close to unity. Typical values are $+10$ G, $+2$ G, and -12 G. A lower spin density on the tervalent carbon is reflected in a proportionate decrease in the anisotropic coupling. Anisotropic coupling to β protons may be analysed in a similar way, but these couplings are usually very small as the interaction is inversely proportional to the cube of the distance between the unpaired electron and the magnetic nucleus.

Whiffen and various associates have made notable contributions to the study of radicals trapped in single crystals during the last decade. An example of their work[25] demonstrates the utility of these studies. The spectra from several X-irradiated alkylmalonic acids were obtained over a range of temperatures extending from 77 K to room temperature. Not only were most of the radicals produced satisfactorily identified, but information was also obtained about the conformation of various substituent groups in radicals where rotation is restricted in the solid lattice at low temperatures. At 77 K lines attributable to the radical RCH$_2$ĊH(COOH), formed by loss of a carboxyl group from the acid, dominate the spectra. Two conformations of this radical can be identified. In both forms characteristic values of the coupling to a single α proton are observed, but differences are found in the couplings to the two β protons. The less stable form of the radical, decomposing or reacting below 195 K, has isotropic couplings to the β protons of about 43 and 11 G. This indicates that the dihedral angle θ is close to zero for one proton and close to 90° for the other. The more stable form of the radical, present at 195 K but disappearing on warming to room temperature, has isotropic couplings to the β protons of about 18 and 7 G. One proton must be very close to the nodal plane of the p orbital, while a value of θ of about 60° is suggested for the other proton. Possible conformations that satisfy these values of θ are shown in XI

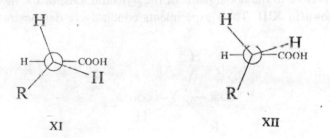

XI   XII

and XII. The anisotropic coupling to the β protons in both conformers is very small. Typical values along the principal directions are $+4$, -1, and -3 G.

A second radical strongly in evidence at 77 K is unstable at 195 K. An isotropic coupling of about 25 G and an anisotropic coupling of about $+2\cdot5$, -1, and $-1\cdot5$ G along the principal directions suggest that the interaction is with a single β proton. In some orientations this doublet shows further small splittings, presumably due to interaction with more remote protons of the alkyl group. Since there are no α protons and only one β proton associated with the unpaired electron, it must be mainly localized in a p orbital on the carbon atom of the carboxyl group. This situation is similar to that observed in irradiated succinic acid at 4 K.[26] At this temperature the spectrum displays a single line with a very anisotropic g value and a group of four lines with an almost isotropic g value close to the free spin value. The singlet is lost on warming to 77 K. The lack of hyperfine interaction and the direction of the g value anisotropy indicates that the unpaired spin is probably localized on the carbonyl oxygen of the carboxyl group. It is believed that the acid cation [HOOC $\cdot$CH$_2$CH$_2\cdot$CÖOH]$^+$ is responsible for this singlet. The four-line spectrum is stable to above 77 K and the magnitude of the hyperfine interaction suggests that there are two β protons and no α protons. Labelling of the carboxyl group with carbon-13 shows that the unpaired electron is mainly localized on this atom and that the atomic framework about this carbon is not quite planar. This spectrum is almost certainly due to the molecular anion of the succinic acid [HOOC$\cdot$CH$_2\cdot$CH$_2\cdot$COOH]$^-$, and the similarities in stability and hyperfine interaction have led Whiffen *et al.* to tentatively ascribe the doublet in irradiated alkyl malonic acids to the molecular anion [RCH(COOH)$_2$]$^-$.

Minor lines are present in irradiated n-propyl and n-butyl malonic acids at 77 K. These have been identified as the radicals formed by abstraction of hydrogen atoms along the alkyl chain. Radicals of this type normally have extensive hyperfine interactions. Thus in CH$_3$ĊHCH$_2$CH(COOH), formed in low yield from irradiated n-propyl malonic acid, interactions with one α and five β protons should be resolvable.

At room temperature all spectra are dominated by the lines associated with radicals of the type RCH$_2$Ċ(COOH)$_2$. As the isotropic and anisotropic couplings to both β protons are almost identical and the value of the isotropic coupling is 23 G, it appears that the alkyl group is almost freely rotating at room temperature. If the crystals are cooled from room temperature back to 77 K, it is found that the couplings to the two protons are no longer identical. One proton has an isotropic coupling of almost 50 G suggesting that the dihedral angle θ is very nearly zero while the isotropic coupling to the other is about 10 G, indicating that this proton is close to the nodal plane of the p orbital. One of the most probable conformations is shown in XIII. These experiments conclusively demonstrate that the alkyl

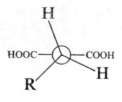

XIII

group is able to rotate at room temperature but is locked in a preferred conformation at 77 K. Studies of this type are making a considerable contribution to the understanding of the course and mechanism of reactions induced by ionizing radiations.

5. STRUCTURAL STUDIES OF ORGANIC TRIPLET STATES

In the last 10 years e.s.r. studies have yielded a great deal of information about the electronic structure of organic triplet states. Most of this work has been concentrated on two types of triplet state: the excited phosphorescent state of aromatic hydrocarbons and the ground state of carbenes and nitrenes. These ground state triplets are usually prepared by the ultraviolet photolysis of a solid solution of the appropriate diazo or azide precursor in a suitable solvent at 77 K. As the magnitude of the dipolar parameter D is strongly dependent on the average distance between the two unpaired electrons some information about the extent of delocalization away from the methylene carbon may be obtained. A D value of 0.70 cm^{-1} is expected when both electrons are entirely localized in the $2p$ orbitals on the methylene carbon, and values of this order are observed for substituted alkyl methylenes such as CF_3CCF_3. In phenyl methylene delocalization into the π electron system of the ring can occur, and this is reflected in the lower D value of 0.52 cm^{-1}, while in diphenyl methylene the D value is only 0.41 cm^{-1}. D values for nitrenes are rather larger than those of carbenes because of the greater electronegativity of the nitrogen atom. A typical value of D for alkyl nitrenes is 1.5 cm^{-1}, but this is reduced to 0.99 cm^{-1} in phenylnitrene. Delocalization on to substituents is also reflected in the size of the hyperfine coupling to a ^{13}C methylene nucleus or a ^{14}N nitrene nucleus. The ^{13}C coupling to the methylene carbon in phenylmethylene is 75 G, but it is only 62 G in diphenyl methylene. Much smaller D values are found in the phosphorescent excited triplet states of the aromatic hydrocarbons as both electrons are extensively delocalized. The observed value for naphthalene is 0.10 cm^{-1}.

The second zero field splitting parameter E is a measure of the deviation of the molecule from linearity at the methylene carbon. Bending of the molecule at the methylene carbon introduces s orbital character into one of the unpaired electron orbitals. The value of E is some measure of this s orbital character and the ratio E/D can be used to estimate the bond angle at the methylene carbon.[27] When there is no delocalization on to substituents, the introduction of s character into one of the unpaired electron orbitals brings the unpaired electrons closer together and hence increases the D value. In CF_3CH the E value is 0.021 cm^{-1} and the D value is 0.71 cm^{-1}. Replacement of the hydrogen atom by a second CF_3 group introduces more s character and increases the E value to 0.044 cm^{-1}. The D value increases to 0.74 cm^{-1}. These values suggest bond angles of about 160° in CF_3CH and 140° in CF_3CCF_3. Extrapolation from these values suggests that the unsubstituted methylene is linear. The E/D ratios indicate bond angles of about 150° in both phenyl methylene and diphenyl methylene. This suggests that only one unpaired electron is delocalized on to the rings. It had been suggested that diphenyl methylene is linear with the two rings at right angles to each other allowing one unpaired electron to delocalize into each ring, but the size of both D and E parameters indicate that this is unlikely. A novel structural effect is encountered in the carbene species derived from cyclopentadienylidene. In such systems one of the unpaired electrons is delocalized around the π orbital system of the ring while the other is localized in an orbital in the plane of the ring system. The D values for fluorenylidene, indenylidene, and cyclopentadienylidene are very similar to that of diphenyl methylene, and the E values suggest that the bond angle at the methylene carbon is about 140°. Since the methylene carbon is part of a five-membered ring, it is unlikely that this bond angle can exceed 120° without the introduction of considerable ring strain. It has been suggested that these results provide evidence of the existence of bent σ bonds in these molecules. A comprehensive account of e.s.r. studies of the triplet state has appeared.[30]

Worked Examples

1. How many lines and of what relative intensity would you expect to find in the isotropic spectrum of the 4-fluoronitrobenzene anion (XIV)? In all we have six magnetic nuclei in this radical: four ring protons, a fluorine nucleus ($I = \frac{1}{2}$), and a nitrogen nucleus ($I = 1$). The four protons are not equivalent but form two groups of two equivalent protons, for although the plane perpendicular to the ring through positions 1 and 4 is a plane of

XIV

symmetry, the plane perpendicular to both this symmetry plane and the ring plane is not. The equivalent protons at positions 2 and 6 will produce a $1 : 2 : 1$ triplet. Each component of this triplet will be further split into a second $1 : 2 : 1$ triplet by the equivalent protons at positions 3 and 5. If one splitting is very much larger than the other, the nine lines produced by the ring protons will occur in three groups of three with relative intensities $1 : 2 : 1$, $2 : 4 : 2$, $1 : 2 : 1$, but if the two splittings are almost equivalent, the nine lines occur in five groups with relative intensities 1, $2 : 2$, $1 : 4 : 1$, $2 : 2$, 1. Each component will be further split into equal triplets by the nitrogen nucleus and then split again into doublets by the fluorine nucleus to give fifty-four lines in all. As there is no indication of the relative sizes of the four different splittings, the precise order of the lines cannot be given, but we do know that unless there is some fortuitous overlapping there will be twenty-four lines of relative intensity 1, twenty-four of relative intensity 2 and six of relative intensity 4.

2. The isotropic couplings in the radical $CH_3\dot{C}HOH$ are 15·0 G to the α proton and 22·6 G to the β protons of the methyl group. Given that the average value of Q_β is 29·3 G, what is the apparent value of Q_α? If the spin withdrawing power of an α proton is zero and of an α methyl group is 8·1%, calculate the spin-withdrawing parameter for an α hydroxy group.

From eqn. (21) the spin density $\varrho(C)$ on the tervalent carbon is 22·6/29·3. This spin density produces an α proton coupling of 15·0 G. Thus, from eqn. (20),

$$15.0 = Q_\alpha \times \frac{22.6}{29.3},$$

$$Q_\alpha = 19.5 \text{ G}.$$

The Fischer equation (22) for this radical is

$$\varrho(C) = (1 - \Delta X_H)(1 - \Delta X_{CH_3})(1 - \Delta X_{OH}),$$

where $\Delta X_H = 0$, $\Delta X_{CH_3} = 0.081$ and ΔX_{OH} is to be determined. Thus

$$\varrho(C) = 1 \times 0.919 \times (1 - \Delta X_{OH}) = \frac{22.6}{29.3},$$

$$(1 - \Delta X_{OH}) = \frac{22.6}{29.3 \times 0.919} = 0.840.$$

Therefore the spin withdrawing parameter for an α hydroxy group is 0.160, that is to say the α hydroxy group can withdraw 16.0% of the spin density from the tervalent carbon atom.

3. The spectrum shown in Fig. 25 is a reconstruction of that obtained when the acidified titanium/peroxide reagent is mixed with diethyl ether and flowed through the cavity of an e.s.r. spectrometer. Measure the splittings and identify the radical producing this spectrum.

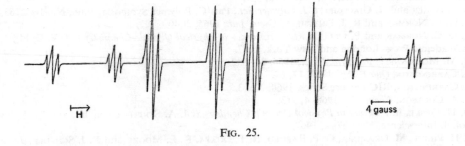

FIG. 25.

There are, in all, twenty-four lines in this spectrum, which we may label 1 to 24 from the low field side, split into eight main groups. Each group consists of a $1:2:1$ triplet, indicating a small interaction with two equivalent protons that are relatively remote from the radical centre. The splitting between lines 1 and 2 corresponds to a coupling of 1.2 G. The relative intensities of the eight main groups are $1:1:3:3:3:3:1:1$, and this immediately suggests a main quartet splitting due to a coupling to three equivalent protons with a subsidiary doublet splitting to an isolated proton. The quartet splitting may be obtained from, say, the distance between lines 1 and 7, and is about 21.2 G. The splitting between lines 1 and 4 gives a doublet coupling of 13.2 G. These two main couplings suggest a radical of the general type $CH_3\dot{C}HX$. The group X must contain two equivalent protons, producing the minor coupling, that are not immediately adjacent to the radical centre. Reaction of the titanium/peroxide reagent with saturated compounds normally results in hydrogen atom abstraction. With diethyl ether the two possible results of such a reaction are the radicals $\cdot CH_2CH_2OCH_2CH_3$ and $CH_3\dot{C}HOCH_2CH_3$. The latter, having one α, three β, and two γ protons, fits the requirements of the spectrum satisfactorily. Not unexpectedly the coupling to the three equivalent δ protons is too small for resolution. Using the measured splittings we may derive the Q_α value for the radical and the spin-withdrawing parameter for the OCH_2CH_3 group by the method given in the previous example. Groups of the general type OR are some of the most effective at withdrawing spin density from the tervalent carbon atom.

References

1. E. Zavoisky, *J. Phys. USSR*, 1945, **9**, 245.
2. C. P. Poole, *Experimental Techniques in Electron Spin Resonance*, Interscience, New York, 1966.
3. T. H. Wilmshurst, *Electron Spin Resonance Spectrometers*, Hilger, London, 1968.
4. R. S. Anderson, in *Methods of Experimental Physics*, Vol. 3, *Molecular Physics* (ed. D. Williams), Academic Press, New York and London, 1960, Chap. 4.2.
5. G. K. Fraenkel, in *Physical Methods of Organic Chemistry*, Part IV (ed. by A. Weissburger), Interscience, New York, 1960, Chap. 42.
6. E. F. Caldin, *Fast Reactions in Solution*, Blackwell, Oxford, 1964, Chap. 10.
7. A. Carrington, in *Molecular Relaxation Processes*, Academic Press and the Chemical Society, London, 1966.
8. R. W. Fessenden and R. H. Schuler, *J. Chem. Phys.*, 1965, **43**, 2704.
9. R. W. Fessenden and R. H. Schuler, *J. Chem. Phys.*, 1964, **39**, 2147.
10. W. T. Dixon and R. O. C. Norman, *J. Chem. Soc.*, 1963, 3119.
11. R. Livingston and H. Zeldes, *J. Chem. Phys.*, 1966, **44**, 1245.
12a. H. Fischer, *Proc. Roy. Soc.* A, 1968, **302**, 321.
12b. H. Fischer and G. Giacometti, *J. Polymer Sci.*, Part C, Polymer Symposia, 1966, No. 16, 2763.
13. R. O. C. Norman and R. J. Pritchett, *Chem. Ind.*, 1965, 2040.
14. R. O. C. Norman and B. C. Gilbert, in *Advances in Physical Organic Chemistry* (ed. V. Gold), Vol. 5, Academic Press, London and New York, 1967.
15. H. M. McConnell, *J. Chem. Phys.*, 1956, **24**, 632.
16. A. Carrington, *Quart. Rev.* 1963, **17**, 67.
17. A. Carrington, RIC Lecture Series, 1964, No. 3.
18. C. A. Coulson, *Chem. Br.*, 1968, **4**, 113.
19. D. H. Geske, in *Progress in Physical Organic Chemistry* (ed. A. Streitweiser, Jr., and R. W. Taft), Vol. 4, Interscience, New York, 1967.
20. L. H. Piette, M. Okamura, G. P. Rabold, R. T. Ogata, R. E. Moore and P. J. Scheuer, *J. Phys. Chem.* 1967, **71**, 29.
21. L. H. Piette, P. Ludwig, and R. N. Adams, *Analyt. Chem.*, 1962, **34**, 917.
22. G. P. Rabold, P. T. Ogata, M. Okamura, L. H. Piette, R. E. Moore and P. J. Scheuer, *J. Chem. Phys.* 1967, **46**, 1161.
23. J. O. Fitzpatrick, C. Steelink and R. E. Hansen, *J. Org. Chem.*, 1967, **32**, 625.
24. J. R. Morton, *Chem. Revs.*, 1964, **64**, 453.
25a. N. Tamura, M. A. Collins and D. H. Whiffen, *Trans. Faraday Soc.* 1966, **62**, 1037.
25b. N. Tamura, M. A. Collins and D. H. Whiffen, *Trans. Faraday Soc.* 1966, **62**, 2434.
26. H. C. Box, in *Electron Spin Resonance and the Effects of Radiation on Biological Systems* (ed. W. Snipes), National Academy of Science and National Research Council, Washington, 1966.
27. J. Higuchi, *J. Chem. Phys.*, 1963, **39**, 1339.
28. C. Thomson, *Quart. Rev.*, 1968, **22**, 45.

FURTHER READING

Books

A. Carrington and A. D. McLachlan, *Introduction to Magnetic Resonance*, Harper & Row, New York, 1967.
M. Bersohn and J. C. Baird, *An Introduction to Electron Paramagnetic Resonance*, Benjamin, New York, 1966.
P. B. Ayscough, *Electron Spin Resonance in Chemistry*, Methuen, London, 1967.
R. S. Alger, *Electron Paramagnetic Resonance*, Wiley–Interscience, New York, London, and Sydney, 1968.
W. Snipes (ed.), *Electron Spin Resonance and the Effects of Radiation on Biological Systems*, National Academy of Science, National Research Council, Washington DC, 1966.
E. T. Kaiser and L. Kevan, *Radical Ions*, Interscience, New York, 1968.

D. J. E. INGRAM, *Biological and Biochemical Applications of Electron Spin Resonance* Adam Hilger, London 1969.
F. GERSON, *High Resolution ESR Spectroscopy*, Wiley-Verlag, 1970.
J. E. WERTZ and J. R. BOLTON, *Electron Spin Resonance*, McGraw-Hill, New York, 1972.
K. A. MCLAUCHLAN, *Magnetic Resonance*, Oxford U. P., London, 1972.

Reviews

(a) General

M. C. R. SYMONS, in *Advances in Physical Organic Chemistry* (ed. by V. GOLD), Vol. 1, Academic Press, London and New York, 1963.
G. A. RUSSELL, *Science*, 1968, **161**, 423.
F. SCHNEIDER, K. MOBIUS, and M. PLATO, *Angew. Chem., Int. Edn.*, 1965, **4**, 886.
B. SMALLER, in *Advances in Chemical Physics* (ed. J. DUCHESNE) Vol. 7, Interscience, New York, 1964.
R. W. FESSENDEN and R. H. SCHULER, in *Advances in Radiation Chemistry* (ed. M. BURTON and J. L. MAGEE), Vol. 2, Wiley-Interscience, New York, 1970.
E. G. JANZEN, *Accounts Chem. Res.*, 1969, **2**, 279.

(b) Organic Radicals in Solution

R. O. C. NORMAN and B. C. GILBERT, in *Advances in Physical Organic Chemistry* (ed. V. Gold), Vol. 5 Academic Press, London and New York, 1967.
A. CARRINGTON, *Quart. Revs.*, 1963, **17**, 67.
A. CARRINGTON, *Orbital Degeneracy and Spin Resonance in Free Radical Ions*, Lecture Series, 1964, No. 3, Royal Institute of Chemistry, London, 1964.
D. H. GESKE, *Progress in Physical Organic Chemistry* (ed. A. STREITWEISER, Jr. and R. W. TAFT), Vol. 4, Interscience, New York, 1967.
N. M. ATHERTON, *Science Prog.*, 1968, **56**, 179.
C. STEELINK, *Adv. Chem. Ser.*, 1966, **59**, 51.
G. A. RUSSELL, E. T. STROM, E. R. TALATY, K. Y. CHANG, R. D. STEPHENS, and M. C. YOUNG, *Rec. Chem. Prog*, 1966, **27**, 3.

(c) Organic Radicals in Solids

J. R. MORTON, *Chem. Revs.*, 1964, **64**, 453.
D. E. O'REILLY and J. H. ANDERSON, in *Physics and Chemistry of the Organic Solid State* (ed. D. FOX, M. M. LABES, and A. WEISSBURGER), Vol. 2, Interscience, New York, 1965.

(d) Organic Triplet States

C. THOMSON, *Quart. Revs.*, 1968, **22**, 45.
M. WEISSBLUTH, in *Molecular Biophysics* (ed. B. PULLMAN and M. WEISSBLUTH), Academic Press, New York and London, 1965.
E. WASSERMANN, *Progress in Physical Organic Chemistry* (ed. A. STREITWEISER, Jr. and R. W. TAFT), Vol. 8, Wiley-Interscience, New York, 1971.

(e) Biochemical Applications

K. G. ZIMMER and A. MÜLLER, in *Current Topics in Radiation Research* (ed. M. EBERT and A. HOWARD), Vol. 1, North-Holland, Amsterdam, 1965.
C. L. HAMILTON and H. M. MCCONNELL, in *Structural Chemistry and Molecular Biology* (ed. A. RICH and N. DAVIDSON), W. H. Freeman and Co., San Francisco and London, 1968

Electron Spin Resonance Spectroscopy Seminar Problems and Answers

H. W. WARDALE

Department of Chemistry and Applied Chemistry, University of Salford

Seminar Problems in Electron Spin Resonance Spectroscopy

1. Calculate the population ratio $N(-\frac{1}{2})/N(+\frac{1}{2})$ for:

    (i) $v = 9.5$ GHz and $T = 4$ K;

    (ii) $v = 9.5$ GHz and $T = 300$ K;

    (iii) $v = 35$ GHz and $T = 300$ K.

2. How many lines, and of what relative intensity, would you expect to find in the fully resolved isotropic spectrum of the following radicals?

    (i) *p*-benzosemiquinone anion $\{OC_6H_4O\}^-$;

    (ii) *n*-propyl radical $\{CH_3CH_2CH_2\}$;

    (iii) *i*-propyl radical $\{(CH_3)_2CH\}$;

    (iv) anthracene cation $\{C_{14}H_{10}\}^+$;

    (v) the monofluoromethyl radicals $\{CH_2F\}$ and $\{^{13}CH_2F\}$;

    (vi) tetracyanoethylene anion $\{C_2(CN)_4\}^-$;

    (vii) hydrazine cation $\{N_2H_4\}^+$;

    (viii) trichloromethyl radical $\{CCl_3\}$;

    (ix) dimethyl nitroxide radical $\{(CH_3)_2NO\}$

    (x) *p*-phenylenediamine cation $\{H_2NC_6H_4NH_2\}^+$.

3. Sketch the spectrum of:

    (i) $^{13}CH_3$ (data given on p. 165);

    (ii) $CH_3COCHCH_3$ (data given on p. 174);

    (iii) naphthalene anion $\{C_{10}H_8\}^-$ if the proton splitting at positions 1, 4, 5, and 8 is 4·90 G and at positions 2, 3, 6, and 7 is 1·83 G;

    (iv) *p*-dinitrobenzene anion $\{O_2NC_6H_4NO_2\}^-$ if the proton splitting is 1·12 G and the nitrogen splitting is 1·48 G;

    (v) the main primary radical formed during the reaction of acrylonitrile with hydroxyl radicals if the nitrogen splitting is 3·5 G, the α proton splitting is 20·1 G, the β proton splitting is 28·2 G, and the hydroxyl proton splitting is unresolved.

4. The e.s.r. spectrum (Fig. 1) was obtained when a few drops of 100 vol hydrogen peroxide was added to a solution of diethylamine in aqueous diethylformamide at room temperature. A similar spectrum was obtained during the oxidation of diethylhydroxylamine with lead dioxide and during the photolysis of nitrosoethane. Identify this radical and suggest a mechanism for its formation in these three reactions.

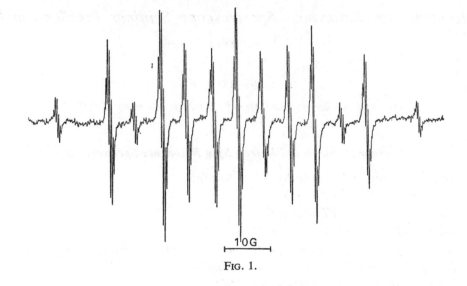

10 G

Fig. 1.

5. Use the McConnell equations and first-order molecular orbital theory to estimate the number and magnitude of all the proton hyperfine lines in:

    (i) *p*-xylene anion;
    (ii) *m*-xylene anion;
    (iii) *o*-xylene anion.

6. An organic solid was γ-irradiated at 77 K. The spectrum at 77 K consists of:
    (i) an anisotropic doublet with principal experimental splittings of 16, 9 and 5 G;
    (ii) an almost isotropic triplet (1:2:1) with a splitting of 8 G;
    (iii) an almost isotropic doublet (1:1) with a splitting of 25 G.

On warming to room temperature the features described in (ii) and (iii) are replaced by an isotropic quartet (1:3:3:1) with a splitting of 12 G while the anisotropic doublet is unchanged. Assuming that all hyperfine interactions are due to protons, deduce as much information as possible about the radical formed during irradiation from this data.

7. The photolysis of a solution of toluene in di-t-butyl peroxide produces a radical having a fifty-four line spectrum. The unpaired electron interacts with seven protons: three sets of two equivalent protons having splittings of 16·30, 5·15, and 1·77 G, and a single proton producing a splitting of 6·18 G. The same radical is obtained as one of the intermediates in the reaction of phenyl acetic acid with the titanium/peroxide reagent. Identify the radical and suggest mechanisms for its formation in these two reactions.

8. A single crystal of zinc acetate was X-irradiated at 77 K. The e.s.r. spectrum obtained at 77 K consists of two distinct parts. An almost isotropic quartet with a splitting of 23 G and relative intensities of $1:3:3:1$ is lost when the crystal is warmed to 150 K. A highly anisotropic part, which for most orientations consists of a pair of doublets having principal experimental hyperfine splitting values of 10·5, 20·6, and 33·4 G, is stable up to 370 K. Above room temperature the anisotropic part becomes modified and the spectrum consists of an essentially isotropic triplet with relative intensities of $1:2:1$ and a splitting of 21·5 G. What radicals are most probably responsible for these spectra? Devise a reaction scheme to account for the presence of these radicals in X-irradiated zinc acetate.

9. How could e.s.r. spectroscopy be used to distinguish between the *cis*, the *trans*, and the unconstrained forms of the 1,4-diacetylbenzene radical anion?

10. Electron spin resonance spectroscopy has been used to follow the course of the oxidation of 2,6-di-t-butyl-4-methyl phenol with lead dioxide in cyclohexane solution. Initially the spectrum consists of twelve lines, a $(1:3:3:1)$ quartet of $(1:2:1)$ triplets with splittings of 10·7 and 1·8 G respectively. After about 10 minutes this is replaced by a $(1:2:1)$ triplet with a splitting of 1·3 G. This decays slowly over several days and is eventually replaced by a ten-line spectrum, a $(1:1)$ doublet of $(1:4:6:4:1)$ quintets with splittings of 5·9 and 1·4 G respectively. Suggest a probable path for this oxidation.

11. The hyperfine coupling constants in the table below refer to radicals of the general type CH_3CHX. The chemical shifts are for the methylene protons in the diamagnetic precursors CH_3CH_2X. Calculate the apparent Q_α value for the radicals and correlate this information with the n.m.r. data. What is the significance of this correlation?

X	$a(H_\alpha)$ gauss	$a(H_\beta)$ gauss	$\delta(CH_2)$
$CH_2.CH_3$	21·8	24·5	1·33
$O.CO.CH_3$	14·8	23·2	4·05
OH	15·0	22·6	3·59
CO.OH	20·2	25·0	2·36

Answers to Seminar Problems in Electron Spin Resonance Spectroscopy

1. Use eqns. (3) and (4) (p. 155) to derive an expression relating population ratio to microwave frequency and temperature. Substitute into this expression using the values $h = 6·626 \times 10^{-34}$ J s, $k = 1·381 \times 10^{-23}$ J K^{-1}, and the values given in the problem. (i) 1·114; (ii) 1·00152; (iii) 1·00559.

2. The formulae required are given in section 4 (p. 163). As the relative magnitudes of the couplings are not given, unambiguous answers cannot always be obtained. With the aid of a little experience it is usually possible to make intelligent guesses about the relative sizes of splittings.

(i) The only magnetic nuclei are the protons. All four are equivalent. The total number of allowed transitions is $2^4 = 16$, but because of the equivalence there is considerable overlapping and only five equally spaced lines are observed with relative intensities of $1:4:6:4:1$.

(ii) Only the seven protons can produce hyperfine interactions. The total number of allowed transitions is $2^7 = 128$, but the total number of lines is reduced by partial equivalence. The two equivalent α protons produce a $1:2:1$ triplet and so do the two equivalent β protons. The three equivalent γ protons produce a $1:3:3:1$ quartet, so that there are in all $3 \times 3 \times 4 = 36$ lines. Generally, β proton splittings are greater than α proton splittings, and both are much greater than γ proton splittings. Possible general shapes of the spectrum can be derived using the coupling diagram shown below. These diagrams are similar to those used in the interpretation of n.m.r. spectra.

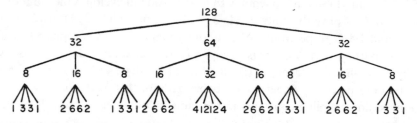

Initially all the allowed transitions are considered to occur at the same field. In this case the single line has an intensity of 128 relative to a line arising from a single allowed transition. The largest hyperfine interaction, with the two β protons, splits this single line into three groups of relative intensity $1:2:1$. Thus the number of allowed transitions in these groups is $32:64:32$. These groups are further split by the similar interaction with the two equivalent α protons producing in all nine groups of transitions. The number of allowed transitions in these groups is $8:16:8$, $16:32:16$, $8:16:8$. Finally, the interaction with the three γ protons splits each group into a $1:3:3:1$ quartet so that the overall situation of the final row is obtained. Variations in the ratios of the three splittings can lead to a different order of lines, but there are always the same number, and the relative intensity of the weakest to the strongest is always the same unless there is a fortuitous overlapping. This is further explored below in parts (iii) and (vii) of this question.

(iii) Again the hyperfine interaction is with seven protons only, and there are 128 allowed transitions. The single α proton produces $1:1$ doublet, but the six β protons are equivalent producing a $1:6:15:20:15:6:1$ septet. In all there are $2 \times 7 = 14$ lines. Assuming first that the β splitting is greater (at least $\times 3$) than the α splitting the central line breaks up into seven equally spaced groups having $2:12:30:40:30:12:2$ allowed transitions in each group. These groups are now split by the smaller α proton interaction to give fourteen lines as shown in the diagram below.

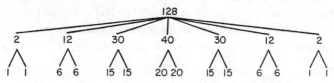

If it is now assumed that the α splitting is very much greater (at least ×7) than the β splitting then the α splitting breaks the central line into two groups each containing 64 transitions. Each is now split into a septet of lines by the β proton interaction to give fourteen lines as shown in the diagram below.

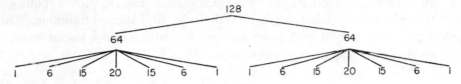

At intermediate ratios of the two couplings these two extremes, a septet of doublets and a doublet of septets, become intertwined. The diagram below correlates these two extremes and shows the effect of fortuitous overlapping at the crossing intersections on the diagram. One such situation, the α splitting is twice the β splitting, is specifically labelled with the expected intensities.

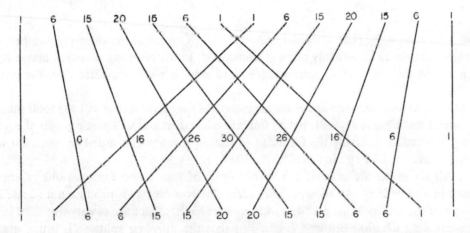

(iv) Only the ten protons can produce hyperfine interactions. The total number of allowed transitions is $2^{10} = 1024$, but the total number of lines is reduced by partial equivalence. Protons at positions 1, 4, 5, and 8 are equivalent and produce a $1:4:6:4:1$ quintet. Similarly, protons at positions 2, 3, 6, and 7 are equivalent and also produce a $1:4:6:4:1$ quintet. The protons at positions 9 and 10 are also equivalent and produce a $1:2:1$ triplet. In all there are $5 \times 5 \times 3 = 75$ lines. The relative intensity of the strongest to the weakest line is $(6 \times 6 \times 2)72 : 1$. In the coupling diagram derived below it is assumed that the smallest interaction is with the protons at positions 9 and 10.

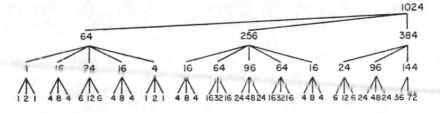

All coupling diagrams for isotropic spectra are symmetrical about the centre, and consequently it is only necessary to show one-half of the complete diagram.

(v) In CH₂F both the protons and the fluorine nucleus can produce hyperfine interactions. All three nuclei have spin of $\frac{1}{2}$, and so the total number of allowed transitions is $2^3 = 8$. If it is assumed that the fluorine splitting is much greater than the proton splitting, then the central line is first split into two groups each having four allowed transitions. The two equivalent protons now split each group into triplets with relative intensities of 1:2:1 as is shown in the coupling diagram below left.

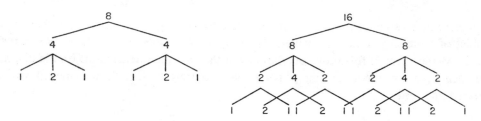

The analysis for ^{13}CH₂F is similar although the ^{13}C nucleus provides yet another spin $\frac{1}{2}$ term and hence twice as many allowed transitions. In the coupling diagram, above right, it is assumed that the carbon splitting, the third step, is slightly greater than the proton splitting.

(vi) Only the four nitrogen nuclei are magnetic. As they have a spin of 1 the total number of allowed transitions is $3^4 = 81$, but all four are equivalent and so a single group of equally spaced lines is found. Using the formulae of section 4 suggests a nine-line spectrum with relative intensities of 1:4:10:16:19:16:10:4:1.

(vii) All six nuclei are magnetic. Two nitrogens of spin 1 and four protons of spin $\frac{1}{2}$ produce in all $3^2 \times 2^4 = 144$ allowed transitions. The two nitrogens produce a 1:2:3:2:1 quintet and the four protons a 1:4:6:4:1 quintet so that there are twenty-five lines in all. In the coupling diagram below it is assumed that the nitrogen splitting is much greater than the proton splitting. As an additional exercise construct the coupling diagram for the situation in which the proton splitting is much greater than the nitrogen splitting. Use this diagram and the previous one to construct a correlation diagram for the two splittings and deduce the nature of the spectrum when the two splittings are equal. Compare the result with the diagram given at the end of this section.

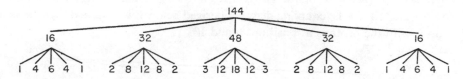

(viii) The chlorine nuclei are magnetic and have spin of $\frac{3}{2}$. The total number of allowed transitions is $4^3 = 64$, but all three chlorine nuclei are equivalent, and so only ten equally spaced lines with relative intensities 1:3:6:10:12:12:10:6:3:1 are obtained.

(ix) There are seven magnetic nuclei, six equivalent protons, and a single nitrogen nucleus. The total number of allowed transitions is $3 \times 2^6 = 192$. The nitrogen nucleus produces a 1:1:1 triplet and the six protons produce a $1:6:15:20:15:6:1$ septet, so that a twenty-one-line spectrum is obtained. The coupling diagram below demonstrates the situation when the nitrogen splitting is much greater than the proton splitting.

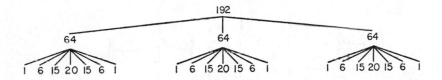

(x) All the nuclei except the six carbons are magnetic. There are four equivalent ring protons, four equivalent amino protons, and two equivalent nitrogen nuclei giving a total of $2^8 \times 3^2 - 2304$ allowed transitions. Two $1:4:6:4:1$ quintets from the protons and a $1:2:3:2:1$ quintet from the nitrogens produce a total of 125 lines with an intensity ratio of strongest to weakest is 108:1. In the coupling diagram below the nitrogen splitting is assumed to be considerably larger than the proton splittings.

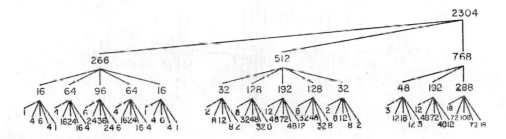

Correlation diagram for additional exercise to (vii) (p. 200):

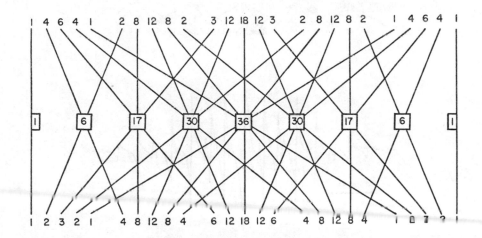

3. (i)

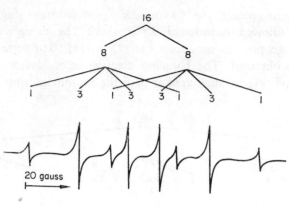

(ii)

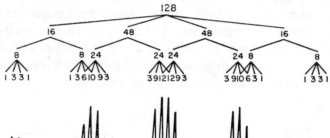

(iii)

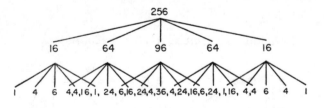

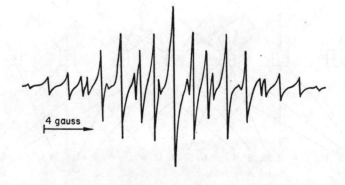

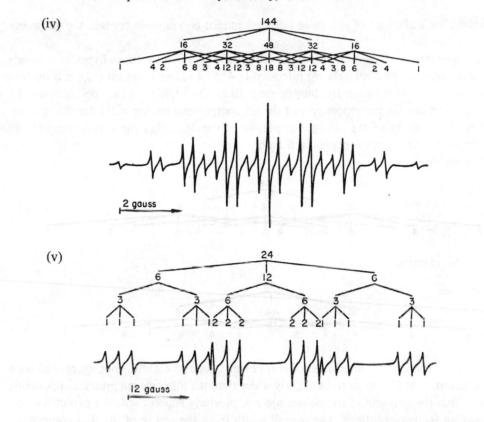

4. Figure 1 is a real, experimental spectrum and not an idealized reconstruction. In such circumstances we may perhaps expect to observe small deviations from the ideal intensity ratios and imperfect resolution of lines associated with very small hyperfine couplings.

The experimental details, particularly the photolytic synthesis, suggest that the only atoms present with magnetic nuclei are hydrogen and nitrogen. Thus the total number of allowed transitions must factorise into $2^m \times 3^n$ where m and n are the total number of protons and nitrogen nuclei producing a resolvable interaction with the unpaired electron.

The spectrum consists of thirteen major components each of which is split by a very small hyperfine interaction. Let us ignore this small splitting for the moment and analyse the major components first. It is not easy to assign relative intensities to these major components as the spectrum is not perfectly symmetrical about the centre; the high field components are slightly smaller than the corresponding ones at low field. Remembering that there should be a centre of symmetry, we must make an intelligent guess at these relative intensities. As a first attempt we might try $1:4:1:6:4:4:6:4:4:6:1:4:1$. This gives a total of forty-six allowed transitions, but as the only factors are 2 and 23 it does not satisfy our conditions. Other unlikely numbers of allowed transitions in this region are $45(= 3^2 \times 5)$, $44(= 2^2 \times 11)$, 43(prime), 47(prime), $49(= 7^2)$, $50(= 2 \times 5^2)$ and $51(= 3 \times 17)$. By far the most probable number of transitions is $48(= 2^4 \times 3)$, for these could arise from an electron hyperfine interaction with four protons and one ^{14}N nucleus. Furthermore, this is reasonable since chemically two of the three different parent compounds contain the grouping

$$\left(\begin{array}{c} | \\ -CH_2-N-CH_2- \end{array}\right).$$ However, this grouping should give rise to fifteen hyperfine components, three quintets each of relative intensity 1:4:6:4:1, so there must be a little for-
tuitous overlapping to reduce it to thirteen main lines. On looking at the spectrum we can
see that we have lost the innermost pair of the six components having unit intensity, and on
measuring the splitting of the nitrogen triplet we can deduce that these two components
are overlapping with components six and eight.

Instead of seeing

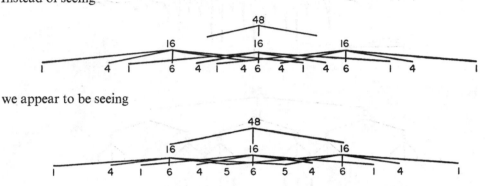

we appear to be seeing

Although the experimental intensities of components six and eight do not agree well with
this conclusion, they do appear to be slightly wider than the other eleven major components,
indication that the two sets of transitions are not precisely superposed. We can check this
by measuring the two splittings. The overall width from the centre of the first component
to the centre of the thirteenth is 79 G, and if our interpretation is correct this will be equal
to twice the nitrogen splitting plus four times the proton splitting. These two splittings can
be measured directly. The proton splitting, obtained from the distance between the first
two components, is 11·3 G. The nitrogen splitting is obtained from the splitting between
components one and three. This is 16·7 G. These splittings would produce an overall
width of 78·6 G in good agreement with our measurement. The fortuitous overlapping
would be precise if twice the nitrogen splitting were equal to three times the proton splitting. Since twice the nitrogen splitting, 33·4 G, is slightly smaller than three times the proton splitting, 33·9 G, the first "missing" unit intensity component occurs just on the low
field side of component six and the second "missing" component is just on the high field
side of component eight. Thus these two components contribute more to the widths than
to the intensities of the main components six and eight.

Having now reached a satisfactory interpretation of the major components in terms of

the group $\left(\begin{array}{c} | \\ -CH_2-N-CH_2- \end{array}\right)$, we must now turn to the hyperfine structure on these
components. Because of the relatively poor resolution we are unlikely to observe the theo-
retical intensity ratios, but by carefully counting the number of lines in the largest major
components we should obtain a strong indication of the origin of this interaction. The
strongest components, numbers 4, 7, and 10, all appear to be septets although the outer-

most pairs of lines are rather poorly defined. On both chemical and spectroscopic grounds the most reasonable explanation of this splitting is that it is the result of a weak hyperfine interaction with the six equivalent protons contained in two equivalent methyl groups. Thus the radical would appear to contain the grouping

$$\left(CH_3\text{---}CH_2\text{---}\overset{|}{N}\text{---}CH_2\text{---}CH_3 \right)$$

This is the diethylamino radical and it is just possible to envisage its formation in these reaction systems. The other major possibility is the diethylnitroxide radical $(C_2H_5)_2NO$. We could easily decide between these two radicals if the g values had been recorded on the spectrum, for the characteristic g value of aliphatic nitroxide radicals is about $2\cdot0055$ while that of aliphatic amino radicals is $2\cdot0035$. Can a firm conclusion be reached from the values of the observed hyperfine splittings?

In the amino radical the unpaired electron is strongly localized on the nitrogen atom; this should result in a large isotropic coupling to the four β protons on the methylene groups of between 20 and 25 G. The unpaired electron in an aliphatic nitroxide radical is mainly localized in a π orbital on the nitroso group, but since this is an antibonding orbital the electron will be slightly more localized on the nitrogen. Consequently the β proton splitting should be little more than half that estimated for the amino radical. The observed splitting of $11\cdot3$ G is in good agreement with such a prediction. Thus the spectrum can be interpreted with some confidence in terms of the diethyl nitroxide radical, and the reaction conditions also favour such an assignment.

Photolysis of nitrosoethane should result in the homolysis of the relatively weak carbon-nitrogen bond

$$C_2H_5NO \rightarrow C_2H_5 + NO$$

The nitroxide may now be formed by addition of the reactive ethyl radical to the parent nitrosoethane

$$C_2H_5 + C_2H_5NO \rightarrow (C_2H_5)_2NO$$

The nitroxide may be produced during the oxidation of the hydroxylamine by the removal of a hydrogen atom.

The precise mechanism for the oxidation of diethylamine with hydrogen peroxide is not known, but an electron balance would require at least 3 moles of peroxide to be reduced for the oxidation of 2 moles of amine. The simplest equation we can write is

$$2(C_2H_5)_2NH + 3\,HO_2^- = 2\,(C_2H_5)_2NO + H_2O + 3\,OH^-$$

but this obviously ignores the probable formation of diamagnetic organic products.

5. The two McConnell equations enable us to relate the size of isotropic proton splittings to spin densities in π orbitals on adjacent atoms. The orbital degeneracy of benzene (Fig. 16) (p. 176) is lifted in substituted benzene anions which have lower than threefold symmetry. If the perturbation due to the substituent is not too large, the unpaired electron distribution

(Fig.17) (p. 176) corresponds fairly closely to that calculated for either ψ_A or ψ_S depending on the electronic effects of the substituents. Since the methyl group is a weakly electron-repelling substituent, the molecular orbital having low spin density on the ring carbons adjacent to the methyl substituents is of lower energy and occupied by the unpaired electron.

The antisymmetric orbital ψ_A, having zero spin density at positions 1 and 4, is stabilized when the two methyl groups occupy these positions. This is the situation in the p-xylene radical anion and should result in a very simple e.s.r. spectrum containing only an interaction with the four equivalent ring protons giving a $1:4:6:4:1$ quintet with a splitting of about $Q_\alpha/4 = 5\cdot7$ G. The observed spectrum contains the expected quintet with a splitting of $5\cdot34$ G, but these components are split by a very small interaction ($0\cdot10$ G) with the β protons in the methyl groups indicating that there is a very small spin density (3%) on the carbons at positions 1 and 4. Thus ψ_A is quite a good approximation to the orbital of the unpaired electron.

The symmetric orbital ψ_S is stabilized when the two methyl groups are either *meta* or *ortho* to each other. The spin density on the adjacent ring carbons is $\frac{1}{12}$ when ψ_S is half filled, and this should result in a $1:6:15:20:15:6:1$ septet from the six protons of the methyl groups with a splitting of about $Q_\beta/12 = 2\cdot4$ G. In the o-xylene radical anion the two ring protons *ortho* to the substituents are equivalent and should give rise to a $1:2:1$ triplet with a splitting of about $Q_\alpha/3 = 7\cdot6$ G. The two remaining ring protons are also equivalent and should produce a further 1:2:1 triplet with a splitting of about $Q_\alpha/12 = 1\cdot9$ G. In all we should anticipate a sixty-three-line spectrum, and the intensity ratio of the strongest to the weakest line should be eighty to one. If ψ_S is a good description of the unpaired electron orbital, then the m-xylene radical anion should produce a similar sixty-three-line spectrum. The experimental spectra are in resonable agreement with these elementary predictions. The splittings in the o-xylene radical anion are $2\cdot20$ G to the methyl protons, and $6\cdot95$ and $1\cdot81$ G to the two types of ring protons. Experimentally the ring protons at positions 2 and 5 are not equivalent in the m-xylene radical anion, the splittings being $6\cdot85$ and $7\cdot72$ G respectively. The ring protons at positions 4 and 6 are equivalent and produce a splitting of $1\cdot46$ G. The observed splitting by the methyl protons is $2\cdot26$ G.

6. The highly anisotropic doublet arises from a hyperfine interaction with a single proton. Assuming that the experimental splittings are all of the same sign, then the isotropic coupling is the mean of the three experimental values. This is 10 G. The values of the anisotropic coupling along the principal axes are then $+6$, -1, and -5 G. The extent of this anisotropy, coupled with the relatively small isotropic coupling, suggests that the proton is in the α position but the spin density on the adjacent carbon atom is rather low.

The triplet must arise from two equivalent protons and the absence of anisotropy suggests that these protons are in the β position. Similarly the large doublet must also arise from a β proton. We can conclude that a methyl group is attached to the radical centre. It appears that this is locked in a preferred conformation at 77 K but on warming to room temperature acquires sufficient energy to enable free rotation about this carbon-carbon bond and the β proton signal is averaged. The radical is of the general type $CH_3\dot{C}HX$. We can deduce something of the nature of X although we have no resolved hyperfine structure associated with it. From the isotropic coupling to the rotating methyl group we may deduce the spin density at the adjacent carbon.

$$\varrho(C) = \frac{a_{iso}(H_\beta)}{Q_\beta} = \frac{14}{29\cdot3} = 0\cdot48.$$

This can be used to calculate Fischer's spin withdrawing parameter for the group X.

$$\varrho(C) = (1 - \Delta CH_3)(1 - \Delta H)(1 - \Delta X),$$

where

$$\varrho(C) = 0\cdot48, \quad \Delta H = 0, \quad \text{and} \quad \Delta CH_3 = 0\cdot08,$$

$$(1 - \Delta X) = \frac{0\cdot48}{0\cdot92} = 0\cdot52,$$

$$\Delta X = 0\cdot48.$$

X is a very powerful spin-withdrawing group, and by strong conjugation to the tervalent carbon reduces the spin density there to 0·48.

Reconsidering the locked methyl group at low temperature in the light of the low adjacent spin density we may conclude that the large doublet indicates that this proton is completely eclipsed with the *p* orbital on the tervalent carbon so that the dihedral angle is zero. The smaller coupling to the two remaining equivalent protons suggests that they must be only 30° out of the nodal plane and that their dihedral angles are about $\pm120°$.

7. The two precursors of the radical both contain the phenyl group. We can make the reasonable assumption that this group remains intact and that five of the seven protons that interact with the unpaired electron are associated with a phenyl group. The splitting of 16·30 G is much too large for protons attached to an aromatic ring, so we may ascribe the other three splittings to the five phenyl group protons: the splitting of 6·18 G to the *para* proton, that of 5·15 G to two equivalent *ortho* protons and the 1·77 G splitting to two equivalent *meta* protons. The two remaining protons are exocyclic and produce a large coupling to the electron. The obvious interpretation is that the spectrum is that of the benzyl radical C_7H_7 and that there is a high spin density on the exocyclic carbon atom that results in the 16·30 G splitting to the two aliphatic protons. Chemically as well as spectroscopically this is a reasonable interpretation. The initial step in the photolysis is the homolysis of the peroxide bond

$$C_4H_9OOC_4H_9 \rightarrow 2\,C_4H_9O$$

This is followed by abstraction of a hydrogen atom from the toluene

$$C_4H_9O + C_6H_5CH_3 \rightarrow C_4H_9OH + C_6H_5CH_2$$

The hydroxyl radical produced by the titanium/peroxide reagent attacks the acidic end of the phenyl acetic acid and the benzyl radical is formed in the subsequent decomposition

$$C_6H_5CH_2COOH + OH \rightarrow C_6H_5CH_2CO_2 + H_2O$$
$$C_6H_5CH_2CO_2 \rightarrow C_6H_5CH_2 + CO_2$$

8. The unstable isotropic quartet clearly shows interaction with three equivalent protons and the size of the coupling suggests that this must be due to the methyl radical. The more stable radical appears to contain two protons which become equivalent above room temperature. The extent of the anisotropy and the size of the isotropic coupling, the mean of the principal values, 21·5 G, suggest that the radical is of the type H_2CR and that free rotation about the C—R bond is restricted below room temperature. The radical $CH_2CO_2^-$, formed by oxidation of the acetate anion, satisfies these criteria.

A possible reaction scheme is

$$CH_3CO_2^- \rightarrow CH_3CO_2^{\cdot} + e$$
$$CH_3CO_2^{\cdot} \rightarrow CH_3^{\cdot} + CO_2$$
$$e + CH_3CO_2^- \rightarrow CH_2CO_2^- + H^{\cdot}$$

or
$$CH_3CO_2^{\cdot} + CH_3CO_2^- \rightarrow CH_3CO_2H + CH_2CO^-$$

or
$$CH_3^{\cdot} + CH_3CO_2^- \rightarrow CH_4 + CH_2CO_2^-$$

Various scavengers might be used to estimate the relative importance of these reactions.

9. There are ten protons in the 1,4-diacetyl benzene radical anion that will interact with the unpaired electron, but it is convenient to consider only the interaction with the ring protons.

When the radical is locked in the *cis* conformation, then the protons at positions 2 and 3 are equivalent and the protons at positions 5 and 6 are equivalent. This results in a nine-line spectrum, a triplet of triplets, with the intensity ratio of the strongest to the weakest line of 4 to 1. When the radical is locked in the *trans* conformation, then the protons at positions 2 and 5 are equivalent and the protons at positions 3 and 6 are equivalent. This produces a similar nine-line spectrum, and hence it is not possible to distinguish between the two conformations without some further experiments. Free rotation about the ring-substituent bond results in the equivalence of the four ring protons, so that the unconstrained form of the radical anion gives rise to a simple 1:4:6:4:1 quintet from the ring protons.

It is possible to distinguish between the *cis* and *trans* isomers by selective deuteration of the ring protons. Substitution of the ring protons at positions 2 and 3 by deuterons ($I = 1$) enables this distinction to be made. These two deuterons are equivalent in the *cis* isomer and consequently split the components of the 1:2:1 triplet from the remaining protons into 1:2:3:2:1 quintets to produce a fifteen-line spectrum with the intensity ratio of the strongest to the weakest line of 6 to 1. The situation is much more complicated in the *trans* isomer as there is no equivalence. In the absence of fortuitous overlapping this results in a thirty-six-line spectrum with each component of unit intensity. In all these cases the effect of the protons on the two acetyl groups is to split the components from nuclei attached to the ring into septets.

10. The initial step in the oxidation of a phenol is the removal of the phenolic hydrogen atom and the formation of a phenoxy radical. The twelve-line spectrum is consistent with the phenoxy radical derived from 2,6-di-t-butyl-4-methyl phenol; the quartet arises from the interaction with the three protons in the methyl group while the triplet is due to the interaction with the two *meta* protons. The nineteen-line multiplet that should result from the interaction with the eighteen equivalent protons on the t-butyl groups is too small to resolve.

The second spectrum is rather more difficult to assign. There is a small interaction with two equivalent protons. The size of this coupling suggests that it is probably the two *meta* protons that are responsible. It has been suggested that the radical responsible is 2,6-di-t-butyl-4-formyl phenoxy radical, but it is not clear why the interaction with the formyl proton is not resolved.

Interaction with five protons is resolved in the third spectrum. The small splitting to four equivalent protons suggests that these are *meta* protons on two identical ring systems. The larger interaction is with a single proton and consequently must be shared by the two ring systems. It is thought that this proton is part of a methine-bridge between the two rings and that the radical responsible is

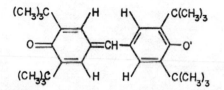

This is a remarkably stable radical and has the trivial name galvinoxyl.

11. The spin density at the radical carbon is calculated from the β proton splitting using the McConnell relationship for β protons and assuming that the Q_β value is constant at 29·3 G. The calculated spin density may then be substituted into the McConnell relationship for α protons and Q_α obtained.
The values obtained are

$$Q_\alpha = \frac{a(H_\alpha)}{a(H_\beta)} Q_\beta.$$

X	Q_α	X	Q_α
$CH_2.CH_3$	26·1	OH	19·5
$O.CO.CH_3$	18·7	$CO.OH$	23·7

When these Q_α values are plotted against the $\delta(CH_2)$ values, a linear relationship is obtained, and this suggests that the factors responsible for the chemical shifts are also responsible for the variation of Q_α. The main factor contributing to the chemical shift is the electron-withdrawing character of the substituent group via the inductive effect. It is suggested that the variation of Q_α is also the result of the electron-withdrawing power of the substituent group. A powerful electron-withdrawing group lowers the total spin density in the C—Hσ bond and hence reduces the spin polarization in the hydrogen 1s orbital and the size of the proton coupling.

Some Recent Developments in Nuclear Magnetic Resonance Spectroscopy

J. A. ELVIDGE

Department of Chemistry, University of Surrey

Introduction

Although nuclear magnetic resonance spectroscopy has been in use by chemists for some 15 years now, it is still a vigorously developing subject.[1] Amongst the recent advances, three have made sufficient impact on structure-determination studies since the appearance of Volume 1 of this book to warrant their mention here. These are the use of lanthanide shift-reagents, applications of the nuclear Overhauser effect (NOE), and the advent of routine carbon-13 nuclear magnetic resonance spectroscopy employing the natural abundance of this isotope. The last constitutes a major advance in technique, and its repercussions in organic chemistry in particular should be very considerable.[2a]

Lanthanide Shift-reagents

EXPERIMENTAL ALTERATION OF CHEMICAL SHIFTS

To facilitate structural deductions, the simplification of complex spectra has been a continuing aim in n.m.r. spectroscopy. Provided chemical shifts are large compared with the coupling constants, then first-order spectra are obtained (Vol. 1, p. 21), capable of ready interpretation by the chemist without recourse to mathematical analysis. Some aids have already been mentioned (Vol. 1, p. 32), amongst which was a method of altering chemical shifts by change of solvent (Vol. 1, p. 36). Solvents of which the molecules are magnetically anisotropic, such as benzene (Vol. 1, pp. 6–7), are effective because they tend to associate with particular structural features in a solute: e.g. for benzene this is with the positive end of polar groupings or multiple bonds. Nearby protons are then selectively shielded or de-shielded, depending on the average mutual orientation. Although the resulting shifts are only 0–1 ppm, they may in favourable cases change a spectrum so that interpretation becomes feasible, or even obvious, as in Fig. 32 of Vol. 1 (p. 38).

Much greater shifts can be expected, with consequent greater simplification of many spectra, from more powerful local magnetic influences (than associated solvent molecules), such as might be provided by paramagnetic ions with large effective moments. Selective shifts of 0–20 ppm in proton resonance spectra have been achieved by the use of certain compounds of europium(III) and other rare earth ions.

EU(DPM)₃, TRISDIPIVALOYLMETHANATOEUROPIUM(III)[†] AND
OTHER LANTHANIDE COMPLEXES

The 8-coordinate europium(III) complex, trisdipivaloylmethanatoeuropium(III) dipyridinate[3] will suffer displacement of pyridine by other compounds having lone-pair atoms. Hinckley[4] first observed that addition of cholesterol to a solution of the europium complex resulted in a loose association and caused substantial downfield shifts in the proton resonance lines of the cholesterol from their normal positions. The magnitude of the induced shifts varied roughly as r^{-3}, where r was the distance from a proton to the europium coordinated at the 3-oxygen of the cholesterol (but see p. 219). Subsequently, Sanders and Williams[5] showed that the pyridine-free 6-coordinate europium complex (1) was superior

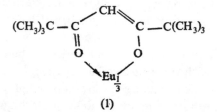

(1)

in coordinating to lone pair functions (as in alcohols, esters, ethers, ketones, amines) and, indeed, produced in cholesterol selective down-field shifts which were some four times larger

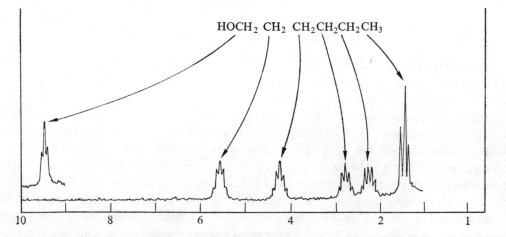

FIG. 1. 100 MHz ¹H spectrum of n-hexanol in CCl₄ after the addition of Eu(DPM)₃ (0·29 m). The inset multiplet is offset 1 ppm. Chemical shifts are in δ ppm from Me₄Si at δ = 0.

† Also named tris(tetramethylheptanedionato)europium, Eu(tmhd)₃.

than observed by Hinckley. The resonances from the protons on the (DPM)₃ ligand groups are shifted upfield, to the region 1–2 ppm higher than tetramethylsilane, and so do not usually interfere. The chemical shift of the internal standard remains constant because tetramethylsilane, being inert and isotropic, does not complex in any way with the europium ion in the added compound (1). However, when the europium compound is present in large amounts its signal may spread sufficiently to obscure the line from the tetramethylsilane. Figure 1 illustrates for n-hexanol the typical effect of the europium reagent. The CH₃ and CH₂ resonance are all separately resolved as a series of first-order multiplets over the region δ 1 to δ 11, downfield of tetramethylsilane, and a complete interpretation is obtained.[5] This is in contrast to the conventional proton spectrum of n-hexanol which shows a complex band in the region δ 1·2 1·7, with only the methylene adjacent to the hydroxyl producing a separately resolved signal, near δ 3·5.

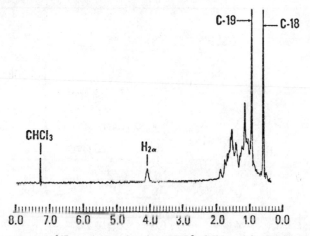

FIG. 2. 100 MHz ¹H spectrum of androstan-2β-ol (20 mg) in CDCl₃ (0·4 ml).

The normal, rather non-informative spectrum of androstan-2β ol (2) obtained at 100 MHz (Fig. 2) was rendered far more informative[6] by the addition of twice the weight of Eu(DPM)₃ to its solution in deuteriochloroform (Fig. 3a). Instead of only the two methyl resonances and that from proton 2α being discerned (Fig. 2), with the rest of the spectrum appearing as a meaningless tight cluster of lines, the shifted spectrum (Fig. 3a) showed

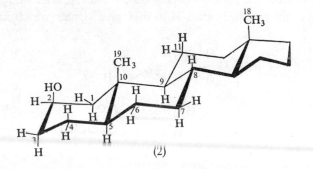

J. A. Elvidge

clearly in addition the resonances (and splitting) from protons 3α and β, 4α and β, and 5α. Further expansion of the still remaining envelope absorption, by operating at 220 MHz, enabled protons 6β, 7α and β, 8β, 9α, and 11α and β to be identified (Fig. 3b).

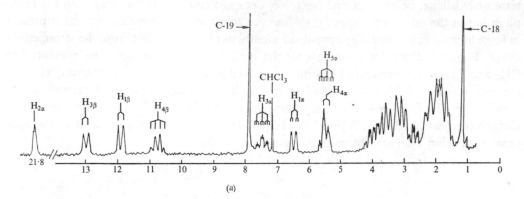

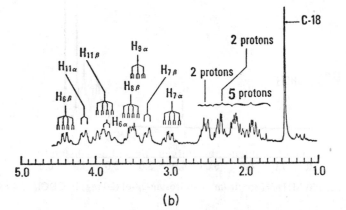

FIG. 3. (a) 100 MHz spectrum of androstan-2β-ol (20 mg) in CDCl₃ (0·4 ml) containing Eu(DPM)₃ (40 mg), and (b) 220 MHz expansion of the region δ 1 to δ 5.

In the spectrum of 1,2:5,6-di-*O*-isopropylidene-α-D-glucofuranose (3) at 100 MHz, only 1-H and 2-H give resonances separately resolved from the complex of signals arising from the other protons on the sugar ring (Fig. 4a). However, after addition of Eu(DPM)₃ to the solution, all of the ring protons 1-H to 6-H give signals which are observed sepa-

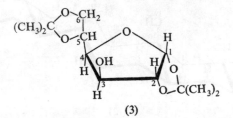

(3)

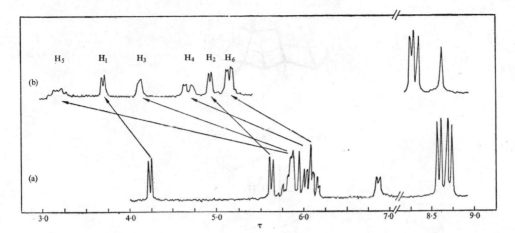

Fig. 4. (a) 100 MHz ¹H spectrum of 1,2:5,6-di-*o*-isopropylidene-α-D-glucofuranose (3) (102 mg) in CCl₄ (0·5 ml), and (b) after addition of Eu(DPM)₃ (33 mg) in CCl₄ (0·55 ml).

rately as first-order multiplets[7] (Fig. 4b). In cases such as this, stereochemical deductions can then at once be made from the directly observed coupling constants, but it must be remembered that the conclusions apply to the complexed molecule and not therefore necessarily to the free molecule in solution.

Many other examples of applications of the europium shift reagent have been reported. Thus this reagent has been used for the determination of the conformation of griseofulvin (4),[8] for proton chemical shift assignments in endrin (5), dieldrin (6), and photodieldrin (7),[9] for checking the methyl group assignments in camphor (8),[10] for determination of

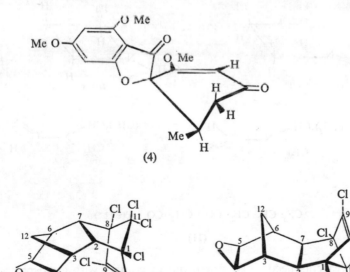

(4)

(5)　　　　　　　　(6)

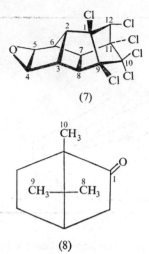

(7)

(8)

Assignments confirmed: τ_{10} 9·14
                           τ_9 9·17
                           τ_8 9·05

the configuration of oximes,[11] and for distinguishing between the *cis*- and *trans*-isomers of azetidine derivatives (9)[12] and between the *E*- and *Z*-isomers of thioamides (10).[13]

More recently, the ligand 1,1,1,2,2,3,3-heptafluoro-7,7-dimethyl-4,6-octanedione (Hfod) (11) has been advocated for forming the europium(III) and praseodymium shift reagents, Eu(fod)₃ and Pr(fod)₃.[14] These chelate more powerfully with weak Lewis bases (e.g. ketones) and are more soluble in nonpolar solvents than Eu(DPM)₃, but because they

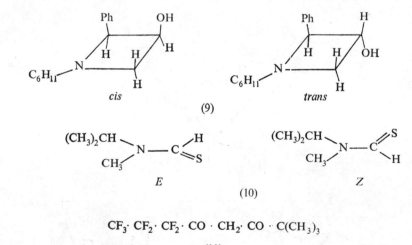

cis   *trans*

(9)

E   *Z*

(10)

$$CF_3 \cdot CF_2 \cdot CF_2 \cdot CO \cdot CH_2 \cdot CO \cdot C(CH_3)_3$$

(11)

chelate water, the reagents need to be stored out of contact with atmospheric moisture. A report on the potentialities of these new shift reagents for ascertaining the micro-structure of stereoregular polymers has appeared.[15]

An interesting finding is that the europium(III) complex (12) of 3-(t-butylhydroxymethy-lene)-D-camphor causes differential shifts in the resonance of enantiomeric amines such as α-phenylethylamine,[16] which should provide the basis of a method for assessing optical purity. An enantiomeric substrate (D+L) forms two complexes (D—D) and (D—L) with a chiral (D) europium compound and because these are diasterioisomeric, the induced shifts differ.

The lanthanide ion (as a water-soluble salt) may be coordinated to a substrate direct if the latter is water soluble.[17a] This approach has been used to investigate,[17b] with the aid of a computer program, the preferred conformation of two representative mononucleo-

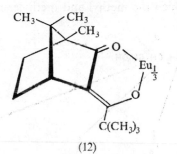

(12)

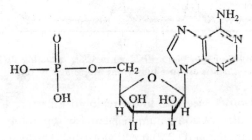

(13)

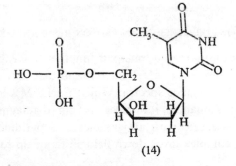

(14)

tides in aqueous solution—adenylic acid (AMP) (13) and deoxythymidylic acid (TMP) (14). The computer searches possible conformations, derived from X-ray crystallographic bond lengths and angles, for best fit with the induced shifts, and the differential line broadening caused by the paramagnetic lanthanide ion, extrapolating these to zero concentration of the ion. The induced shifts are distance and angle dependent (p. 219) and the line broadening is also a function of distance (r^{-6}) from the lanthanide ion. The results so far[17b] indicate that AMP and TMP have preferred conformations in aqueous solution which are slightly different from those in the crystal lattice. Clearly, this is an extremely important development capable of extensive application and modification and of the greatest significance for mechanistic biochemistry.

Whilst formation of loose complexes between substrates and the europium shift reagents in non-polar solvents generally produce marked downfield shifts in proton resonances from their normal positions (but see p. 219), the use of the praseodymium compounds, e.g. $Pr(DPM)_3$, produces up-field shifts.[18] This is illustrated by a spectrum of n-pentanol in carbon tetrachloride, in which the methyl and methylene resonances are all completely

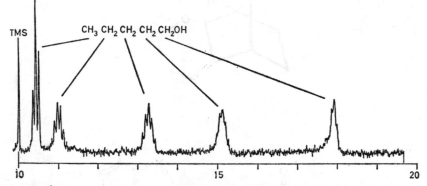

FIG. 5. 100 MHz ^1H spectrum of n-pentanol (0·22 M) in CCl_4 cotaining $Pr(DPM)_3$ (0·053 M). Chemical shifts are in τ ppm from Me_4Si at $\tau = 10$.

separated to give a first-order spectrum in the region 0–8 ppm to high-field of tetramethyl-silane (Fig. 5). Use of the praseodymium reagent for separating the proton resonances from 4-t-butylcyclohexanone and 2,2-dimethyl-4-t-butylcyclohexanone has also been described.[19]

SHIFT REAGENTS FOR OTHER THAN PROTON RESONANCE SPECTROSCOPY

Making use of the fact that nitrogen lone-pair functions coordinate with the lanthanide ion of a shift reagent, Witanowski *et al.*[20] have found that very large chemical shifts can be induced in nitrogen-14 resonance spectra. Whilst $Eu(DPM)_3$ induces down-field (paramagnetic) shifts in proton resonances, it induces up-field (diamagnetic) shifts of nitrogen-14 lines, the largest shifts observed, with alkylamines and pyridine, being 1500 ppm. The corresponding ytterbium complex gives down-field shifts of up to 400 ppm.

REQUIREMENTS FOR AND MODE OF ACTION OF SHIFT REAGENTS

The requirements for a successful shift reagent are rather stringent.[18] The paramagnetic metal ion, besides having a large effective magnetic moment, must have a short electron spin-lattice relaxation time of the order of 10^{-11} s or less.[21] This will preclude coupling of the electron spin with the nuclear spins of a complexed ligand, so that their relaxation times will be largely unaffected (see Vol. 1, p. 20). Nuclear magnetic resonance lines approaching normal width will then be observable. Possible ions are thus restricted to Eu(III) and a few other lanthanides. The substrate—the compound under investigation—must, of course, have one or more functions in its molecule capable of coordinating to the ion. The lanthanide ion may be coordinated to the substrate direct, which is readily accomplished in water or, for example, dimethyl sulphoxide.[17] For compounds soluble in non-polar solvents, it is generally convenient to add to the solution of the substrate the lanthanide ion as a preformed stable complex. This reagent must be sufficiently soluble, stereochemically rigid, and capable of further coordination to the substrate. Resonances from the ligands of the reagent must not interfere with the spectra to be observed. The conditions are met by, for example, Eu(DPM)$_3$. Gradual addition of the shift reagent is advisable, with repeated observation of the n.m.r. spectrum, to avoid ambiguities from signal crossover (see Fig. 4): some resonances will be moved much further than others, and so their relative position in the spectrum may change.

The shifts induced are quite large because the electron magnetic moment is some 10^3 times that of nuclear moments. As the unpaired electrons of a lanthanide are in the deep $4f$ shell, they are effectively confined to the ion: it is as if the ion were a small oriented magnet. The shifts are called pseudo-contact shifts because there is essentially no probability of unpaired electron spin density being found at ligand nuclei. The chemical shifts induced in ligand nuclear resonances are then essentially dipolar in nature, and consequently their magnitude shows both a distance and angular dependence. The induced shift of a proton is inversely proportional to the cube of the distance from the lanthanide ion[4, 19] and also varies as $(3 \cos^2 \theta - 1)$, where θ is the angle between the crystal field axis (or principal symmetry axis) and the lanthanide–proton direction.[21, 22] Because of this last relationship, induced shifts can be positive or negative.[23, 24] Inevitably the paramagnetic lanthanide ion gives rise to some broadening of the nearby proton resonances, and this effect varies with distances as (r^{-6}).[17, 21]

Nuclear Overhauser Effect

A prediction originally made by Overhauser[25] for metals was that if coupling between the nuclear spins and unpaired electron spins provided the principle means of nuclear spin relaxation, then saturation of the electron spin resonance would lead to increased signal strength in the simultaneously observed nuclear resonance signal. Subsequently this was verified, and somewhat analogous effects were observed in purely n.m.r. spectroscopy. In general, changes in signal strength encountered as a result of double resonance experiments, either homo- or hetero-nuclear, are now termed nuclear Overhauser effects (NOE).

Anet *et al.*[26] were the first to employ a nuclear Overhauser effect for making proton resonance spectral assignments. They examined organic compounds in solvents devoid of constitutional hydrogen, such as carbon tetrachloride and deuteriochloroform, and observed that it was generally important, too, to exclude oxygen because its molecule has unpaired spins and so is paramagnetic. The procedure and precautions ensure that intermolecular proton–proton and electron–proton spin interactions are minimized and that the main mechanism for proton spin relaxation amongst the solute molecules is dipole–dipole interaction across space between neighbouring protons on the same molecule. Irradiation of one particular proton at its resonance frequency then increases somewhat the relaxation rate of a neighbouring proton and the observed n.m.r. signal from this proton increases in intensity. This is effectively because it is then less easy for that proton resonance to become saturated (cf. Vol. 1, p. 2). More precise explanations have been given of the original Overhauser effect observed in metals or with organic free radicals by Pople *et al.*[27] and of the nuclear Overhauser effect by Lynden-Bell and Harris[28] and Abragam.[29]

EXAMPLES

The proton resonance spectrum of $\beta\beta$-dimethyl-acrylic acid (15) includes two doublets at τ 8·58 and 8·03 (with J 1·3 Hz) from the methyl groups. These are coupled to the olefinic proton which gives a septet at lower field. Anet *et al.*[26] observed that irradiation of the

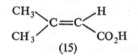

(15)

higher field methyl group caused an increase of 17% in the integrated intensity of the signal from the olefinic proton, now collapsed by the decoupling to a quartet. When the lower field methyl group was irradiated at its resonance frequency, the intensity of the olefinic proton signal was practically unaffected. Thus it was deduced that the higher field methyl group was *cis* to the olefinic proton, which was in line with earlier deductions from the chemical shifts. It is important to note that the reverse experiment failed. Irradiation of the olefinic proton had no effect on the integrated intensities of either of the methyl signals. This was evidently because the methyl protons relax one another: outside protons, even nearby ones, contribute little by comparison, so that saturation of them causes no change in the observed methyl signal intensities.

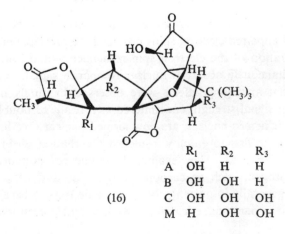

	R_1	R_2	R_3
A	OH	H	H
B	OH	OH	H
C	OH	OH	OH
M	H	OH	OH

(16)

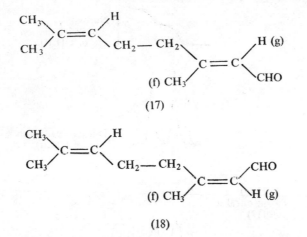

(17)

(18)

Use of the NOE has been made in elucidating the structures of the ginkgolides (16).[30] In another study, the proton resonance lines obtained from citrals *a* and *b* have been assigned with complete certainty.[31] Citral *a* had previously been assigned[32] the *trans*-configuration (17) and citral *b* the *cis*-configuration (18). Irradiation of $CH_3(f)$ in a double resonance experiment with citral *b* (18) decoupled this group of protons from adjacent ones, but at the same time the integrated intensity of the signal from H(g) was increased by 18%

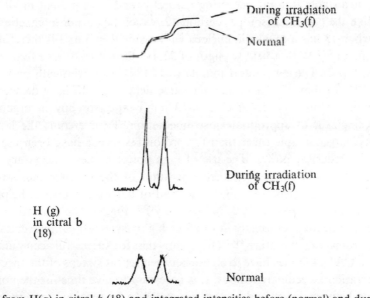

During irradiation
of $CH_3(f)$

Normal

During irradiation
of $CH_3(f)$

H (g)
in citral b
(18)

Normal

Fig. 6. Signals from H(*g*) in citral *b* (18) and integrated intensities before (normal) and during irradiation of $CH_3(f)$.

(Fig. 6). Repetition of the experiment with citral *a* (17) effected no significant change (Fig. 7). Thus it was confirmed that in citral *b*, $CH_3(f)$ was *cis* to H(g) and necessarily therefore *trans* to the aldehyde function, which substantiated the configuration (18).

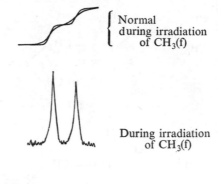

Normal
during irradiation
of $CH_3(f)$

During irradiation
of $CH_3(f)$

H (g)
in citral a
(17)

Normal

FIG. 7. Signals from H(g) in citral a (17) and integrated intensities before (normal) and during irradiation of $CH_3(f)$.

Carbon-13 Nuclear Magnetic Resonance

Carbon-13 is a stable isotope, occurring naturally, and so is present in all carbon compounds. Unlike the common isotope, carbon-12, which has a non-magnetic nucleus with zero spin, carbon-13 has a magnetic nucleus with a spin $I = \frac{1}{2}$ and is therefore capable of detection by n.m.r.[33] With a field strength of 23,487 G, the resonance frequency is 25·14 MHz (cf. Vol. 1, p. 2). The low natural abundance (1·1%) and intrinsically poor sensitivity to detection (1·59% of that of a proton at the same field strength) have delayed widespread investigation of the possibilities of carbon-13 n.m.r. spectroscopy in organic chemistry until the development of appropriate instrumentation. To overcome the low abundance and sensitivity, larger sample tubes than for proton resonance have been used and higher values of radio frequency power. The use of high power makes it necessary to sweep the signals rapidly to avoid saturation; then operation of the spectrometer with the phase control adjusted to give dispersion signals instead of adsorption lines is helpful[34] as is a modulation technique employed by J. N. Shoolery[35] (Fig. 8). The relaxation times of carbon-13 nuclei can be long, especially in small or highly symmetrical molecules, with values of 10–50 s at ordinary temperature.[36] This means that for successful accumulation of weak spectra (with a CAT), minutes have to elapse between repeat sweeps of the spectrum making the whole operation exceedingly tedious. It is necessary to give time for the normal population difference between lower and upper spin states to be regained, otherwise the signals saturate and disappear. However, in larger molecules with intrinsically slower reorientation times in solution, such as cholesteryl chloride, sucrose, adenosine-5'-monophosphate, many of the carbon-13 nuclei have short relaxation times (< 1 s)[37] and only some side-chain carbons or those not bearing a proton have longer relaxation times of the order of 2–8 s, but this still makes conventional spectral accumulation a lengthy operation.

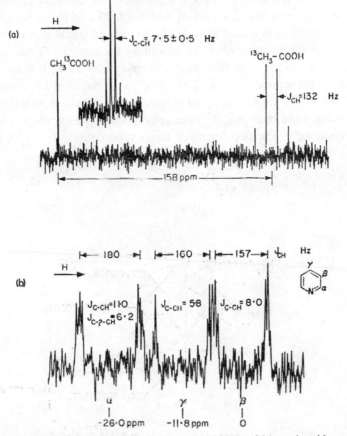

FIG. 8. Carbon-13 (natural abundance) spectrum at 15·085 MHz of (a) acetic acid, and (b) pyridine.

Of the methods for enhancing signal-to-noise, that involving a double resonance technique has perhaps been most widely used.[38] All of the protons in the compound under investigation are irradiated at their resonance frequencies (this being achieved with an audiofrequency-modulated r.f. source—so-called noise or broad-band decoupling). Not only is the signal-to-noise ratio of the carbon-13 signals improved because the intensity of collapsed multiplets is then concentrated into single lines, but there is added enhancement because of a NOE (p. 219). Even quaternary carbons (which lack attached hydrogen) show some enhancement of their signals,[37] but this is relatively small and, indeed, might not have been expected.[36b] All of the carbon atoms (as natural carbon-13) can be detected in cholesterol by this method after only 100 accumulations,[39] but the time involved is still long (*ca.* 1–2 days).

The development of pulse spectrometers with computerized Fourier transform facilities has enabled the time factor to be reduced to minutes or hours, but these facilities are at present very costly. A very short duration (e.g. 40 μs) high-power r.f. pulse is given (instead of slowly sweeping the spectrum). This is followed by the rapid free decay of all of the resonances from the sample (in up to 3 s or so). Repetition can therefore be rapid, and a large number of decay trains can be accumulated within 15–30 min (or at most several hours). By Fourier transformation in a computer, the accumulated signal is converted into a conventional absorption spectrum which is traced out by the recorder. Thus in obtaining the carbon-13 Fourier transform spectrum of cholesteryl chloride with simultaneous proton spin decoupling,[37] the recycle time was 21·7 s and so the required 128 accumulations took only 47 min. Carbon-13 Fourier transform spectra, again obtained with simultaneous proton spin decoupling, have been used[40] to assign configurations amongst N-nitroso-amines (19) and to investigate[41] the two forms of aquocyano-cobyric acid (20, a, b). Whereas

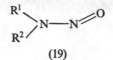

(19)

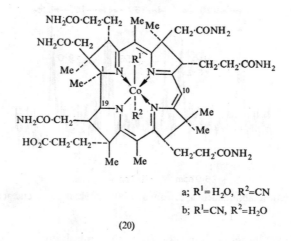

a; $R^1 = H_2O$, $R^2 = CN$

b; $R^1 = CN$, $R^2 = H_2O$

(20)

only one very small shift difference (0·05 ppm) results (for 10-H) in the proton resonance spectra, there are many appreciable differences between the carbon-13 shifts in the two isomers.[41] Such decoupled spectra give no ^{13}C—^1H spin–spin coupling constants of course. These may be derived from the less rapidly obtained, plain Fourier transform spectra, or from the very tediously accumulated conventional absorption spectra (Fig. 8). So far these coupling constants have been derived, most often perhaps, from observation of the ^{13}C—H satellite signals present in conventional proton resonance spectra (see Vol. 1, pp. 35–36, 44). A technique of off-centre double resonance,[2c] in which the appropriate proton signals are irradiated near to (rather than at) their resonance frequencies, allows the carbon-13 signals (accumulated by the pulse technique) to appear as recognizable multiplets, so that assignments do not depend solely on carbon-13 chemical shifts but can be made also on the basis of the numbers of directly attached hydrogen atoms. Because the carbon-13 multiplets appear with diminished ^{13}C—H couplings, the true coupling constants are not directly obtained. It appears, however, that this is seldom a disadvantage in structural work.

An example which illustrates unique advantages of carbon-13 resonance comes from biosynthesis studies.[42] Carbon-13 labelled acetate, first as 1-C and then as 2-C enriched material, was supplied to cultures of *Aspergillus versicolor*, which produces the metabolite sterigmatocystin. From the carbon-13 resonance spectra of the labelled products from the two experiments, the complete pattern of incorporation of acetate was determined directly,

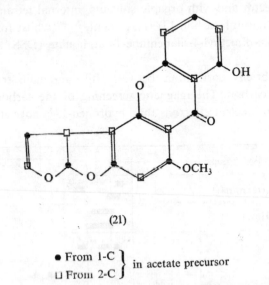

(21)

● From 1-C ⎤
⎦ in acetate precursor
⊔ From 2-C ⎦

with the result as shown in formula (21). Thus the use of radioactive (^{14}C) acetate precursor, and tedious stepwise degradation of the metabolite and counting of the fragments, were obviated.

CHARACTERISTICS OF CARBON-13 SPECTRA

Because carbon-13 has a spin of one-half, the same as that of the proton (and, for example ^{19}F and ^{31}P), the spin multiplets in the undecoupled spectra have a similar appearance, and their interpretation is exactly analogous to that of proton multiplet signals (see Vol. 1, pp. 21–32). Because of the low natural abundance of carbon-13, any one molecule of an organic compound is very unlikely to contain more than one such isotopic atom.[43] Direct carbon-13 to carbon-13 splittings will not therefore be observed. In the bulk sample, all of the possible positions in the molecule will be represented, so that the carbon-13 spectrum comprises a series of signals, one from each carbon position, which are split only by the attached adjacent hydrogen or other magnetic nuclei. Most of the structural information to be gained comes then from the carbon-13 chemical shifts (Fig. 9 and Tables 1–5). The carbon-13 to proton coupling constants are large (*ca.* 0–300 Hz) compared with proton to proton coupling constants (0–20 Hz), and the carbon-13 chemical shifts cover a much larger range (up to 300 ppm) than proton shifts (*ca.* 20 ppm). For calibrating carbon-13 spectra, three main references have been used which may or may not be enriched in carbon-13: these are carbon disulphide, aqueous potassium carbonate, and the carboxyl carbon of acetic acid. In the past the reference substances have often been used externally, the bulk susceptibility correction (Vol. 1, p. 9) being ignored because it is relatively small compared with many carbon-13 chemical shifts. Now that carbon-13 resonance is being applied to the solution of structural problems and because it can provide configurational and conformational information where small shifts can be diagnostic,[2b] the use of good (isotropic) internal standards leading to accurate chemical shifts becomes important. For proton decoupled spectra and with organic solvents, internal tetramethylsilane (which by symmetry has 4·4% natural carbon-13) is excellent,[2c, 40] whilst for aqueous solutions it seems that sodium 4,4-dimethyl-4-silapentane-1-sulphonate (DSS) should prove satisfactory.

Figure 9 provides, briefly, correlations between shifts and main structural types involving sp^3 to sp hybridized carbon. The magnetic screening of the carbon-13 nucleus, with its larger number of extra-nuclear electrons than hydrogen-1, is not yet well understood, and

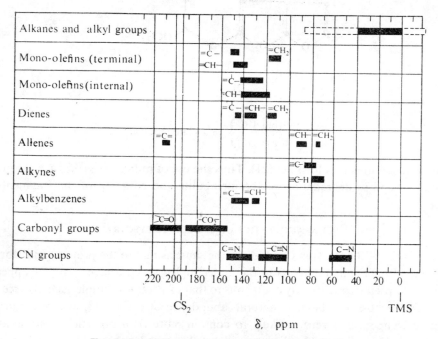

FIG. 9. Correlation chart for carbon-13 chemical shifts.

direct comparisons with arguments used in proton resonance spectroscopy are fraught with danger. Often the screening of a carbon-13 nucleus varies with the electronegativity of an attached substituted (Table 1), rather as is the case for protons (cf. Vol. 1, p. 6), but additional factors operate when adjacent groupings have multiple bonds or are highly polarizable. For example, when carbon-13 is part of the carbonyl group, the nucleus is very strongly de-shielded, and within the range of carbonyl compounds (Table 2) the nucleus is more shielded in esters than in ketones, whereas the reverse might have been expected. Iodine causes a large upfield shift in the resonance of the attached carbon-13 in both alkyl (Table 1) and aryl compounds (Table 5), whereas most other substituents produce downfield shifts. The carbon-13 shifts for sp^2 hybridized carbon in alkenes are very similar to those in aromatic compounds (Table 4), ring current effects which are so prominent in proton resonance (see Vol. 1, pp. 7, 54–55) being negligible. Substituents on a benzene ring cause moderate shifts (± 35 ppm) in the resonance position of the carbon-13 nucleus at the

TABLE 1

CARBON-13 CHEMICAL SHIFTS FOR CH_3, CH_2, CH, AND C
ATTACHED TO GROUPS X AND ALKYL (METHYL) GROUPS
R [34, 44–47]

X	CH_3X	RCH_2X	$RR'CHX$	$RR'R''CX$
—I	−22·0	1·7	22·9	43·8
—H	−1·8			
—CN	5·0			
—Br	9·7	28·9	46·0	62·5
—R	ca. 12	ca. 29		
—≡		ca. 27–37		
—NH₂	13·0			
—SMe	19·8			
—CO₂H	21·0			
—Ph	22·2			
—COMe	25·0			
—Cl	25·2	41·0	55·7	67·1
—CHO	29·9			
—CMe	31·9			
—S(O)Me	43·8			
—NMe₂	47·8			
—OH	48·0	59·4	64·6	69·6
—OPh	55·6			
+				
—NMe₃	57·0			
—NO₂	57·6	73·2	80·9	87·0
—OMe	59·7			
—F	75·7			

TABLE 2
CARBON-13 CHEMICAL SHIFTS FOR THE CARBONYL GROUP [34, 48, 49]

Compound	δ	Compound	δ
Dimethyl acetylenedi-carboxylate	154	Ethyl lactate	177
		Acetic acid	178
Diethyl carbonate	155	Butyrolactone	179
Dimethyl formamide	164	Acetophenone	198
Methyl benzoate	167	Acetaldehyde	200
Acetic anhydride	167	Acetone	206
Vinyl acetate	169	2-Nonanone	207
Methyl acetate	171	Cyclohexanone	210
Acrylic acid	174	Cyclopentanone	220

TABLE 3
CARBON-13 SHIFTS IN CYCLO-
HEXANES, 4-t-Bu-C_6H_{10}-OR, FOR
THE CARBON AT THE POINT OF
ATTACHMENT OF GROUPS OR[50]

R	Axial	Equatorial	$\Delta\delta$
H	67·6	72·5	4·9
Ac	70·4	74·4	4·0
Me	76·5	81·4	4·9

TABLE 4
CARBON-13 CHEMICAL SHIFTS FOR SOME
UNSATURATED FUNCTIONS, OLEFINS,
ACETYLENES, AND AROMATICS[2b, 49, 51, 52]

Function	δ
—C≡N	121
>C=NOH	161
CN⁻	178
Olefins	
=CH₂ (α)	106–117
—CH=(CH₂) (β)	138–150
—C=(CH₂) (β)	144–153
=CH— (otherwise, in chain)	119–144
=C— (otherwise, in chain)	124–144
Allenes	
=CH₂	74
=CH—	86
=C=	207–211
Dienes (conjugated and unconjugated)	
=CH₂	112–118
=CH—	*ca.* 139
=C—	*ca.* 146
Acetylenes	
—C≡C—	*ca.* 82
—C≡CH (α)	70–79
(β)	80–90
Aromatics	
=CH— (benzene)	129·0
=CH— (alkylbenzenes)	127–132
=C— (alkylbenzenes)	133–138
=CH (naphthalene)	128·1
=C— (naphthalene)	137·4

TABLE 5
CARBON-13 SHIFTS FOR MONOSUBSTITUTED BENZENES,
IN PPM RELATIVE TO BENZENE AT δ 129 PPM FROM
TMS[44, 52-54]

Substituent	1-C	o-C	m-C	p-C
NO_2	19·6	−5·3	0·8	6·0
COMe	9·3	0·2	0·2	4·2
CHO	9·0	1·2	1·2	6·0
COCl	5·8	2·6	1·2	7·4
CO_2Me	2·2	0·4	0·4	4·4
Ph	10·0	0·4	0·4	0·4
Me	9·1	0·3	0·3	2·8
CH_2OH	12·3	−1·4	−1·4	−1·4
$SiMe_3$	13·4	4·4	−1·1	−1·1
F	35·1	−14·3	0·9	−4·4
Cl	6·4	0·16	1·0	−2·0
Br	−5·4	3·3	2·2	−1·0
I	−32·3	9·9	2·6	0·4
OPh	29·2	−9·4	1·6	−5·1
OMe	30·2	−14·7	0·9	−8·1
OH	27·6	−11·8	2·6	−6·1
O^-	39·6	−8·2	1·9	−13·6
O.COMe	23·0	−6·4	1·3	−2·3
NH.COMe	11·1	−9·9	0·2	−5·6
NH_2	19·2	−12·4	1·3	−9·5
NMe_2	22·4	−15·7	0·8	−11·8

point of attachment, but there seems no correlation with electronic character (Table 5). In contrast, the smaller shifts of the *para*-carbon do reflect in some measure the electronic character of the substituent, but the shifts at the *ortho*-carbon do not necessarily do so. Thus the nitro group in nitrobenzene, which causes a marked downfield shift in the resonance of the *ortho*-protons, actually causes an upfield shift in the *ortho*-carbon-13 resonance.

In conclusion it may be stated that there are now many useful though often purely empirical correlations between carbon-13 chemical shifts and structure,[2, 33, 49] as the Fig. 9 and Tables 1–5 briefly indicate. Indeed, carbon-13 shifts can be very sensitive probes of structure —see, for example, Table 3. It is certain that increasingly wide use will be made of carbon-13 magnetic resonance spectra for solving structural problems, especially now that the spectra are reasonably conveniently provided by pulse spectrometers.

CARBON-13 CHEMICAL SHIFTS

Tables 1–5 and Fig. 9 give approximate shifts for carbon-13 as δ values in ppm *downfield* from internal tetramethylsilane (TMS). Approximate conversions to other standards, where the upfield shifts have been regarded as positive (in contradistinction to the convention for ¹H shifts), are:

$$\delta(\text{TMS}) = -[\delta(\text{benzene})_{\text{ext}} - 129]$$
$$= -[\delta(\text{CS}_2) - 194]$$
$$= -[\delta(\text{CH}_3 \cdot \overset{*}{\text{C}}\text{O}_2\text{H})_{\text{ext}} - 179]$$

References

1. *Progress in NMR Spectroscopy* (ed. J. W. EMSLEY, J. FEENEY, and L. H. SUTCLIFFE), Pergamon Press; *Advances in Magnetic Resonance* (ed. J. S. WAUGH), Academic Press; *Annual Review of NMR Spectroscopy* (ed. E. F. MOONEY), Academic Press (later, *Annual Reports on NMR Spectroscopy*).
2a. J. FEENEY, Carbon-13 resonance, in *Annual Reports on the Progress of Chemistry*, 1968, **65A**, 75.
2b. E. F. MOONEY and P. H. WINSON, Carbon-13 NMR spectroscopy: carbon-13 chemical shifts and coupling constants, in *Annual Review of NMR Spectroscopy*, 1969, **2**, 153.
2c. E. W. RANDALL, Carbon-13 magnetic resonance, in *Chemistry in Britain*, 1971, **7**, 371.
3. K. J. EISENTRAUT and R. E. SIEVERS, *J. Amer. Chem. Soc.*, 1965, **87**, 5254.
4. C. C. HINCKLEY, *J. Amer. Chem. Soc.*, 1969, **91**, 5160.
5. J. K. M. SANDERS and D. H. WILLIAMS, *Chem. Comm.*, 1970, 422.
6. P. V. DEMARCO, T. K. ELZEY, R. B. LEWIS, and E. WENKERT, *J. Amer. Chem. Soc.*, 1970, **92**, 5737.
7. I. ARMITAGE and L. D. HALL, *Chem. Ind.*, 1970, 1537.
8. S. G. LEVINE and R. E. HICKS, *Tetrahedron Letters*, 1971, 311.
9. L. H. KEITH, *Tetrahedron Letters*, 1971, 3.
10. C. C. HINCKLEY, *J. Org. Chem.*, 1970, **35**, 2834.
11. Z. W. WOLKOWSKI, *Tetrahedron Letters*, 1971, 825.
12. T. OKUTANI, A. MORIMOTO, T. KANEKO, and K. MASUDA, *Tetrahedron Letters*, 1971, 1115.
13. W. WALTER, R. F. BECKER, and J. THIEM, *Tetrahedron Letters*, 1971, 1971.
14. R. E. RONDEAU and R. E. SIEVERS, *J. Amer. Chem. Soc.*, 1971, **93**, 1522.
15. J. E. GUILLET, I. R. PEAT, and W. F. REYNOLDS, *Tetrahedron Letters*, 1971, 3493.
16. G. M. WHITESIDES and D. W. LEWIS, *J. Amer. Chem. Soc.*, 1970, **92**, 6979.
17a. F. A. HART, G. P. MOSS, and M. L. STANIFORTH, *Tetrahedron Letters*, 1971, 3389.
17b. C. D. BARRY, A. C. T. NORTH, J. A. GLASEL, R. J. P. WILLIAMS, and A. V. XAVIER, *Nature*, 1971, **232**, 236.
18. J. BRIGGS, G. H. FROST, F. A. HART, G. P. MOSS, and M. L. STANIFORTH, *Chem. Comm.*, 1970, 749.
19. P. BÉLANGER, C. FREPPEL, D. TIZANÉ, and J. C. RICHER, *Chem. Comm.*, 1971, 266.
20. M. WITANOWSKI, L. STEFANIAK, H. JANUSZEWSKI, and Z. W. WOLKOWSKI, *Tetrahedron Letters*, 1971, 1653.
21. H. J. KELLER and K. E. SCHWARZHANS, *Angew. Chem., Int. Edn.*, 1970, **9**, 196.
22. D. R. EATON and W. D. PHILLIPS, NMR of paramagnetic molecules, in *Advances in Magnetic Resonance* (ed. J. S. Waugh), Academic Press, New York, 1965, **1**, 103; R. R. FRASER and Y. Y. WIGFIELD, *Chem. Comm.*, 1970, 1471; J. BRIGGS, F. A. HART, and G. P. MOSS, *Chem. Comm.*, 1970, 1506.
23. T. H. SIDDALL, *Chem. Comm.*, 1971, 452.
24. B. L. SHAPIRO, J. R. HLUBUCEK, G. R. SULLIVAN, and L. F. JOHNSON, *J. Amer. Chem. Soc.*, 1971, **93**, 3281.
25. A. W. OVERHAUSER, *Phys. Rev.*, 1953, **91**, 476.
26. F. A. L. ANET, A. J. R. BOURN, P. CARTER, and S. WINSTEIN, *J. Amer. Chem Soc.*, 1965, **87**, 5249; F. A. L. ANET and A. J. R. BOURN, *J. Amer. Chem. Soc.*, 1965, **87**, 5250.
27. J. A. POPLE, W. G. SCHNEIDER, and H. J. BERNSTEIN, *High-resolution Nuclear Magnetic Resonance*, McGraw-Hill, New York, 1959, pp. 211–212.
28. R. M. LYNDEN-BELL and R. K. HARRIS, *Nuclear Magnetic Resonance Spectroscopy*, Nelson, London, 1969, p. 143.
29. A. ABRAGAM, *The Principles of Nuclear Magnetism*, Oxford, 1961, p. 367.
30. M. C. WOODS, I. MINURA, Y. NAKADAIRA, A. TERAHARA, M. MARUYAMA, and K. NAKANISHI, *Tetrahedron Letters*, 1967, 321.
31. M. OHTSURU, M. TERAOKA, K. TORI, and KEN'ICHI TAKEDA, *J. Chem. Soc.* (B), 1967, 1033.
32. P. B. VENUTO and A. R. DAY, *J. Org. Chem.*, 1964, **29**, 2735; J. W. K. BURRELL, R. F. GARWOOD, L. M. JACKMAN, E. OSKAY, and B. C. L. WEEDON, *J. Chem. Soc.* (C), 1966, 2144.

33. J. W. EMSLEY, J. FEENEY, and L. H. SUTCLIFFE, *High Resolution Nuclear Magnetic Resonance*, Pergamon, Oxford, 1966, **2**, 988.
34. P. C. LAUTERBUR, *Ann. NY Acad. Sci.*, 1958, **70**, 841.
35. Ref. 33, p. 994.
36a. E. D. BECKER, J. A. FERRETTI, and T. C. FARRAR, *J. Amer. Chem. Soc.*, 1969, **91**, 7784.
36b. K. F. KUHLMANN, D. M. GRANT, and R. K. HARRIS, *J. Chem. Phys.*, 1970, **52**, 3439.
36c. R. FREEMAN and H. D. W. HILL, *J. Chem. Phys.*, 1970, **53**, 4103.
36d. J. R. LYERLA, D. M. GRANT, and R. K. HARRIS, *J. Phys. Chem.*, 1971, **75**, 585.
37. A. ALLERHAND, D. DODRELL, and R. KOMOROSKI, *J. Chem. Phys.*, 1971, **55**, 189.
38. E. G. PAUL and D. M. GRANT, *J. Amer. Chem. Soc.*, 1964, **86**, 2977, 2984.
39. J. FEENEY, High resolution nuclear magnetic resonance spectroscopy, in *Annual Reports on the Progress of Chemistry*, 1968, **65A**, 64–65.
40. P. S. PREGOSIN and E. W. RANDALL, *Chem. Comm.*, 1971, 399.
41. D. DODDRELL and A. ALLERHAND, *Chem. Comm.*, 1971, 728.
42. M. TANABE, T. HAMASAKI, H. SETO, and L. JOHNSON, *Chem. Comm.*, 1970, 1539.
43. R. FREEMAN, *J. Chem. Phys.*, 1964, **40**, 3571.
44. H. SPIESECKE and W. G. SCHNEIDER, *J. Chem. Phys.*, 1961, **35**, 722.
45. C. H. HOLM, *J. Chem. Phys.*, 1957, **26**, 707.
46. P. C. LAUTERBUR, *J. Amer. Chem. Soc.*, 1961, **83**, 1846.
47. G. E. MACIEL, *J. Phys. Chem.*, 1965, **69**, 1947.
48. P. C. LAUTERBUR, *J. Chem. Phys.*, 1957, **26**, 217.
49 P. C. LAUTERBUR, Chapter 7 in *Determination of Organic Structures by Physical Methods* (ed. F. C. Nachod and W. D. PHILLIPS), Academic Press, 1962, **2**, 465.
50. J. J. BURKE and P. C. LAUTERBUR, *J. Amer. Chem. Soc.*, 1964, **86**, 1870.
51. R. A. FRIEDEL and H. L. RETCOFSKY, *J. Amer. Chem. Soc.*, 1963, **85**, 1300.
52. P. C. LAUTERBUR, *J. Amer. Chem. Soc.*, 1961, **83**, 1838.
53. G. E. MACIEL and J. J. NATTERSTAD, *J. Chem. Phys.*, 1965, **42**, 2427.
54. K. S. DHAMI and J. B. STOTHERS, *Can. J. Chem.*, 1967, **45**, 233.

The Elucidation of Structural Formulae of Organic Compounds by Application of a Combination of Spectral Methods

F. SCHEINMANN

Department of Chemistry and Applied Chemistry, University of Salford

THE previous chapters have reviewed the power of the various individual spectroscopic methods as aids in elucidating structural formulae of organic compounds. Often one spectroscopic method by itself is inadequate for a complete structure analysis, and a combination of physical and chemical methods is used. All the structural information obtained from the various sources should fit together like a jig-saw puzzle to give in many cases an unambiguous picture of the structural formula.

The change in emphasis from chemical to physical methods has been quite dramatic, especially during the last decade. The writer recalls, as a first-year undergraduate, learning about the skill and patience of Körner[1] who, in 1874, orientated the three dibromobenzenes

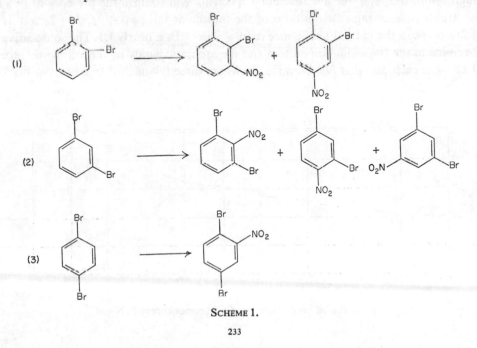

SCHEME 1.

(1), (2), and (3) by preparing all the possible mononitro derivatives of each isomer. Thus, as shown by Scheme 1, *ortho*-dibromobenzene (1) can give two nitro derivatives, *meta*-dibromobenzene (2), three nitro derivatives, and *para*-dibromobenzene (3) only one nitro derivative.

This commendable pioneering effort was not only time consuming but required large amounts of material because it must be assumed that at least one of the three mononitro derivatives of *meta*-dibromobenzene may only be formed in low yield, and that indirect methods of synthesis may be necessary. Today the same orientation problem could be solved within a couple of hours using physical methods and relatively small quantities of material, most of which could be recovered. Thus an examination of the infrared spectra (as potassium bromide discs or liquid films or in carbon disulphide) below 900 cm^{-1} may show out of plane C—H bending frequencies characteristics of either a 1,2-, 1,3-, or a 1,4-disubstituted benzene derivative (see Vol. 1, p. 174). These data must not be regarded as definitive since these frequencies cannot always be reliably assigned. The proton magnetic resonance spectra will, however, provide an independent check for tentative assignments. Thus the *para*-dibromobenzene will show only one aromatic resonance peak consistent with an A_4 system: the spectrum of *ortho*-dibromobenzene is classified as an AA′BB′ system, and therefore the complex aromatic absorptions will be symmetrical about the centre (see Vol. 1, p. 28). The spectrum of *meta*-dibromobenzene, which as an AB$_2$X system, can be expected to be more complex than the other two isomers, can be identified by comparison with a recorded spectrum or computed data for this spin system, or the AB$_2$X system in favourable cases can be simplified by double irradiation experiments to resemble an AB$_2$ system.

If the three isomers are presented as unknown compounds, the mass spectrum in each case will give the molecular formula by accurate mass measurement of the molecular ion at high resolution; even the low resolution spectrum will confirm the presence of two bromine atoms by showing that the ratio of the molecular ion peaks, M, $M+2$, and $M+4$, is 1:2:1 because the natural abundance ratio of ^{79}Br: ^{81}Br is nearly 1:1. The consecutive loss of bromine in the fragmentation pattern and the aromatic bands in the ultraviolet spectra will show in each case that two bromine atoms are directly attached to a benzene nucleus.

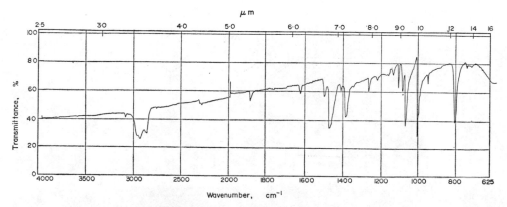

FIG. 1. The infrared spectrum of *p*-dibromobenzene in Nujol.

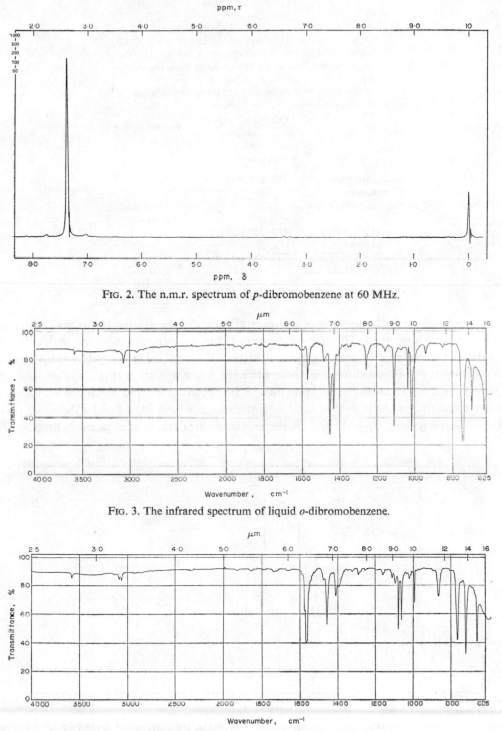

FIG. 2. The n.m.r. spectrum of *p*-dibromobenzene at 60 MHz.

FIG. 3. The infrared spectrum of liquid *o*-dibromobenzene.

FIG. 4. The infrared spectrum of liquid *m*-dibromobenzene.

TABLE 1
OUT OF PLANE BENDING FREQUENCIES FOR THE
Ortho- AND *Meta-*DISUBSTITUTED BENZENES QUOTED
FROM VOL. 1, P. 174, COMPARED WITH THE ABSORPTION
FREQUENCIES FOUND IN *Ortho-* AND *Para-*DIBROMO-
BENZENE
Relative band intensities are given in brackets

	Quoted absorption range (cm^{-1})	Found frequency (cm^{-1})
o-Disubstituted benzenes	770–735 (vs)	
o-Dibromobenzene		744 (vs) and 700 (m→s)
m-Disubstituted benzenes	810–750 (vs) and 725–680 (m→s)	
m-Dibromobenzene		770 (s) and 725 (s)

To check these predictions about the spectral nature of the dibromobenzenes commercial samples were examined. The infrared spectrum for *para*-dibromobenzene in Nujol (Fig. 1) shows the presence of a strong band at 810 cm^{-1} representing the out of plane bending frequency of a 1,4-disubstituted benzene while the n.m.r. spectrum (Fig. 2) shows only one peak (at τ 2·6). The infrared spectra (Figs. 3 and 4) of *ortho-* and *meta*-dibromobenzenes do not lead to unambiguous structural assignments since both Figs. 3 and 4 show two strong bands in the region 700–800 cm^{-1} characteristic of three and four adjacent hydrogens. The

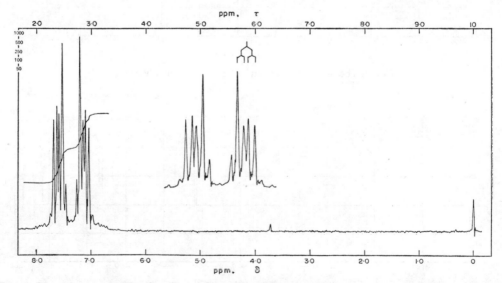

FIG. 5. The n.m.r. spectrum of *o*-dibromobenzene at 60 MHz. Inset shows the signals at 250 Hz sweep width.

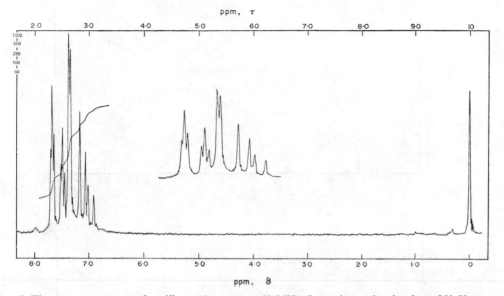

FIG. 6. The n.m.r. spectrum of *m*-dibromobenzene at 60 MHz. Inset shows the signals at 250 Hz sweep width.

difficulty in interpreting the infrared data is illustrated in Table 1. However, the n.m.r. spectrum of *ortho*-dibromobenzene (Fig. 5) is symmetrical about its centre point, while the spectrum of *meta*-dibromobenzene (Fig. 6) is more complex but is similar to a measured and computed spectrum of *meta*-dinitrobenzene reported by Abraham *et al.*[2] In examination of the commercial samples it was found that *ortho*-dibromobenzene had been mistakenly labelled *meta*-dibromobenzene and that *meta*-dibromobenzene had the label for *ortho*-dibromobenzene. This error was immediately apparent on inspection of the proton magnetic resonance spectra but, in addition, the infrared spectra of the samples were compared with the spectra recorded for *ortho*- and *meta*-dibromobenzenes contained in the *Documentation of Molecular Spectroscopy* (catalogue numbers 3048 and 3331 respec-

TABLE 2
ULTRAVIOLET SPECTRAL DATA FOR THE DIBROMO-
BENZENES IN HEXANE

Compound	λ_{max} (nm)	log ε
o-Dibromo-	271[a]	2·4
benzene	226	3·7
m-Dibromo-	271[a]	2·5
benzene	235	3·6
p-Dibromo-	273[a]	2·5
benzene	228	4·2

[a] Fine structure also present.

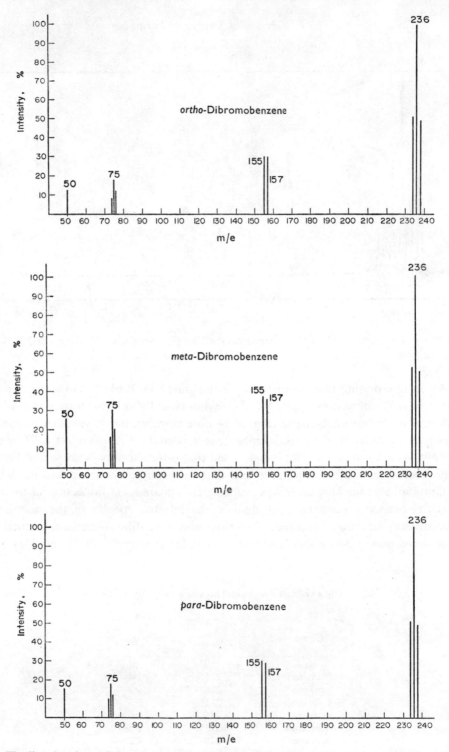

FIG. 7. The line drawings of the mass spectra of *o*-, *m*-, and *p*-dibromobenzenes measured on the AEI MS12 spectrometer.

tively).[3] The literature data readily confirmed that the samples of *ortho-* and *meta*-dibromobenzene had been incorrectly labelled, and later an apology was received from the suppliers. The ultraviolet data (Table 2) and mass spectra (Fig. 7) show the expected aromatic feature for the isomeric dibromobenzenes but these methods alone do not show the orientation of the bromine atoms. The mass spectral fragmentation pattern for *ortho-*, *meta-*, and *para*-dibromobenzene (Fig. 7) shows that in each the molecular ion prefers to fragment first by loss of a bromine atom to give peaks in a 1:1 ratio at m/e 155 and 157 followed by loss of HBr to give a peak at m/e 75. However, from their mass spectra it is not possible to orientate the three dibromobenzenes.

The elucidation of the structures of the dibromobenzenes by physical methods illustrates that often two physical methods, usually infrared and proton magnetic resonance spectroscopy, will suffice to solve the structure of an organic compound. This is especially true in synthetic work when it is necessary to confirm the structure of an expected reaction product or to distinguish between the two or three likely products. However, it is with unknown structures such as natural products, or photochemical reaction products, or valence tautomers, or when an unpredictable reaction occurs, that a combination of all spectroscopic methods is most valuable; often because only milligram quantities of material may be available and because analysis by chemical methods on unstable structures may be too vigorous to give meaningful information. The remainder of this chapter will consider a general approach to the elucidation of structural formulae with a combination of spectral methods and then deals with the case history of some representative examples.

A General Approach

Our experience with natural products suggests the following sequence of spectral measurements for elucidating the structures of unknown compounds.

MASS SPECTRUM FOR DETERMINATION OF THE MOLECULAR FORMULA

A mass measurement of the molecular ion can immediately check the molecular formula established by more classical methods. This is a worthwhile exercise because the data obtained from a combustion analysis and a Rast molecular weight determination often cannot distinguish between two possible homologues nor can the number of hydrogen atoms be accurately established. This can be illustrated by the percentages of carbon and hydrogen in $C_{19}H_{18}O_6$ (scriblitifolic acid, see p. 241) and its higher homologue $C_{20}H_{20}O_6$:

$$C_{19}H_{18}O_6 \text{ requires } C, 66\cdot8 \text{ and } H, 5\cdot7\%$$
$$C_{20}H_{20}O_6 \text{ requires } C, 67\cdot4 \text{ and } H, 5\cdot7\%$$

Thus a combustion analysis which gives C, $67\cdot1$ and H, $5\cdot5\%$, is in good agreement for both homologues. Ambiguities of this kind have led to errors, but care must also be taken with the mass spectral method because in some cases the ion at highest mass may not represent the molecular ion (see p. 14). Thus a tertiary alcohol may thermally lose water in the probe, and the derived olefin after loss of an electron will be seen as the ion of highest mass. Fortunately, in many cases the ion of highest mass does correspond to the molecular ion. It is often possible, however, to check the molecular formula by counting the relative number of protons in the proton magnetic resonance spectrum from integration of the peak areas, and, of course, the results of a combustion analysis must also support the conclusions from the spectral methods.

From the molecular formula the number of rings and double bonds present can be calculated. Thus for a saturated acyclic hydrocarbon with a carbon atoms there are $(2a+2)$ hydrogen atoms. For each double bond introduced or for each ring formed from the same carbon skeleton, two hydrogen atoms are deducted. The Equation (1) gives the number of double bonds and rings present, i.e. double-bond equivalents (DBE). When oxygen or other divalent elements are present in a molecule, eqn. (1) is still valid.

$$\text{For} \qquad C_aH_b \quad \text{or} \quad C_aH_bO, \quad DBE = \frac{2a+2-b}{2}. \tag{1}$$

To apply eqn. (1) when halogens are present, each halogen must be replaced by an atom of hydrogen. For nitrogen and trivalent phosphorus a hydrogen must be subtracted from b for each trivalent atom present.

$$\text{For} \qquad C_aH_bO_cN_d, \quad DBE = \frac{2a+2-(b-d)}{2}. \tag{2}$$

For each pentavalent phosphorus atom three hydrogen atoms must be subtracted from the total number of hydrogen atoms present. It should be remembered that from eqns. (1) and (2) a triple bond will be represented as two double-bond equivalents.

A calculation of the number of double-bond equivalents may suggest the type of information that can be obtained from an ultraviolet spectrum. For complicated molecules it is often more prudent to interpret other regions of the mass spectrum after more structural information has accumulated from other methods.

THE ULTRAVIOLET SPECTRUM

The chief value of ultraviolet spectrum is to provide information about the nature of a conjugated system. If the molecular formula shows some double-bond equivalents, comparison of the ultraviolet spectrum with reference data may identify the chromophore. Sometimes the infrared and the proton magnetic resonance spectrum will assist in the search for the chromophore by showing, for example, aromatic absorptions and the number and environment of the various aromatic protons. Precise structural information can also be obtained from ultraviolet spectra by application of the Scott rules (see p. 110) for simple aromatic systems and the Woodward–Fieser rules (see p. 110) for polyenes and enones. The ultraviolet spectrum is often useful in differentiating between isomeric structures, e.g. between conjugated and non-conjugated systems, or between *cis*- and *trans*-stilbenes, or between anthracene or phenanthrene structures. Heterocyclic systems can often be detected by their characteristic chromophores or by comparison with the carbocyclic analogue. Thus, as expected, the ultraviolet spectrum of pyridine resembles that of benzene, and the spectra of quinoline and isoquinoline are similar to the spectrum of naphthalene.

THE INFRARED SPECTRUM

The infrared spectrum will reveal the presence of functional groups not detected by the ultraviolet spectrum, and is particularly valuable in the carbonyl region (1600–1800 cm^{-1}) and for X—H stretching frequencies where X is oxygen, nitrogen, sulphur, phosphorus, and saturated and unsaturated carbon.

Triple bonds as found in terminal acetylenes, nitriles, azides and diazonium salts have strong absorptions in the region of 2100–2300 cm^{-1}. However, for symmetrical disubstituted acetylenes, where there is no net change in dipole moment during the stretching vibration, absorption in this region is not observed (see Vol. 1, p. 133).

In addition, infrared spectrum should be able to confirm the data from the ultraviolet spectrum about the unsaturation. Thus aromatic compounds will have a C—H stretching frequency just above 3000 cm^{-1} and C=C stretching frequencies at about 1600 cm^{-1} and 1500 cm^{-1}.

THE PROTON MAGNETIC RESONANCE SPECTRUM

This spectrum will indicate the environment of each proton present in the molecule. In addition, the integration of the spectrum will give the relative number of protons present in each environment, and therefore these data can be used to check the empirical formula. The information from the spectrum must be consistent with the nature of the ultraviolet and infrared spectra. Thus an aromatic system will have its aromatic protons in the region τ 1·5–4, and the presence of hydroxyl and amine functions can be readily detected by immediate exchange on addition of deuterium oxide.

ANALYSIS OF THE FRAGMENT IONS IN THE MASS SPECTRUM

The structural information obtained already can be verified by consideration of the fragment ions found in the mass spectrum, and if ambiguities still exist, unique fragmentations or rearrangements may only be relevant to one structure. The advantage of considering the mass spectrum at this stage is that the functional groups, alkyl, aryl, and heterocyclic moieties will have characteristic changes and rearrangements associated with them. Thus it may be possible to map out the chief fragmentation patterns for a given structure, and the mass spectrum will thus provide an independent check for the previous deductions and give additional structural information.

Scriblitifolic Acid

Some time ago a new project was initiated at Salford to examine the metabolites from tropical timbers of the Guttiferae family. The little previous work that had been done on this large plant family suggested that some of the metabolites may be of interest for their insecticidal and antibacterial properties. The availability of chromatographic methods for the isolation of metabolites and the spectral methods for identification enabled rapid progress to be made with the structure elucidation of phenolic pigments from *Calophyllum*, *Symphonia*, and *Garcinia* species.

Scriblitifolic acid was isolated from the heartwood of *Calophyllum scriblitifolium*, Henderson and Wyatt-Smith, which came from the Setapok forest in Sarawak.[4] The case history of the studies on the structure of scriblitifolic acid will be given in detail to illustrate the value of using together data from the various spectral methods.

The molecular formula of scriblitifolic acid from elemental and mass spectrometric analyses is $C_{19}H_{18}O_6$. Calculation of the double-bond equivalent from eqn. (1) gives a value of 11, suggesting that the molecule is highly unsaturated and that rings may be present. Previous phenolic extractives from Guttiferae have included derivatives of xanthone (4) coumarin (5) and anthracene (6) and the ultraviolet spectrum of scriblitifolic acid was examined specifically with these systems in mind.

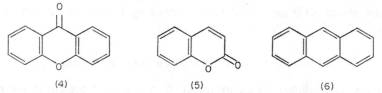

(4)   (5)   (6)

The ultraviolet spectrum (Fig. 8) immediately favoured the presence of a xanthone nucleus, and comparison with the recorded spectra of 1-hydroxyxanthone and 1-methoxyxanthone (Fig. 9) suggested that the presence of two peaks in the region of 230–260 nm may be indicative of a 1-hydroxy-group. A closer comparison of the absorption maxima for scriblitifolic acid and 1-hydroxyxanthone shows that they are not identical, and therefore scriblitifolic acid may have other auxochromic groups.

The infrared spectrum (Fig. 10) shows two carbonyl peaks; one due to the xanthone carbonyl at 1650 cm⁻¹ and the other due to the presence of an aliphatic carboxylic acid at 1710 cm⁻¹. The broad hydroxyl band between 2500–3000 cm⁻¹ confirmed the presence of the carboxylic acid (as a dimer). The eleven double-bond equivalents are now all accounted for with ten for the xanthone nucleus and one for the carboxylic acid group.

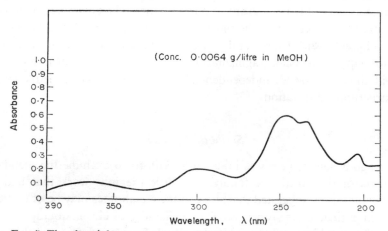

FIG. 8. The ultraviolet spectrum of scriblitifolic acid measured in methanol.

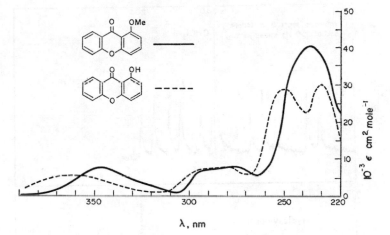

FIG. 9. The ultraviolet spectra of 1-hydroxyxanthone and 1-methoxyxanthone measured in methanol.

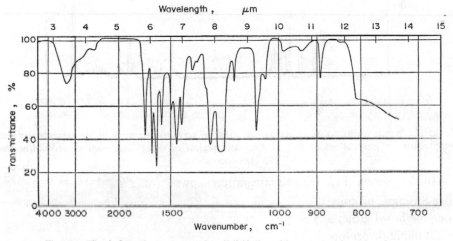

FIG. 10. The infrared spectrum of scriblitifolic acid measured in chloroform.

Methylation experiments confirmed the presence of one phenolic and one carboxylic acid group consistent with partial formula (7).

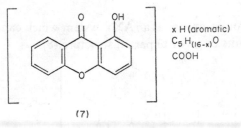

(7)

Thus at this stage it remains to account for one oxygen atom, the substitution pattern of the xanthone, and the nature of the side chain.

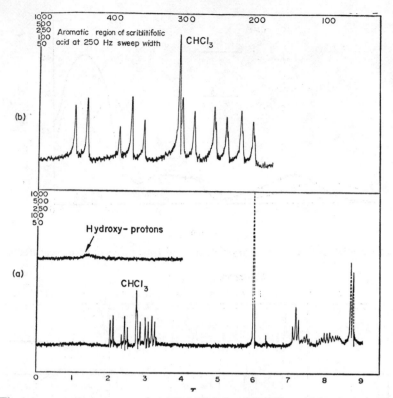

FIG. 11. (a) The n.m.r. spectrum measured at 100 MHz of scriblitifolic acid in deuteriochloroform with tetramethylsilane as internal reference. (b) Aromatic region of n.m.r. spectrum of scriblitifolic acid at 250 Hz sweep width.

The n.m.r. spectrum (Fig. 11) and integration shows:

(a) five aromatic protons;

(b) one methoxyl group at τ 6·0;

(c) eight aliphatic protons;

(d) two hydroxylic protons undergoing exchange at τ −2·6.

The aromatic protons (Fig. 11b) appear as two groups:

(a) Two *ortho* split doublets (J 8 Hz) at τ 2·8 and 2·0 which must be due to protons at C-7 and C-8 since only H-8 can appear in the region of τ 2 from de-shielding by the adjacent ring carbonyl group.

(b) Three adjacent protons appear as an AXY system which can be assigned to C-2, C-3, and C-4. This now permits us to write partial structure (8)

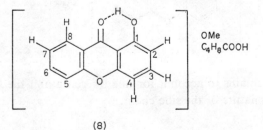

(8)

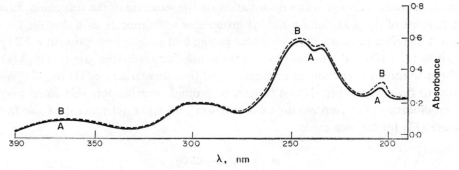

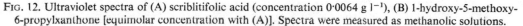

FIG. 12. Ultraviolet spectra of (A) scriblitifolic acid (concentration 0·0064 g l⁻¹), (B) 1-hydroxy-5-methoxy-6-propylxanthone [equimolar concentration with (A)]. Spectra were measured as methanolic solutions.

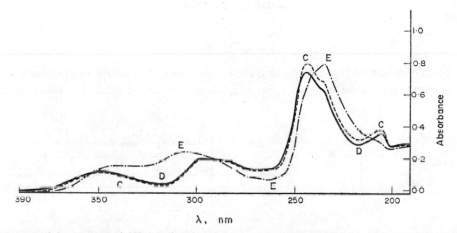

FIG. 13. Ultraviolet spectra of (C) 1,5-dimethoxy-6-propylxanthone [equimolar concentration with (D)], (D) O-methylscriblitifolic acid (concentration 0·004 g l⁻¹), (E) 1,6-dimethoxy-5-propylxanthone [equimolar concentration with (D)]. All spectra determined as methanolic solutions.

At this stage the oxygenation pattern is still uncertain, and model compounds were synthesized for comparison between 1,5- and 1,6-oxygenation patterns (9) and (10). The ultraviolet spectra (Figs. 12 and 13) clearly indicate that scriblitifolic acid is a 1-hydroxy-5-methoxyxanthone derivative (Fig. 13).

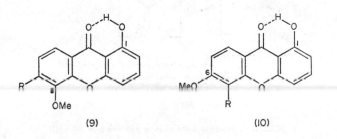

(9)   (10)

The n.m.r. spectrum also provides information on the structure of the side chain. From the eight protons of the side chain a methyl group at τ 8·75 appears as a doublet (J, 7 Hz) and must therefore be adjacent to a methine group, and a methylene group at τ 7·2 appears as a triplet (J, 7 Hz) and must be placed next to another methylene group (Fig. 11a). This spin–spin splitting pattern can be accommodated by either structure (11) or (12). We were inclined to favour structure (12) on biogenetic grounds because this side chain may arise from modification of an isoprene unit. Comparison with n.m.r. reference data also favoured structure (12) for the side chain.

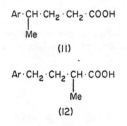

To confirm structure (12) for the side chain, the mass spectrum was examined. All the expected peaks due to the fragmentation of a 2-methylbutanoic acid side chain were ob-

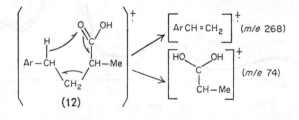

SCHEME 2. Fragmentation of side chain (12)

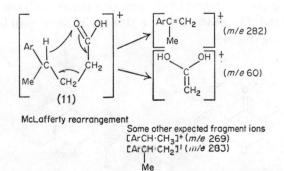

SCHEME 3. Fragmentation of other possible side chain (11)

served (Fig. 14 and Scheme 2). Quite clearly most of the peaks expected from the alternative side chain were absent (Scheme 3).

Finally, as an independent proof deuterium exchange was studied, by n.m.r. spectroscopy, at the site adjacent to the carboxyl group. Scheme 4 (A and B) illustrates exchange by

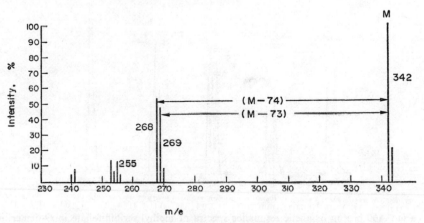

FIG. 14. Part of the line drawing of the mass spectrum of scriblitifolic acid.

enolization for side chains 11 and 12. No exchange occurred with the parent acid even under forcing conditions but with the methyl ester (12a) exchange occurred catalysed by

(A) $Ar \cdot CH_2 \cdot CH_2 \cdot \overset{Me}{\underset{|}{CH}} \cdot COOR$ $\underset{\rightleftharpoons}{\overset{D_2O}{}}$ $Ar \cdot CH_2 \cdot CH_2 \cdot \overset{Me}{C} = \overset{OD}{C} \cdot \overset{}{OR}$

       12 ;   R = H

      12a ; R = Me

                                                $\updownarrow$ Me

                                    $Ar \cdot CH_2 \cdot CH_2 \cdot \overset{|}{CD} \cdot COOR$

(B) $Ar \cdot \overset{Me}{\underset{|}{CH}} \cdot CH_2 \cdot CH_2 \cdot COOR$ $\underset{\rightleftharpoons}{\overset{D_2O}{}}$ $Ar \cdot \overset{Me}{\underset{|}{CH}} \cdot CH_2 \cdot CD_2 \cdot COOR$

      11 ;   R = H

      11a ; R = Me

SCHEME 4.

sodium methoxide in deuteriomethanol at room temperature by the mechanism shown in Scheme 4(A) leading to collapse of the methyl doublet and thus excluding structure (11) for the side chain. Collapse of the methyl doublet was followed for 9 days (Fig. 15). It follows that scriblitifolic acid is a 1-hydroxy-5-methoxyxanthone with a 2-methylbutanoic acid located at C-6 in accordance with structure (13). That scriblitifolic acid has structure

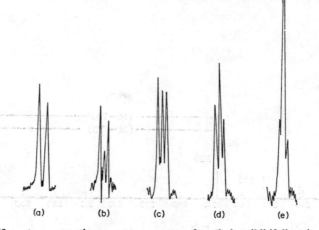

FIG. 15. Varian A60 proton magnetic resonance spectra of methyl scriblitifoliate in *O*-deuteriomethanol containing sodium methoxide showing the effect of deuteration of the methine proton–$CH(CH_3)CO_2$ Me upon the methyl doublet at τ 8·75 after (a) zero time, (b) 1 day, (c) 2 days, (d) 3 days, and (e) 9 days. The spectrum amplitude was not held constant for the consecutive determinations.

(13) has been confirmed by an unambiguous synthesis of its methyl ether (14).

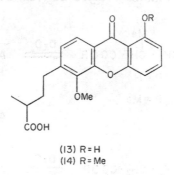

(13) R = H
(14) R = Me

Alpinumisoflavone

The elucidation of the structure of alpinumisoflavone illustrates the use of data from mass, infrared, ultraviolet, and nuclear magnetic resonance spectra to derive two possible isomeric structures (in this case (15a) and (16a)) for the metabolite.[5] The availability of both linear and angular pyranoisoflavones then allows unambiguous structure assignment for both isomers from proton magnetic resonance spectral data involving the nuclear Overhauser effect (NOE) and the effect of acetylation on chemical shift.[5]

The laburnum (family Leguminosae, subfamily Papilionaceae, the pea-flower family) is a well-known ornamental tree, but after the yew tree it is the most poisonous tree grown in Britain.[6] All parts of it are toxic, and numerous cases are recorded of Laburnum poisoning of human beings who have eaten the flowers or seeds or who have carried twigs or bunches of flowers in their mouths. A poisonous principle in *L. alpinum* J. Presl., the alkaloid cytisine (17) was reported in 1943, but it is expected that further studies using modern methods may reveal other metabolites.

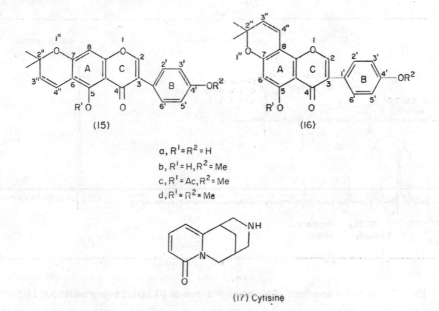

(15)   (16)

a, $R^1 = R^2 = H$
b, $R^1 = H, R^2 = Me$
c, $R^1 = Ac, R^2 = Me$
d, $R^1 = R^2 = Me$

(17) Cytisine

The twigs of *L. alpinum* J. Presl. were extracted with hot chloroform. Column and thin-layer chromatography yielded a new pyranoisoflavone which was named alpinumisoflavone. The molecular formula $C_{20}H_{16}O_5$ follows from the mass spectrum and accurate measurement of the molecular ion. The infrared spectrum (Fig. 16) shows the presence of hydroxyl groups with bands at 3410 cm^{-1} and an intramolecular hydrogen-bonded carbonyl group at

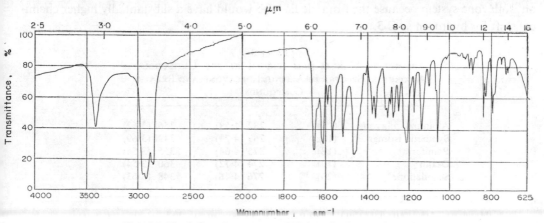

FIG. 16. The infrared spectrum of alpinumisoflavone in Nujol.

1655 cm^{-1}. The n.m.r. spectrum (Fig. 17) is in good agreement with the pyranoisoflavone structures (15a or 16a). Thus the hydrogen-bonded hydroxy-group at C-5 resonates at $\tau - 3\cdot14$ and the other hydroxyl at C-4' is a broad band centred at τ 4·68; both signals are removed on addition of deuterium oxide. The protons of ring B give rise to an AA'BB' spin

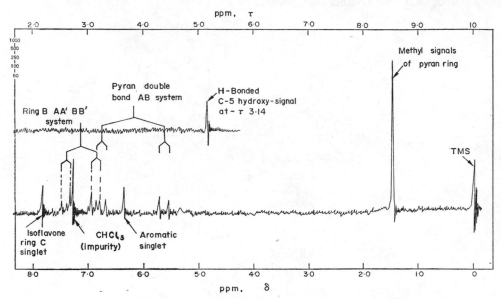

FIG. 17. The n.m.r. spectrum of alpinumisoflavone at 60 MHz: sweep offset 500 Hz.

system approximating to two doublets (J, 9 Hz) at τ 2·61 (2H) and 3·15 (2H). The single aromatic proton of ring A gives rise to a signal at τ 3·64 consistent with a phloroglucinol ring proton of a flavonoid system. The dimethylpyran ring gives a methyl signal at τ 8·53 (6H) and the olefinic proton signals as two doublets (J, 10 Hz) centred at τ 3·23 (1H) and 4·37 (1H).[5] The signal at τ 2·18 is due to the hydrogen at C-2 and is of diagnostic value for an isoflavone system because the isomeric flavone would have a substantially higher chemical shift for hydrogen at C-3.

TABLE 3
ABSORPTION MAXIMA [N.M. (log ε)] IN THE ULTRAVIOLET
ABSORPTION SPECTRA OF ALPINUMISOFLAVONE AND RELATED
COMPOUNDS

Alpinumisoflavone	(15a)	282 (4·58)	356i (3·62)
Synthetic isomer	(16a)	269 (4·73)	349i (3·69)
Pomiferin[a]	(18)	275 (4·64)	355 (3·54)
Osajin[a]	(19)	274 (4·72)	360 (3·54)
Scandenone[b]	(20)	276 (4·46)	348 (3·63)

[a] *The Chemistry of the Flavanoid Compounds* (ed. T. A. Geissman), p. 367.
[b] A. PELTER and P. STAINTON, *J. Chem. Soc.* (C), 1966, 701.
i denotes inflection.

The ultraviolet spectrum supports a 4',7-trioxygenated pyranoisoflavone structure with absorption maxima at 282 nm (log ε 4·58) and **an** inflection at 356 nm (log ε 3·62) in good agreement with maxima for pomiferin (18), osajin (19), and scandenone (20) (Table 3).

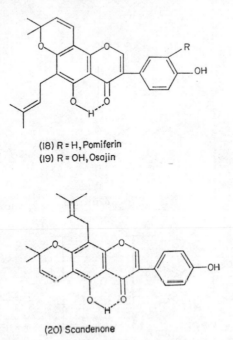

(18) R = H, Pomiferin
(19) R = OH, Osajin

(20) Scandenone

Mass spectral fragmentation (Fig. 18) is consistent with a dimethylpyranoflavonoid system since the molecular ion readily loses 15 mass units to give the base peak at *m/e*

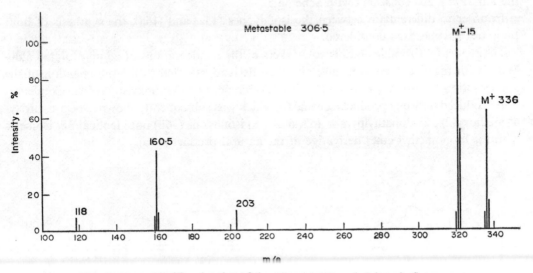

FIG. 18. Part of the line drawing of the mass spectrum of alpinumisoflavone.

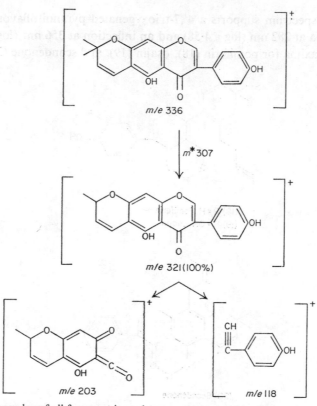

m/e 336

m*307

m/e 321(100%)

m/e 203   m/e 118

SCHEME 5. The formulae of all fragment ions shown are supported by accurate mass measurements

321 and the doubly charged ion at m/e160·5. A retro-Diels–Alder process follows to give the ion at m/e 203 consistent with Scheme 5.

In order to differentiate between the isoflavones (15a) and (16a), the synthesis of both pyranoisoflavones was considered.

The report by Crombie and his co-workers of the condensations of $\alpha\beta$-unsaturated aldehydes with reactive phenols to give chromen derivatives[7] led to the investigation of the condensation of 3-methylbut-2-enal with 5,7-dihydroxy-4′-methoxyisoflavone. Condensation produced only one product (scheme 6), which was subsequently shown to be 5-hydroxy-4′-methoxy-2″, 2″-dimethylpyrano (5″, 6″, 7, 8) isoflavone (16b) not identical but isomeric with the monomethyl ether derivative of the natural product (15b).

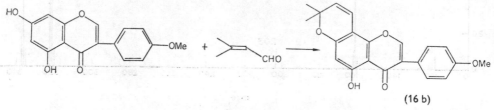

(16 b)

SCHEME 6.

The availability of both isomeric pyranoisoflavones (15b) and (16b) enabled us to examine further aspects of proton n.m.r. in order to distinguish between the linear and angular isomers. Studies of the effect of acetylation[8] and of the NOE[9] were both successful in solving the problem.

The acetylation of the 5-hydroxy-group of a chromen has been shown to cause a marked diamagnetic shift ($\Delta_\tau = +0.3$ to $+0.4$) of the peri-proton [4″—H in (15a)] and a small paramagnetic shift ($\Delta_\tau = -0.1$ of the second olefinic proton [3″—H in (15a)]. One would thus predict that, for compound (15b) acetylation of the 5—OH would cause a shift to higher field of the 4″—H signal and a slight downfield shift of the 3″—H signal. For compound (16b) there should be no such movement of the signals. Table 4 shows that acetyl-

TABLE 4

COMPARISON OF CHEMICAL SHIFTS (τ) OF THE OLEFINIC PRO-
TONS IN THE 2,2-DIMETHYLCHROMEN RING OF ALPINUMISO-
FLAVONE, ITS ANGULAR SYNTHETIC ISOMERS, AND THEIR 5-
ACETOXY DERIVATIVES

Natural compound	Olefinic proton resonances		Synthetic compound	Olefinic proton resonances	
	4″-H	3″-H		4″-H	3″-H
15b	3·47	4·58	16b	3·46	4·55
15c	3·72	4·45	16c	3·40	4·46
Δ_τ acetyl	+0·25	−0·13	Δ_τ acetyl	−0·06	−0·09

ation of the natural product derivative shows the shifts expected for a linear structure. Thus the natural product has structure (15a). No such diamagnetic shift is observed for the acetylated synthetic product which is assigned structure (16c).

NUCLEAR OVERHAUSER EFFECT

Irradiation with a moderately strong audiofrequency at the frequency of the C–5 methoxy-group in the synthetic compound (16d) resulted in a marked increase (33%) in the intensity of the signal of the phloroglucinol proton but no appreciable increase for the signal due to the olefinic (4″—H) proton (Table 5). Such an internal NOE is observable only when two or more nucleii are in close proximity, so that their spin-lattice relaxation times are strongly influenced by internuclear dipole–dipole interactions (see p. 219).

Irradiation at the frequency of the C-5 methoxy-group in the derivative (15d) of the natural compound did not cause any appreciable enhancement of the phloroglucinol proton signal; however, an increase of 10% in the intensity of the 4″-H signal was observed. Thus compound (16d) must have its C-5 methoxy-group adjacent to an aromatic proton, i.e. structure (16d) must be correct for the angular isomer and structure (15d) for the linear isomer.

These results prove that alpinumisoflavone has the linear pyranoisoflavone structure (15a).

TABLE 5

NUCLEAR OVERHAUSER EFFECT MEASUREMENTS; ENHANCE-
MENT OF INTENSITIES AFTER IRRADIATION AT THE 5-*O*-Me
FREQUENCY

	Relative intensities[a]			
	Natural isomer (15d)		Synthetic isomer (16d)	
	4″-H	8-H	4″-H	6-H
No irradiation	780[b]	74[c]	49[c]	147[d]
Irradiation	858[b]	74[c]	50[c]	195[d]
Enhancement on irradiation (%)	10			33

[a] The relative intensities are average values. The deviations in these values cause an uncertainty of about 3% in the enhancement factor.
[b] Measurement by weighing of areas.
[c] Measurement by integration of areas.
[d] Measurement of areas by planimeter.

The Structure of Triazolinonorbornanes

As part of a synthetic programme to prepare norbornanes with fused heterocyclic rings, phenyl azide was reacted with 5-aminomethylnorborn-2-ene (21).

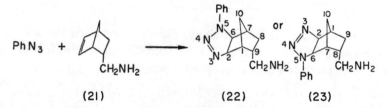

(21)   (22)   (23)

Assuming *exo*-attack two possible adducts (22 and 23) were anticipated. However, only one product (22) was isolated and it was necessary to establish unambiguously its structure.[10]

The mass spectrum under the usual operating conditions failed to give a molecular ion for a phenyl azide-aminomethylnorbornene adduct; instead, the ion of highest mass appears at $[M-28]^+$. The combustion analysis, however, supports the molecular formula $C_{14}H_{18}N_4$ as shown by the data: Found C, 69·0; H, 7·5; N, 22·7. $C_{14}H_{18}N_4$ requires C, 69·4; H, 7·4; N, 23·1%. Therefore the $[M-28]^+$ peak in the mass spectrum is due to loss of a nitrogen molecule from the thermally labile triazoline ring. In an attempt to obtain the molecular ion M^+ by lowering the temperature of the injection block to 150° and also by increasing the sample pressure, an $[M+1]^+$ peak was obtained. The intensity ratio $[M+1]^+/[M-28]^+$ was found to vary with sample pressure (Fig. 19) indicating a molecule —molecular ion interaction of the type:

$$ABCD^{+\cdot} + ABCD \longrightarrow [ABCDABCD]^{+\cdot}$$
$$\downarrow$$
$$ABCDA^+ + BCD^{\cdot}$$

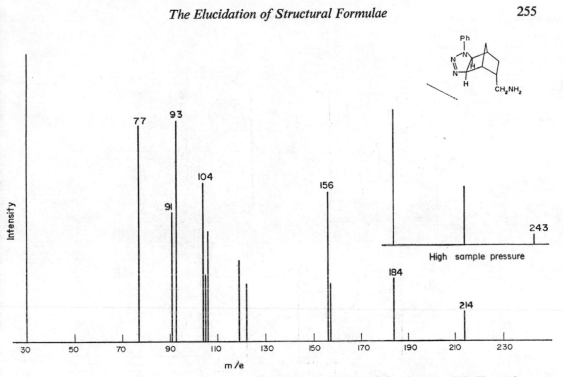

Fɪɢ. 19. The line drawing for the mass spectrum of the aminomethyltriazolinonorbornane (22). The molecular ion is not seen under normal operating conditions but at high sample and a probe temperature of 150° an $M+1^+$ peak is seen at m/e 243 shown as inset.

Replacement of the aminomethyl by cyanomethyl or by hydroxymethyl leads to similar mass spectral features. In fact the first fragment lost from the $[M-28]^+$ ion is CH_2R in all cases ($R = NH_2$, CN, or OH). Thus the mass spectra of all three triazolines are identical below m/e 184. This ion ($C_{13}H_{14}N$) appears to be directly responsible for the ions at m/e 91 (C_7H_7), 104 (C_7H_6N), and 93 (C_6H_7N), and the expected retro-Diels–Alder fragment at m/e 156 ($C_{11}H_{10}N$). Although no obvious stereochemical information is provided by the mass spectral study, the rationalization of the data as shown in Scheme 7 supports the ring structure of the triazolinonorbornanes.

The infrared spectrum (Fig. 20) shows the symmetrical and antisymmetrical N—H stretching frequencies at 3300 and 3380 cm^{-1} to confirm the presence of a primary aliphatic amine group. A monosubstituted phenyl group is indicated by the stretching frequencies at 3030 ν (C—H), 1606 and 1500 cm^{-1}, ν (C=C), and out-of-plane bending frequencies at 695 and 753^{-1}. The ultraviolet spectrum (Fig. 21), with maxima at 227, 290, and 309 nm, as expected, shows some chromophoric features also seen in aniline and its derivatives. Thus aniline has absorption maxima at 230 and 280 nm and N-methylaniline at 245 and 295 nm.

Studies in n.m.r. spectroscopy provide stereochemical information and prove that the triazoline ring is in the exo-position, and by use of the NOE and of the shift reagent tris-(dipivaloylmethanato) europium(III) Eu(DPM)$_3$ (see p. 212), it has been possible to locate the phenyl and aminomethyl substituents as shown in structure (22).

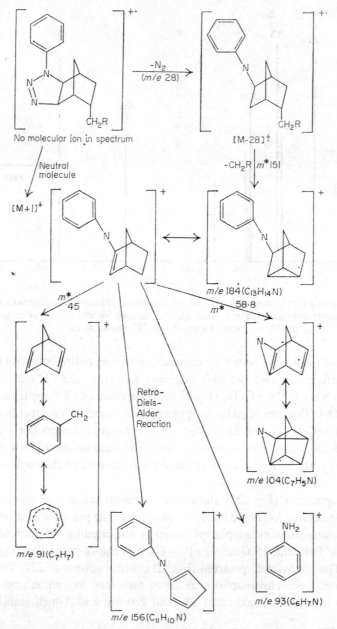

SCHEME 7. Accurate mass measurements were obtained for all fragments R = CN, NH$_2$, or OH

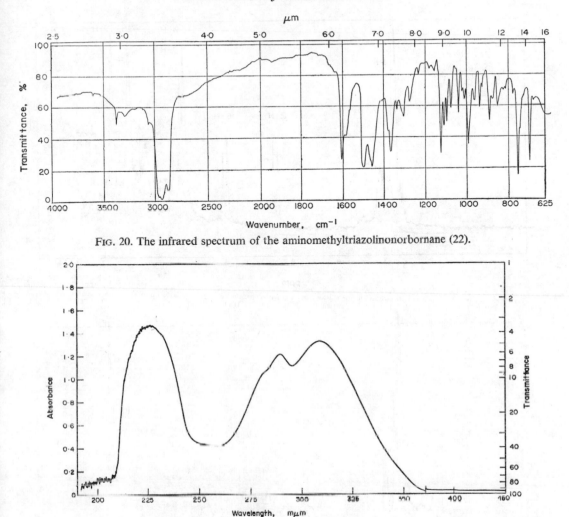

FIG. 20. The infrared spectrum of the aminomethyltriazolinonorbornane (22).

FIG. 21. The ultraviolet spectrum of the aminomethyltriazolinonorbornane (22) measured at a concentration of 2 mg in 100 ml ethanol.

Examination of the n.m.r. spectrum of the triazolinonorbornane (Fig. 22a) from low- to high-field shows the aromatic protons in the region of τ 2·8. The protons at C-2 and C-6 form an AB system with doublets centred at 5·35 and τ 6·40 (*J*, 9·0 Hz). This necessitates these protons assuming an *endo*-position, where there is virtually no coupling to the adjacent protons at C-1 and C-7 because the dihedral angle approximates to 90°. Since the protons at C-2 and C-6 are in *endo*-positions, the heterocyclic ring must assume an *exo*-location. The *endo*-methylene protons appear with the signals at τ 7·4, and the other signal of diagnostic value is due to the C-8 *endo*-proton which is at high-field and appears as a multiplet centred at τ 9·2. Previous work[11] suggests that this abnormal shielding is caused by the anisotropy of the bond from C-9 to the *endo*-methylene group. It now remains to diffe-

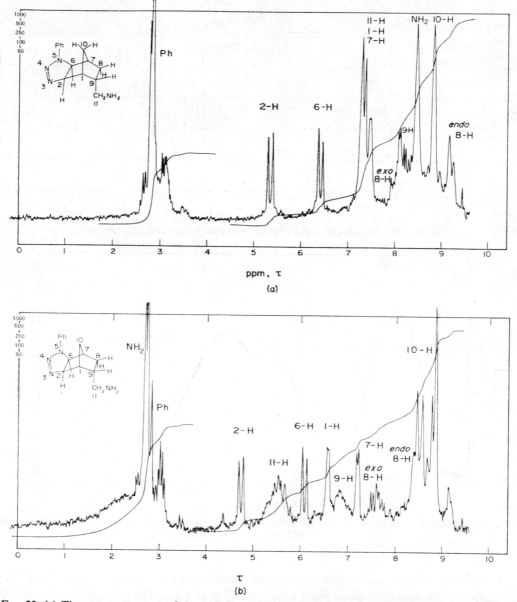

FIG. 22. (a) The n.m.r. spectrum of the aminomethyltriazolinonorbornane (22) measured at 100 MHz in deuteriochloroform prior to addition of Eu(DPM)$_3$. (b) The n.m.r. spectrum of (22) after addition of Eu(DPM)$_3$.

rentiate between the structures (22) and (23). The rigid nature of the tricyclic ring system enables the NOE experiments to locate the *endo*-substituent, the *endo*-hydrogen atoms, and the phenyl group relative to one another. Empirical data suggest that the high-field doublet (τ 6·40) is due to the *endo*-proton at C-6, which is adjacent to the phenyl substituent, and NOE experiments support this assignment. Thus irradiation at the frequency

TABLE 6

NUCLEAR OVERHAUSER EFFECT EXPERIMENTS
ON THE TRIAZOLINONORBORNANE (22)

Signal(s) irradiated	Proton affected	% Enhancement
Phenyl	6 – H	8
endo-9-CH₂	2 – H	16
endo-8-H	6 – H	9–10

Peak enhancement was deduced using electronic integration by comparing integrals of the C–2 and C–6 *endo*-proton signals before and after irradiation. Area measurements of the peaks with a planimeter gave slightly higher values.

of the most intense peak in the phenyl region (τ 2·8) causes an increase of 8% in the integral of the high-field doublet (Table 6). That the phenyl group is on the opposite side of the molecule to the aminomethyl, as in structure (22), was confirmed by double irradiation at the signal of the *endo*-methylene protons. A positive NOE of 16% was observed in the integral of the low-field doublet due to the *endo*-proton at C-2. Final proof in support of structure (22) was obtained by observing the NOE caused by triple irradiation of the two major peaks due to the C-8 *endo*-proton. This gave rise to approximately 10% enhancement of the integral of the high-field doublet corresponding to the C-6 *endo*-proton (Fig. 23). The data are summarized in Table 6.

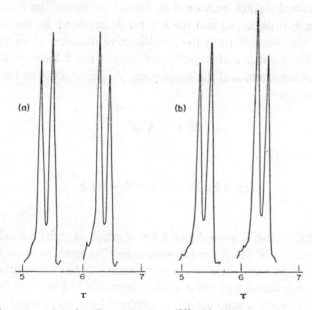

(a)   (b)

FIG. 23. NOE experiments on the triazolinonorbornane (22): (a) AB quartet due to the C-2 and C-6 *endo*-protons with no double irradiation; (b) AB quartet due to the C-2 and C-6 *endo*-protons showing the NOE caused by simultaneous irradiation of the two major signals due to the C-8 *endo*-proton.

The results in Table 6 could only be observed when the deuteriochloroform solutions of the triazolinonorbornane were carefully de-gassed. It was experimentally easier to obtain similar stereochemical information using the pseudo-contact shift reagent Eu(DPM)$_3$.

APPLICATION OF THE SHIFT REAGENT EU(DPM)$_3$ TO

ASSIGNMENT OF STEREOCHEMISTRY

The effect of trisdipivaloylmethanatoeuropium(III) Eu(DPM)$_3$ on the chemical shift behaviour of the protons (see p. 212) can be used to establish the stereochemistry of the aminomethyl triazolinonorbornane (22) because the distance of each proton from the shift reagent or the group to which it is complexed bears a logarithmic relationship to the change in chemical shift produced on addition of the reagent.

This relationship arises from the pseudocontact interaction equation of McConnell and Robertson (see also p. 219) [eqn. (1)],

$$\Delta Eu_i = k(3\cos^2\theta_i - 1)r_i^{-3}, \tag{1}$$

where ΔEu_i is the paramagnetic shift induced in proton i, r_i the distance from the europium ion to the ith proton H_i, θ_i is the angle between the N–Eu axis and the Eu–H$_i$ axis, and k is a constant. In our studies approximations concerning θ_i and r_i have been made which are justified in structural studies because it is difficult to locate the europium atom with precision. The angle θ_i is neglected and the r_i term is replaced by the distance vector R_i where R_i measures the distance from the coordinating nitrogen atom to the ith proton. By ignoring the N–Eu distance and angle θ_i, not only is the k term modified but also the power of distance vector alters and the subsequent graphical analysis leads to the approximation given in eqn. (2):

$$\Delta Eu_i \approx K/R_i^2. \tag{2}$$

The logarithm of eqn. (2) is

$$\log \Delta Eu_i \approx -2\log R_i + \log k, \tag{3}$$

and a plot of log ΔEu_i vs. log R_i should give a linear graph of slope *ca.* -2.

In solving the structure of the 5-aminonorbornene (24) application of eqn. (3) led to a slope of -1.92. The aminomethyltriazolinonorbornane (22) has four nitrogen atoms where complex formation may occur, but, since steric, electronic and basicity factors are important, the primary aliphatic amino-group is preferentially complexed with Eu(DPM)$_3$. Figure 22b shows the n.m.r. spectrum of the aminomethyltriazolinonorbornane after addition of Eu(DPM)$_3$.

As expected, the most dramatic shift is seen for the amino-group which has moved 6 ppm down-field. The adjacent *endo*-methylene group has moved almost 2 ppm down-field. It is now possible to locate all the alicyclic protons since there is a linear relationship between the changing chemical shift and the amount of Eu(DPM)₃ that has been added. This is illustrated by the graph in Fig. 24 which shows the variation in chemical shift (τ) for each proton with concentration of Eu(DPM)₃ for the triazolinonorbornane (22). Furthermore, a logarithmic plot of ΔEu or, better, the slope of the various graphs in Fig. 24

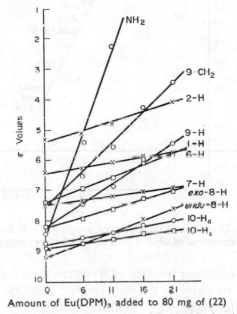

FIG. 24. Variation in chemical shift (τ) with concentration of Eu(DPM)₃ for the triazolinonorbornane (22).

against the vector distance R of the various protons from the primary amino-group, shows a linear relationship of slope $-1\cdot92$ (Fig. 25). Thus complex formation must occur exclusively with the primary amino-group, since other work [12] shows that substantial association at another site would result in a difference between the measured shift and the straight line of Fig. 25. Although the aminomethyl group of compound (22) may rotate freely, the europium complex appears to show a conformation preference. Thus to obtain the best linear relationship in Fig. 25 it was necessary to measure the vector distance R of the various protons in the norbornane system from the nitrogen in an almost eclipsed conformation with the hydrogen at C-9 illustrated in the projection formula (25).

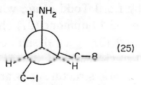

(25)

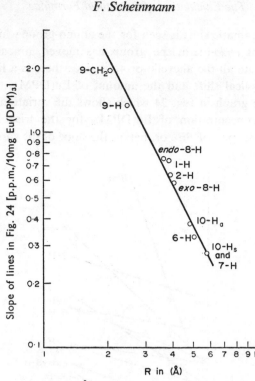

FIG. 25. Slope of lines in Fig. 24 vs. R in (Å) (the primary amino N ... H distance) for (22). The slope of lines in Fig. 24 vs. R is given instead of ΔEu vs. R because the slopes embrace several values of ΔEu.

These results, together with those from the NOE experiments, confirm the stereochemistry given for structure (22). In addition, the shift reagent Eu(DPM)$_3$ provides information on the conformation of the complexed aminomethyl group.

The Structure of Elsinochrome A

The bright red photosensitizing pigments produced by moulds of the genus *Elsinoe* and named the elsinochromes have been studied by Weiss and his colleagues. In elucidating the structure of elsinochrome A, elegant use has been made of ultraviolet, infrared, mass, and n.m.r. spectroscopy on the parent metabolite and its derivatives. In addition, polarographic reduction gives a radical anion which, on examination by electron spin resonance (e.s.r.) spectroscopy, provides confirmation of the structure of elsinochrome A.[13, 14]

The molecular formula $C_{26}H_{12}O_6(OMe)_4$ of elsinochrome A follows from elemental analysis, mass spectrometry, and a Zeisel determination for methoxy-groups. Figure 26 shows the striking similarity of the ultraviolet spectrum of elsinochrome A with that of erythroaphin-fb (26) reported by Lord Todd *et al.*, which indicates that elsinochrome A also has the 4,9-dihydroxyperylene-3,10-quinone (27) chromophore.[14] The striking colour change (red → emerald green) and the spectral shift on addition of base, confirmed the presence of phenolic groups. The evidence now described shows that elsinochrome A has the rapidly interconverting tautomeric structures (28 and 29). The infrared spectrum does

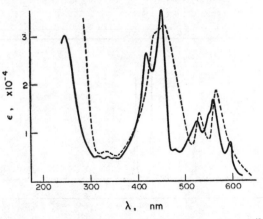

FIG. 26. Absorption spectra of erythroaphin-fb (-----) and elsinochrome A (-----).[14] (Reproduced by permission).

not show a hydroxy-band at *ca.* 3300 cm^{-1}, which suggests that the phenolic hydroxy-groups are strongly hydrogen-bonded to the carbonyl group. In the carbonyl region the band at 1623 cm^{-1} is consistent with that of an extended chelated quinone system. The band at 1715 cm^{-1} is due to aliphatic ketone group.

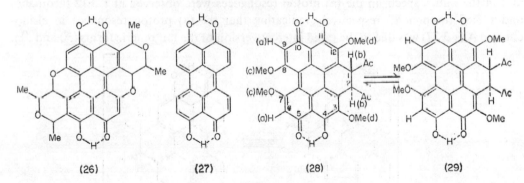

(26)   (27)   (28)   (29)

The n.m.r. spectrum of elsinochrome A (Fig. 27) consists of six single, unsplit peaks which are consistent only with a symmetrical arrangement of the substituents on the perylene-quinone ring system. Integration of the spectrum gave the ratios for the twenty-four protons in the molecular formula as $6:6:6:2:2:2$. At lowest field hydrogen-bonded hydroxy-groups appear together at τ −6·20. The two aromatic protons at τ 3·37 and four aromatic methoxy-groups (two at τ 5·68 and two at τ 5·92) are also readily assigned. Thus there are only two positions to account for the remaining four substituents, two H(b) protons at τ 4·8 and two methyl ketone groups at τ 7·95. From the infrared absorption band at 1715 cm^{-1} it follows that the methyl ketone groups are not directly attached to the aromatic nucleus. To satisfy the requirements for a symmetrical structure the protons H(b) are attached to an alicyclic system of the tautomeric structures (30 and 31).

Evidence that elsinochrome A has two tautomeric structures comes from the chemical shift of the aromatic protons which appear between the normally accepted range of aromatic

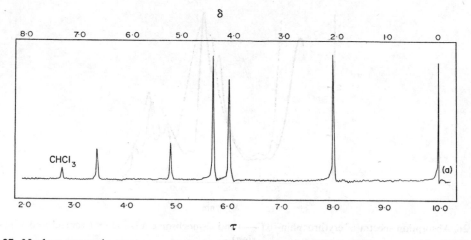

FIG. 27. Nuclear magnetic resonance spectrum of elsinochrome A (28) ⇌ (29) measured in CDCl$_3$.[14] The signal due to the hydrogen-bonded hydroxy groups at τ −6·20 is not shown. (Reproduced by permission.)

and quinonoid protons. This assignment is supported by methylation studies on elsinochrome A which give two isomeric dimethyl ethers. One ether is red and the other is yellow, and in the n.m.r. spectrum the (a) proton resonances were observed at τ 3·12 (aromatic) and τ 3·87 (quinonoid) respectively, indicating that the (a) proton resonance in elsinochrome A (τ 3·37) was due to the rapid interconversion of the tautomeric forms (30 and 31).

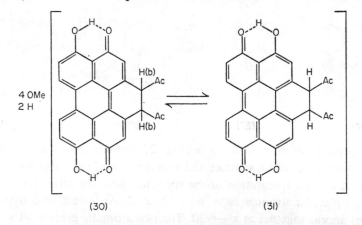

(30)   (31)

For the complete structure determination of the optically active elsinochrome A it is necessary to locate the methoxy groups and to give the stereochemistry of the H(b) protons. The configuration of the methyl ketone groups could be either *cis* or *trans*. However, for the *cis*-configuration one of the benzylic protons will always be in the axial and the other in the equatorial position; they will thus be magnetically non-equivalent. The sharp single peak of the (b) protons in the n.m.r. spectrum of elsinochrome A measured in either deuteriochloroform or deuteriobenzene favours the *trans*-configuration for the (b) protons. This assignment is supported by equilibration studies. Previous work has shown that

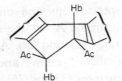

equilibration of optically active forms (*cis* and *trans*) of cyclohexyl methyl ketones in alkaline solution leads to the *trans*-isomer with the methyl ketone group in the equatorial position. Equilibration of elsinochrome A in pyridine-deuterium oxide solution caused disappearance of the n.m.r. signals of the benzylic protons. Equilibration in the presence of pyridine-water gives back the original compound which shows that elsinochrome A has both ketone groups in the equatorial position and that the two (b) protons have the *trans*-diaxial configuration. For a *rigid trans*-diaxial configuration the two (b) protons would be at an angle of 180° to each other and in such a case the vicinal spin coupling constant would be in the range 10–16 Hz. Since the benzylic protons are equivalent, it was necessary to obtain the coupling constant from the n.m.r. spectrum of the ^{13}C-satelite.

The natural abundance of ^{13}C is only of the order of 1·1%, and therefore repeated scanning of the weak signal and summation of all scans by a computer time averaging technique (see Vol. 1, p. 35) was used to give the coupling between ^{13}C and the benzylic proton and also to show the further splitting due to the *trans* (b) protons. The expected coupling between ^{13}C and benzylic proton of 130 Hz was observed, but the signal was split into a doublet with a coupling of only 5 Hz for protons Hb, and this indicates that the *trans*-configuration of the side chain is not rigid.

There are three possible arrangements (32, 33, and 34) that must be considered in locating the four methoxy-groups and the two aromatic protons in accordance with a symmetrical arrangement for the substituents.

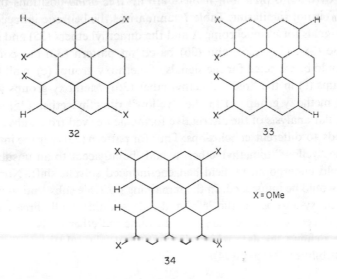

32   33   34

X = OMe

Two methods based on n.m.r. spectroscopy and one method involving e.s.r. studies have been used to prove that elsinochrome A has the substitution pattern given in (32), and therefore the tautomeric structures (28 and 29). As part of this work, the dimethyl ethers of elsinochrome A, dimethyl ethers "yellow" and "red", have been examined and the results are in agreement with structures (35 and 36)

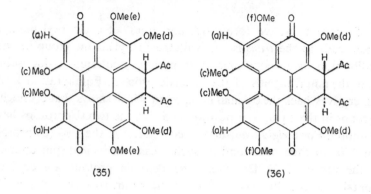

(35)   (36)

Previously it was shown that the "yellow" ether (35) has the nuclear proton (a) in a quinonoid environment, whereas the "red" ether (36) has proton (a) in a benzenoid region.

NUCLEAR MAGNETIC RESONANCE SOLVENT STUDIES

Studies on the solvent dependence of methoxy-group resonances have contributed considerably to the elucidation of a variety of aromatic structures including natural coumarins, flavones, and quinones. A marked solvent shift $\Delta_\tau(C_6H_6\text{—}CDCl_3)$ is usually observed if the methoxy-group has at least one free *ortho*-position. The methoxy-resonance shifts up field in benzene by 0·5–1·0 ppm, but if there are no free *ortho*-positions the solvent shift $\Delta_\tau(C_6H_6\text{—}CDCl_3)$ is not significant. Table 7 summarizes the benzene-induced solvent shifts of the methoxy-signals for elsinochrome A and the dimethyl ethers (35) and (36).

If the formulae (28), (29), (35), and (36) based on pattern (32) are correct, benzene-induced shifts would be expected for the signals of methoxy-groups (c) in all three cases and for methoxy-groups (f) in the "red" dimethyl ether (36). Methoxy-groups (d) in the three compounds and methoxy group (e) in the "yellow" dimethyl ether (35) should not be influenced. A similar analysis of the alternative formulae derived from substitution patterns (33) and (34) leads to different conclusions. Thus for pattern (33) *only* the introduced methoxy-group in the "yellow" dimethyl ether would be adjacent to an unsubstituted *ortho*-position and would undergo an up-field benzene-induced solvent shift. None of the other methoxy-groups would be influenced. In the remaining possible substitution pattern (34) the original two methoxy-groups on the left-hand side should in all three cases be shifted whereas the introduced methoxy-groups in both dimethyl ether derivatives should not be influenced. Table 7 shows the data which clearly support the substitution pattern (32) and eliminate the possibilities (33) and (34).

TABLE 7

METHOXY-RESONANCES (SOLUTIONS IN $CDCl_3$ AND C_6D_6; 60 MHz
SPECTRA; Me_4Si AS INTERNAL REFERENCE)

Compound	OMe	$\tau(CDCl_3)$	$\tau(C_6D_6)$	$\Delta(C_6D_6-CDCl_3)$ (p.p.m.)
(28) ⇌ (29)	c	5·95 (L)*	6·77 (L)	+0·82
	d	5·64	5·70	+·006
(35)	c	6·09 (L)	6·85 (L)	+0·76
	d, e	5·75 and 5·87	5·90 and 5·95	+0·15 and +0·08
(36)	c, f	5·90 and 5·81 (L)	6·59 and 6·40 (L)	+0·69 and +0·59
	d	5·70	5·67	−0·03

(L) Peak with larger half-linewidth, i.e. lower height, than comparable resonances in the spectrum.
(Reproduced by permission from reference [13].)

NUCLEAR OVERHAUSER EFFECT STUDIES

Table 7 shows that the methoxy-groups (c) could be easily located since they appeared as signals of larger half-line width and lower height than the resonances of methoxy-groups (d). This reproducible difference in peak heights (without change in area under the methyl peaks) is clearly shown in Fig. 27. This effect is due, at least in part, to long-range coupling between the protons of the methoxy-group and the one on the ring *ortho* to it. It was desirable to confirm that the methoxy-groups (c) were adjacent to nuclear protons (a) by NOE measurements. Indeed, irradiation at the methoxy-groups (c) in elsinochrome A (28 and 29) and its two dimethyl ethers (35 and 36) caused considerable enhancement in the integrated areas of the signals due to the nuclear protons. Furthermore, irradiation at the resonance of the introduced methoxy-groups (f) of the "red" dimethyl ether (36) likewise resulted in an increase of the integrated areas of the signal due to protons (a) (Table 8). Thus elsinochrome

TABLE 8

DOUBLE-RESONANCE EXPERIMENTS[a]

Compound	Irradiation at OMe	Relative intensity of H(a)		Relative intensity of H(b)		Enhancement of H(a) (%)
		No irr.	irr.	No irr.	irr.	
(28) ⇌ (29)	c	113	140	118	119	24±3
(35)	c	100	121	112	112	21±3
(36)	c	103	121	100	100	17±2
	f	102	120	100	100	17±2

* The relative intensities are average values of ten measurements. The deviations in these values cause an uncertainty of *ca.* 3% in the enhancement factor. The units are arbitrary. (Reproduced by permission from reference [13].)

A must have the tautomeric structures (28) and (29) since these are the only possibilities arising from arrangements (32), (33), and (34) which give a dimethyl ether (36) having two pairs of methoxy groups (c) and (f) adjacent to protons (a).

ELECTRON SPIN RESONANCE STUDIES

Electrolytic reduction of 1,4-naphthoquinone derivatives gives radical anions which can be identified from their e.s.r. spectra (see p. 179). The spin density at the quinonoid hydrogens is usually greater than that at the aromatic hydrogens, and for methoxy-substituents, coupling of the unpaired electron with the methoxy-protons it is necessary to have $p\pi$ orbital overlap between the oxygen and the π system of the chromophore. In addition, for effective splitting there must be sufficient spin density on the oxygen or on the oxygen and the carbon atom of the system to which the oxygen is attached. The e.s.r. spectrum of the radical anion of elsinochrome A appears to consist of eleven poorly resolved lines (Fig. 28a) with a single splitting of 0·49 G. Exchange of the hydroxy-protons with deuterium oxide led to a spectrum consisting of nine lines. Thus the hydroxy-protons contribute to the splitting in Fig. 28a. The e.s.r. spectrum of 1,2-dideuterioelsinochrome A (Fig. 28b) still consists of eleven lines, showing that the two aliphatic protons H(b) only cause broadening of the lines.

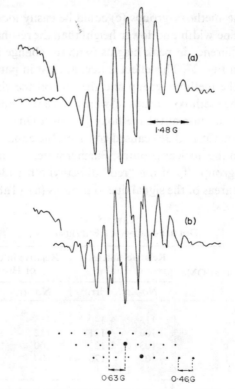

FIG. 28. The e.s.r. spectra of (a) the radical anion of elsinochrome A (28) $\rightleftharpoons$ (29); (b) the radical anion of 1,2-dideuterioelsinochrome A. The insets show amplifications of the outermost low-field signals. (Reproduced by permission from reference [13].)

There are ten protons that are responsible for the eleven-line spectrum: these are the two hydroxy-protons, the two ring protons (a) at C-6 and C-9 and six methoxy-protons, and this has been confirmed by the spectra of derivatives.

The most satisfactory analysis of the eleven-line spectrum is that the two-ring protons (a) give rise to a 1 : 2 : 1 triplet with a splitting of 0·63 G and that each component is further split into a nonet with a splitting of 0·46 G (Fig. 28b) due to the six methoxy- and two hydroxy-protons which fortuitously show the same splitting. This should give rise to a twenty-seven line spectrum, but because of the small difference between the two splittings, complete resolution is not obtained. Figure 28 shows each nonet as nine dots and the overlap between the three sets gives rise to the eleven lines shown in Fig. 28. The fact that there are two different couplings (0·63 and 0·46 G) in the eleven lines shown in Fig. 28b rules out equal coupling of ten protons with one unpaired electron.

The consequence of this analysis in structure elucidation is that *two identical methoxy-groups must be capable of pπ orbital overlap with the ring system* for the six methoxy-protons to be involved in coupling with an unpaired electron. The methoxy-groups on the left-hand side in both substitution patterns (32) and (34) would be consistent with the e.s.r. data, but substitution pattern (33) can be ruled out because steric hindrance prevents all four methoxy-groups from attaining the necessary position for good pπ overlap. For similar reasons the methoxy-groups on the right-hand side of structures (32) and (34) cannot be involved in coupling because the adjacent groups prevent the methoxy-groups from lying in the same plane as the benzene rings.

In order to decide between systems (32) and (34) the e.s.r. spectra derived from the radical anions of the two dimethyl ethers of elsinochrome A were studied. If system (32) and thus structures (35) and (36) are correct, compound (35) will have two quinonoid protons (a) which will have a larger coupling constant than the aromatic protons (a) in (36). Only one pair of methoxy-groups in structure (35) viz. those at C-7 and C-8, can assume the steric arrangement for coupling to an unpaired electron. One can therefore account for the observed spectrum of a triplet (1·48 G) of septets (0.47 G). Compound (36) has the four methoxy-groups (c) and (f) which are capable of coupling. Since the coupling constant of the benzenoid proton is lower than that of the quinonoid proton, one would expect a complex spectrum resulting from the coupling of fourteen almost equally coupled protons. This is in agreement with the observed complex fifteen-line spectrum derived from the radical anion of the red dimethyl ether.

In contrast, if the red dimethyl ether was derived from system (34), only one pair of methoxy-groups (6H) and two aromatic protons would be available for coupling wth the unpaired electron. Thus the spectrum expected on this latter basis would be quite different from the observed one. Hence these results provide independent support for the arrangement of the methoxy-groups as given in pattern (32), and consequently for the tautomeric structures (28) and (29) for elsinochrome A. These conclusions were further supported by analysis of the e.s.r. spectrum derived from 6,9-dibromoelsinochrome A (37). After introduction of bulky two bromine atoms, the adjacent methoxy-groups at C-7 and C-8 will be forced out of plane so that the necessary pπ overlap for coupling with an unpaired electron is no longer possible. Since the two ring protons (a) at C-6 and C-9 are no longer present, only coupling with the two hydroxy-protons occurs. This in accord with the simple observed three-line spectrum of 6,9-dibromoelsinochrome A (37) with a coupling constant of approximately 0·5 G which on deuteration was transformed to a single, broad-line spectrum.

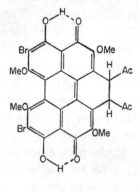

(37)

References

1. W. Körner, obituary notice by J. B. Cohen, *J. Chem. Soc.*, 1925, 2975.
2. R. J. Abraham, E. O. Bishop, and R. E. Richards, *Mol. Phys.*, 1960, **3**, 485.
3. *Documentation of Molecular Spectroscopy, Infrared Absorption Spectra.*, Butterworths, London, catalogue numbers 3048, and 3331.
4. B. Jackson, H. D. Locksley, and F. Scheinmann, *J. Chem. Soc.* (C), 1967, 785.
5. B. Jackson, P. J. Owen, and F. Scheinmann, *J. Chem. Soc.* (C), 1971, 3389 and cited references.
6. A. A. Forsyth, *British Poisonous Plants*, Ministry of Agriculture, Fisheries and Food Bulletin No. 161, HMSO, 1968, p. 60.
7. W. M. Bandaranayake, L. Crombie, and D. A. Whiting, *Chem. Comm.*, 1969, 58 and 970.
8. A. Arnone, G. Cardillo, L. Merlini, and R. Mondeli, *Tetrahedron Letters*, 1967, 4201.
9. F. A. L. Anet and A. J. R. Bourn, *J. Amer. Chem. Soc.*, 1965, **87**, 22 and 5250.
10. D. Barraclough, J. S. Oakland, and F. Scheinmann, *J. Chem. Soc.*, Perkin I, 1972, 1500.
11. R. G. Foster and M. C. McIvor, *Chem. Comm.*, 1967, 280.
12. C. C. Hinckley, M. R. Klotz, and F. Patil, *J. Amer. Chem. Soc.*, 1971, **93**, 2417.
13. R. J. J. Ch. Lousberg, L. Paolillo, H. Kon, U. Weiss, and C. A. Salemink, *J. Chem. Soc.* (C), 1970, 2154 and references quoted therein.
14. R. J. J. Ch. Lousberg, C. A. Salemink, U. Weiss, and T. J. Batterham, *J. Chem. Soc.* (C), 1969, 1219.

Seminar Problems and Answers in the Identification of Organic Compounds Using a Combination of Spectral Methods[†]

B. J. HOPKINS and F. SCHEINMANN

Department of Chemistry and Applied Chemistry, University of Salford

Spectral Measurements

Nuclear magnetic resonance spectra were measured on either Varian A60 or HA100 spectrometers.

Infrared spectra were measured with a Perkin-Elmer 257 grating spectrometer.

Ultraviolet spectra were measured on the Unicam SP800 spectrometer.

Mass spectra were measured with an AEI MS12 instrument at 70 eV.

† A discussion of the answers is provided on p. 305.

PROBLEM 1 — USING A COMBINATION OF SPECTRAL METHODS

Deduce the structure of the compound $C_3H_2O_2$.

U.V. spectrum: λ_{max}^{hexane} 235 nm (ε 2500).

I.R. spectrum: liquid film.

N.M.R. spectrum in $CDCl_3$ at 60 MHz (500 Hz sweep width).

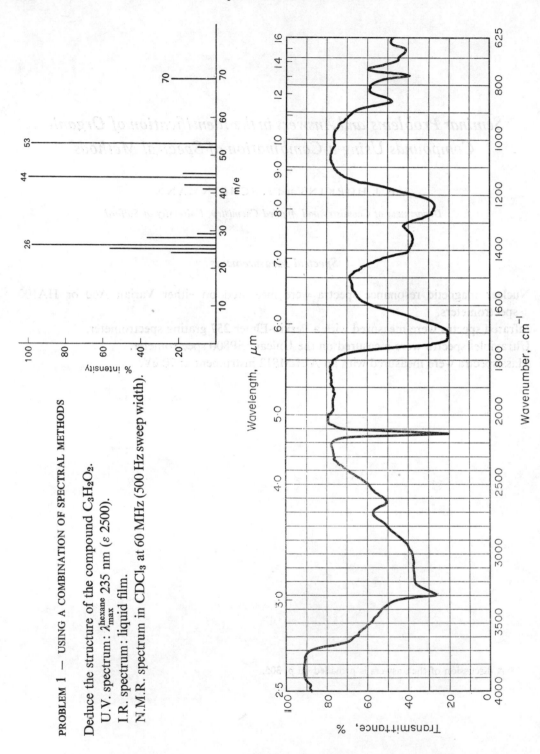

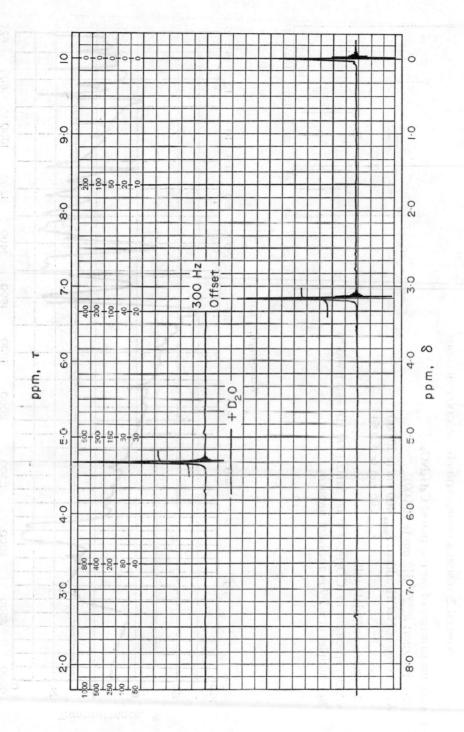

PROBLEM 2—USING A COMBINATION OF SPECTRAL METHODS

Deduce the structure of the compound $C_6H_5NO_3$.

U.V. spectrum: $\lambda_{max}^{ethanol}$ 312 nm (ε 10,000).

With 1 drop N NaOH added: λ_{max} 400 nm (ε 20,000).
λ_{max} 305 nm (ε 8,500).

I.R. spectrum: Nujol mull.

N.M.R. spectrum in $(CD_3)_2CO$ at 60 MHz (500 Hz. sweep width).

The multiplet at τ 7·9 is a solvent impurity.

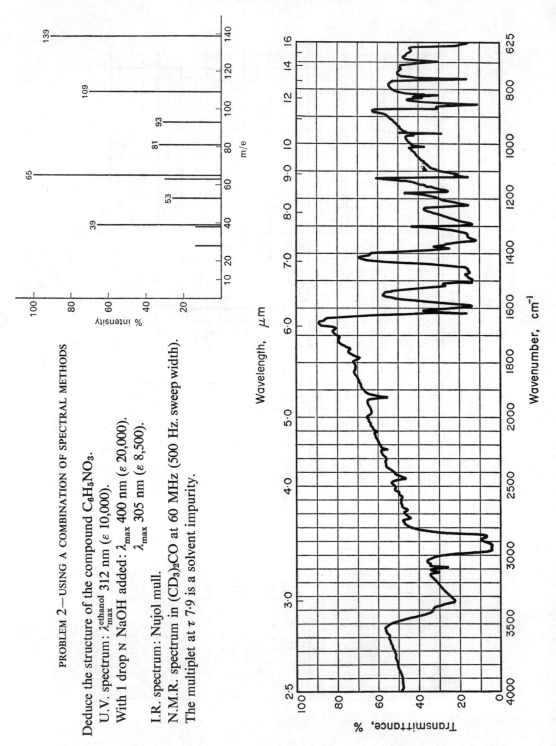

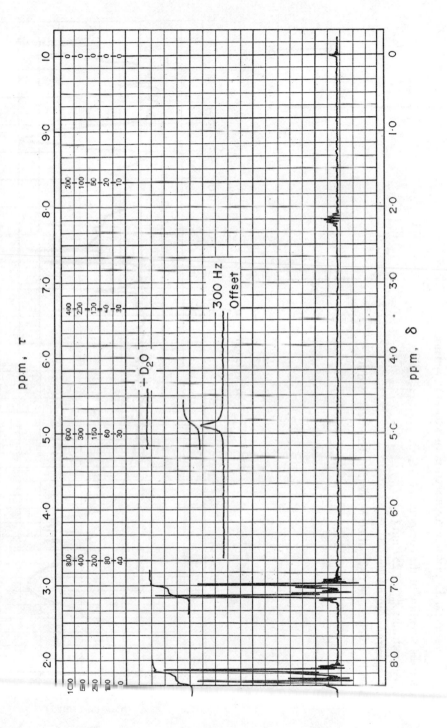

PROBLEM 3—USING A COMBINATION OF SPECTRAL METHODS

Deduce the structure of the compound $C_8H_{10}O$.

U.V. spectrum: λ_{max}^{hexane} 257 nm (ε 122).

I.R. spectrum: liquid film.

N.M.R. spectrum in CCl_4 at 60 MHz (500 Hz sweep width).

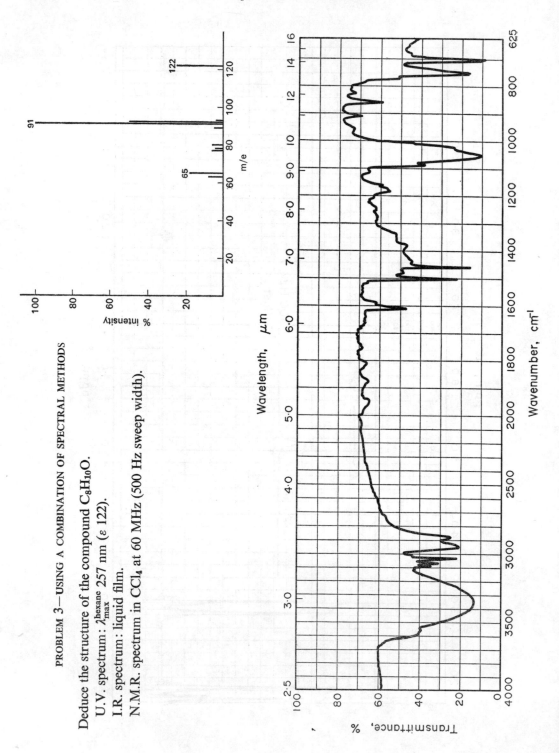

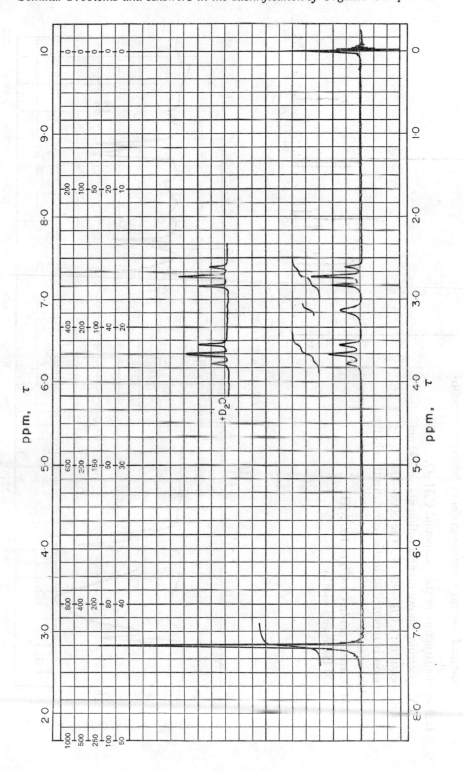

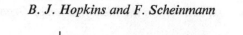

PROBLEM 4—USING A COMBINATION OF SPECTRAL METHODS

Deduce the structure of the compound $C_5H_{12}O$.

U.V. spectrum: shows no absorption above 200 nm.

I.R. spectrum: CCl_4 solution. The 3000–4000 cm^{-1} region is repeated at higher concentration in the lower trace.

N.M.R. spectrum in $CDCl_3$ at 100 MHz (500 Hz sweep width).

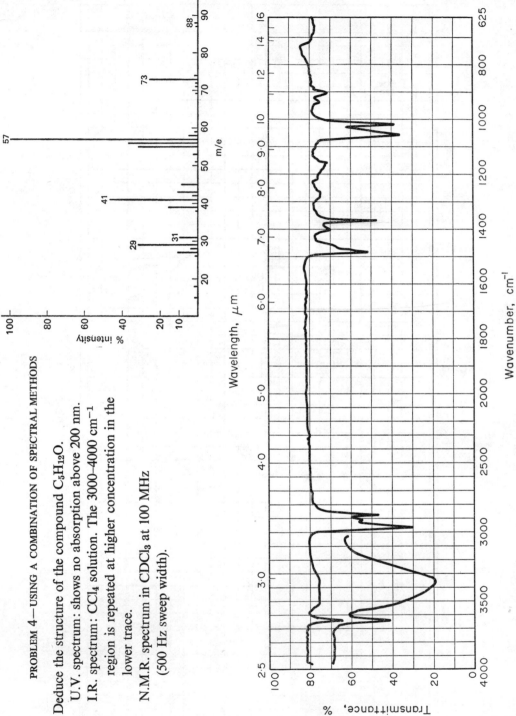

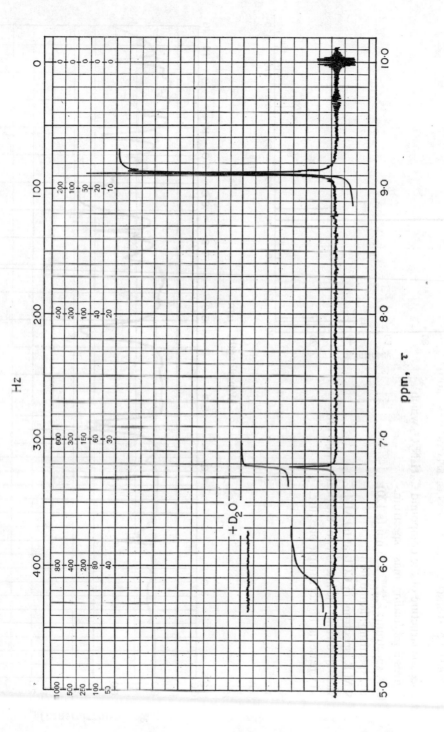

PROBLEM 5—USING A COMBINATION OF SPECTRAL METHODS

Deduce the structure of the compound $C_3H_5N_3O_2$ which gives a misleading mass spectrum.

U.V. spectrum: λ_{max}^{hexane} 235 nm (ε 120).

I.R. spectrum: liquid film.

N.M.R. spectrum in CCl_4 at 60 MHz.

Mass spectrum: the intensity of the peak at m/e 116 increases at higher sample pressure.

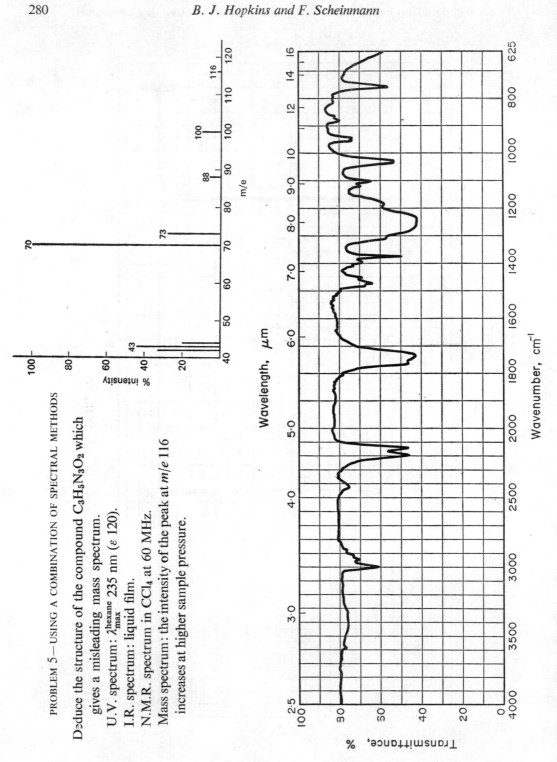

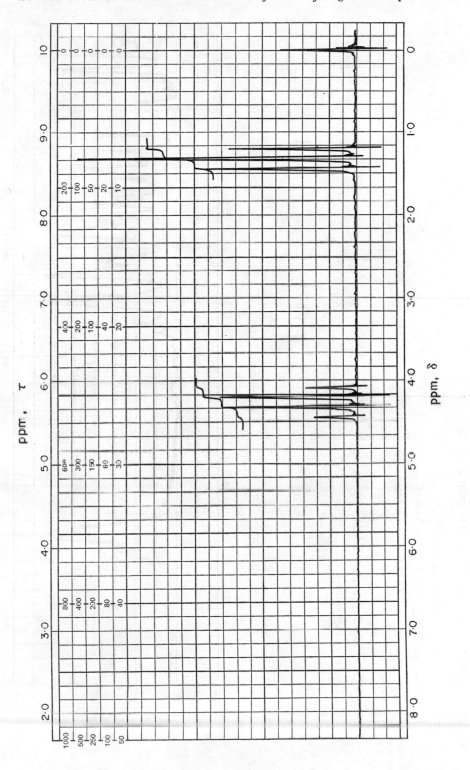

PROBLEM 6—USING A COMBINATION OF SPECTRAL METHODS

Deduce the structure of the base $C_{14}H_{19}N$.

U.V. spectrum: λ_{max}^{hexane} 252 nm (ε 20,400)
               and   210 nm (ε 20,000).

I.R. spectrum; liquid film.

N.M.R. spectrum in CCl_4 at 60 MHz
         (500 Hz sweep width).

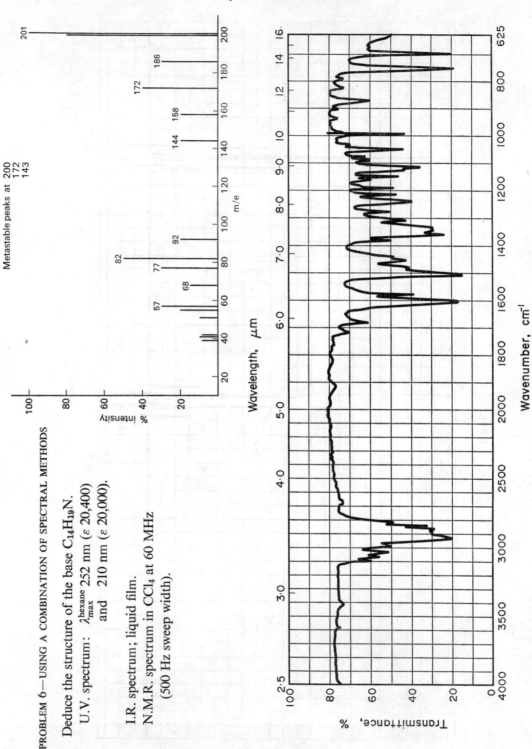

Metastable peaks at 200, 172, 143

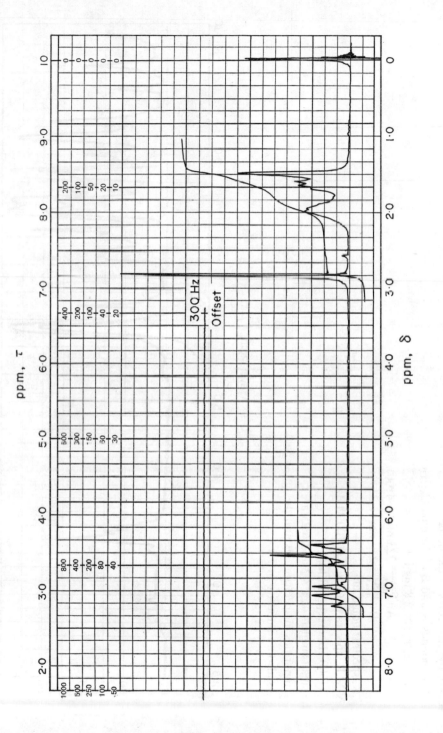

PROBLEM 7— USING A COMBINATION OF SPECTRAL METHODS

Deduce the structures of the isomeric compounds A and B of formula $C_8H_{12}O_4$.

U.V. spectra: at 2·8 mg l⁻¹:

    A: λ_{max}^{hexane} 223 nm (ε 4100).
    B: λ_{max}^{hexane} 219 nm (ε 2300).

I.R. spectra: liquid film.

N.M.R. spectra in $CDCl_3$ at 100 MHz (1000 Hz sweep width).

The small singlets at τ 3·2 and 6·25 are impurities in spectrum 7B and the latter peak is also an impurity in 7A.

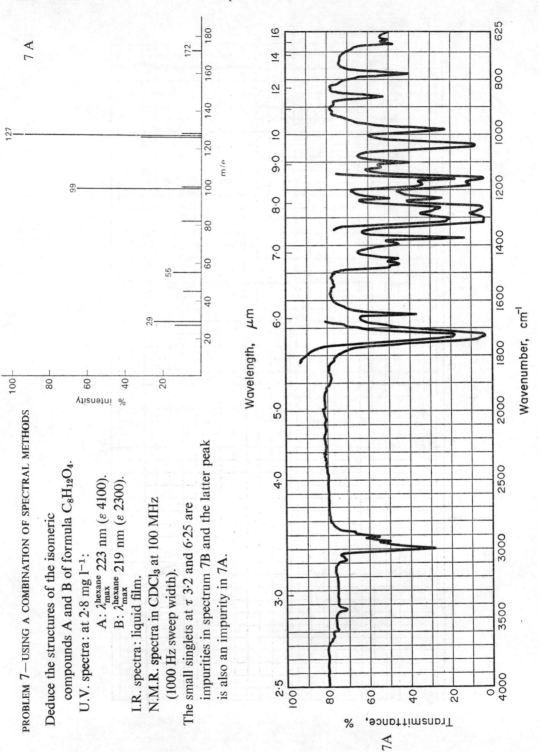

7 A

7A

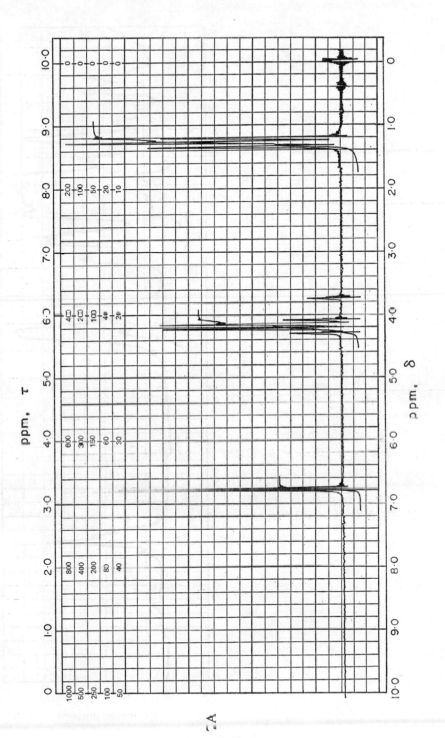

PROBLEM 7 – USING A COMBINATION
OF SPECTRAL METHODS (cont.)

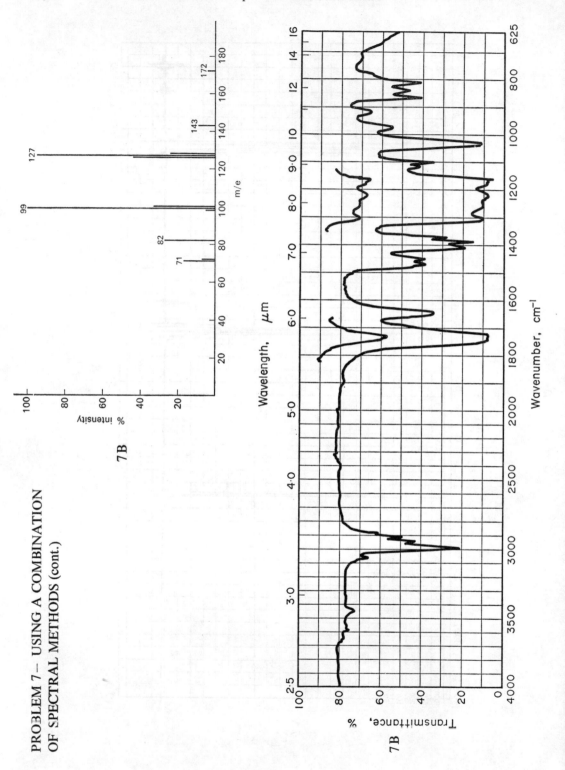

7B

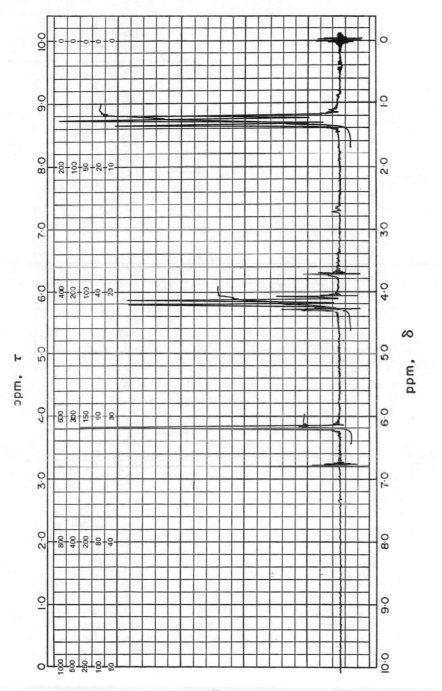

7 B

PROBLEM 8—USING A COMBINATION OF SPECTRAL METHODS

Deduce the structure of the compound $C_{10}H_{14}O_3$.

U.V. spectrum: λ_{max}^{hexane} 231 nm (ε 10,000)
and 272 nm (ε 10,000) pH dependent.

I.R. spectrum: liquid film.

N.M.R. spectrum in CCl_4 at 60 MHz
(500 Hz sweep width).

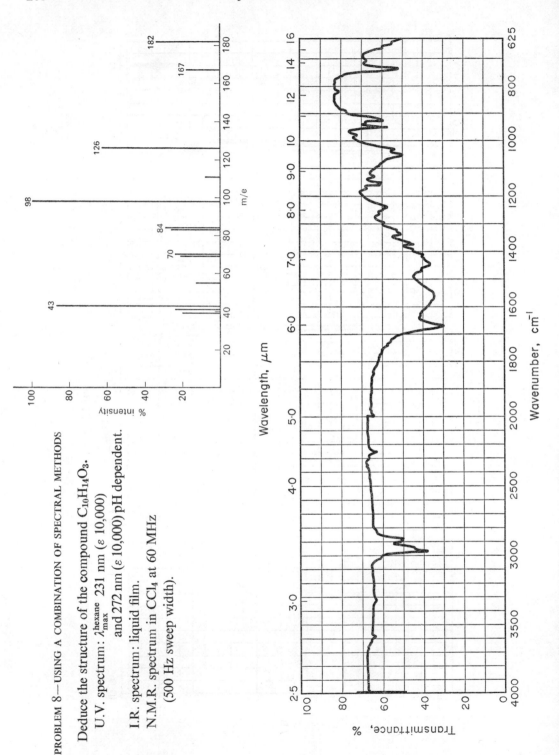

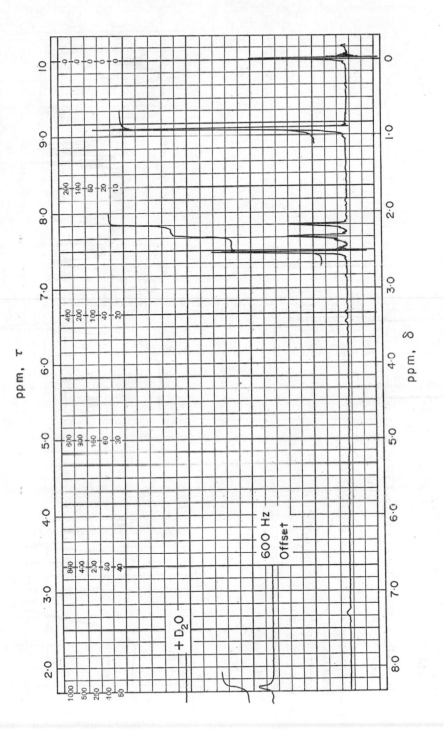

PROBLEM 9—USING A COMBINATION OF SPECTRAL METHODS

Deduce the structure of the product derived by
the treatment of 2,2,5,6,6-pentamethyl-3-
oxoheptan-1,7-dial with concentrated
hydrochloric acid at room temperature.

$$OHC . CMe_2 . CHMe . CH_2 . CO . CMe_2 CHO \xrightarrow{\text{conc. HCl}} ?$$

The mass spectrum of the product is difficult to
rationalize.

U.V. spectrum: λ_{max}^{hexane} 208 nm (ε 207).
I.R. spectrum as potassium bromide disc.
N.M.R. spectrum in deuteriobenzene at 60 MHz
(500 Hz sweep width).

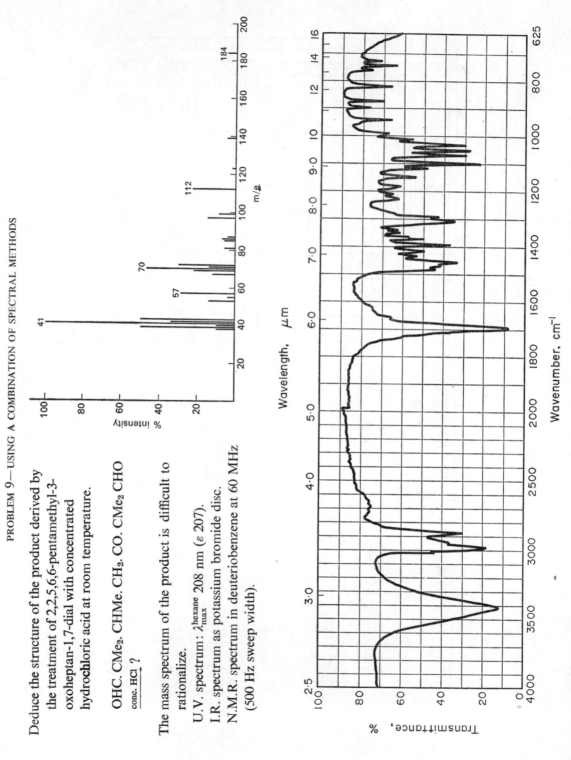

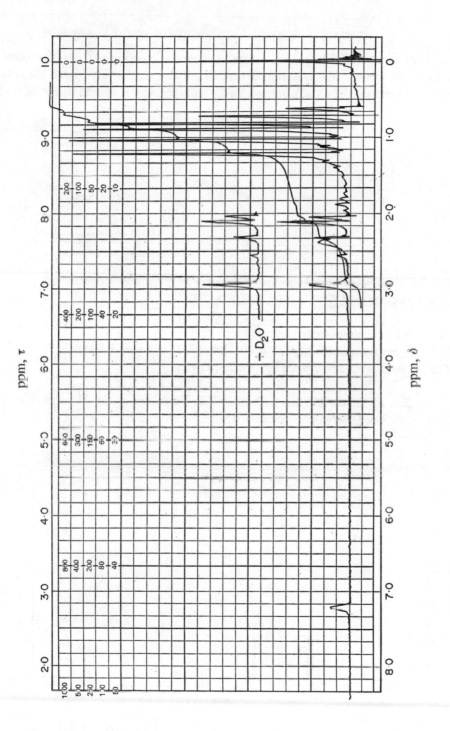

PROBLEM 10—USING A COMBINATION OF SPECTRAL METHODS

Deduce the structure of the yellow naturally occurring compound $C_{15}H_{10}O_5$ found commonly in higher plants. If it is assumed that the fifteen carbon atoms are biogenetically derived by combination of a C_6C_3 unit with a C_6 unit, flavonoids should be considered:

e.g.

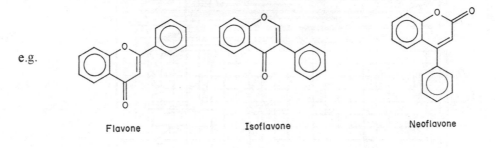

Flavone   Isoflavone   Neoflavone

The u.v. spectrum resembles the reported data for polyhydroxyflavones.
  U.V. spectrum: $\lambda_{max}^{methanol}$ nm (ε): 268 (23,700),
              335 (25,700).

  I.R. spectrum: Nujol mull.
  N.M.R. spectrum in $(CD_3)_2$ SO at 60 MHz (500 Hz sweep width).
  The peaks at τ 6·5 and 7·4 are solvent impurities.

  For u.v. data on flavonoids, see L. Jurd in *The Chemistry of the Flavonoid Compounds*, (ed. T. A. Geissman), Pergamon, 1962, p. 151.

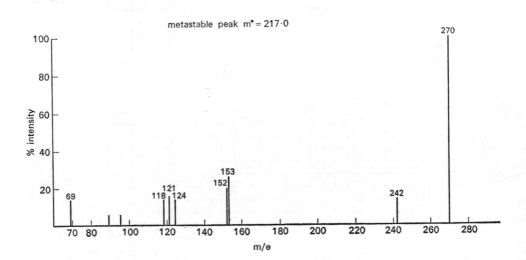

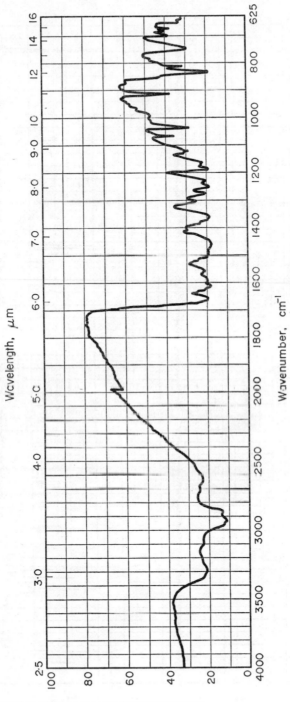

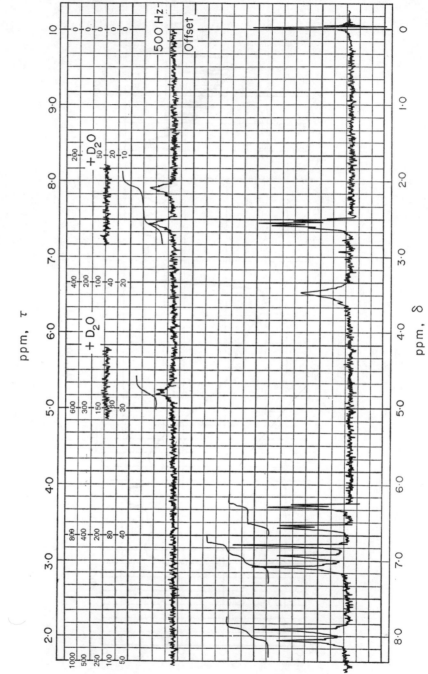

PROBLEM 10 (cont.)

PROBLEM 11 — USING A COMBINATION OF SPECTRAL METHODS

The 3,3-dimethylallyl oestrone ether was heated in boiling diethylaniline for 20 h. A methyl ether of the phenolic product was then prepared and subjected to spectral measurements as below. Determine the structure of the product and comment on its formation.

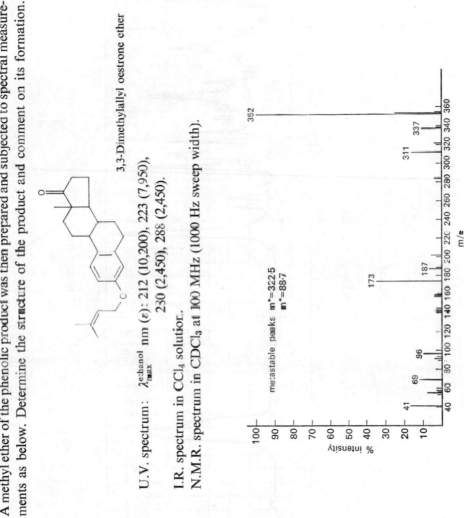

3,3-Dimethylallyl oestrone ether

U.V. spectrum: $\lambda_{max}^{ethanol}$ nm (ε): 212 (10,200), 223 (7,950), 230 (2,450), 288 (2,450).

I.R. spectrum in CCl_4 solution.

N.M.R. spectrum in $CDCl_3$ at 100 MHz (1000 Hz sweep width).

metastable peaks $m^* = 322.5$
                    $m^* = 88.7$

% Intensity

m/e

PROBLEM 11 (cont.)

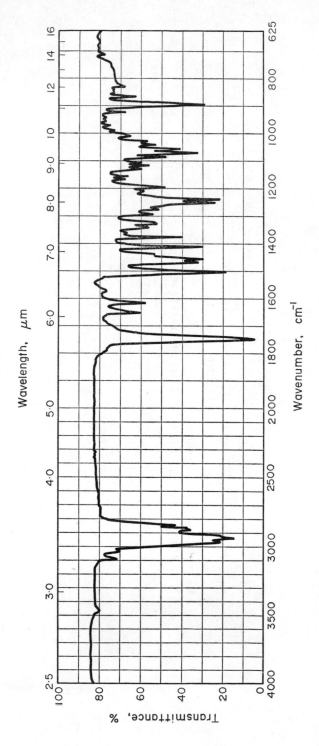

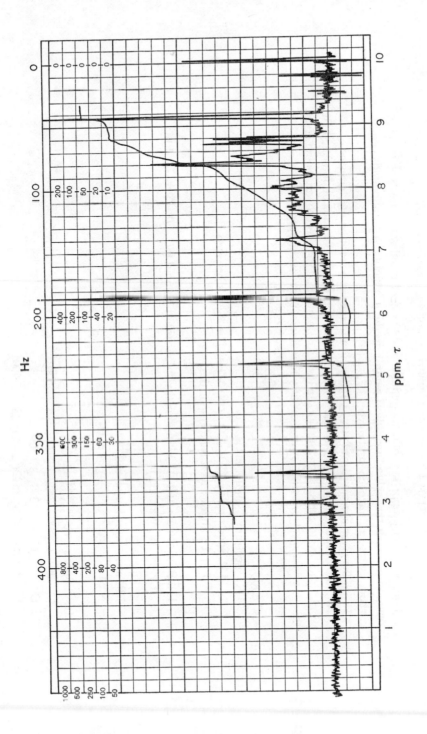

PROBLEM 12—USING A COMBINATION OF SPECTRAL METHODS

Equimolecular proportions of compounds C and D were treated with base. The resulting reaction proceeds through a free radical anion E to the product F. All compounds contain carbon, hydrogen, nitrogen, and oxygen.

From the given spectra identify C, D, E, and F and suggest a reaction mechanism.

An initial analysis of the possible hyperfine interactions in the radical intermediate would lead one to expect fifty-four lines in the e.s.r. spectrum. Fortuitous overlapping gives the observed spectrum with fewer than fifty-four lines: interpret the spectrum.

C and D
I.R. spectrum: Nujol mull.
N.M.R. spectrum in $(CD_3)_2CO$ at 60 MHz. The signals between τ 7 and 8 are due to solvent impurities.

E
E.S.R. spectrum at 10^{-2} M in dimethylsulphoxide-t-butanol (80 : 20). (Provided by Dr. H. W. Wardale.)

F
I.R. spectrum: Nujol mull.
N.M.R. spectrum in CCl_4 at 60 MHz.

U.V. Spectrum: $\lambda_{max}^{ethanol}$ nm (ε).
C   303, (7,500); 281, (8,800); 218, (7,500).
D   281, (1,500); 239, (10,500)
F   323, (15,800); 261, (7,900); 231, (7,900).

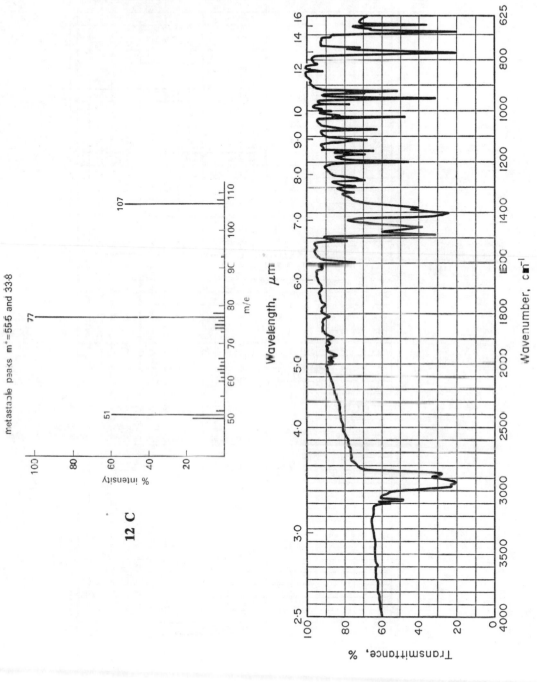

metastable peaks m⁺=55·5 and 33·8

107

77

51

% intensity

100

80

60

40

20

50 60 70 80 9C 100 110

m/e

12 C

Wavelength, μm

2·5 3·0 4·0 5·0 6·0 7·0 8·0 9·0 10 12 14 16

% Transmittance,

100 80 60 40 20 0

4000 3500 3000 2500 2000 1800 1600 1400 1200 1000 800 625

Wavenumber, cm⁻¹

12 C

PROBLEM 12 (cont.)

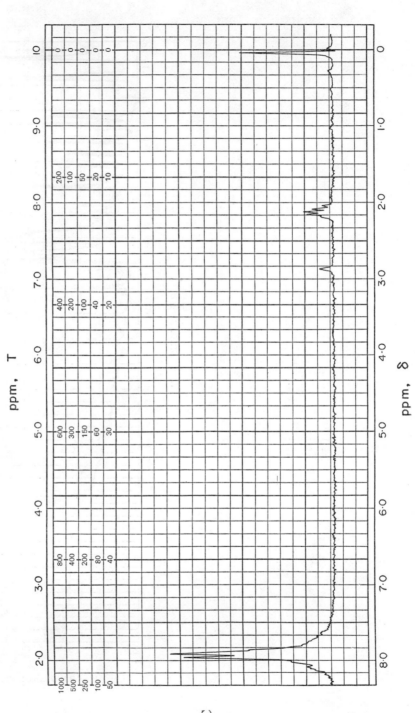

12 C

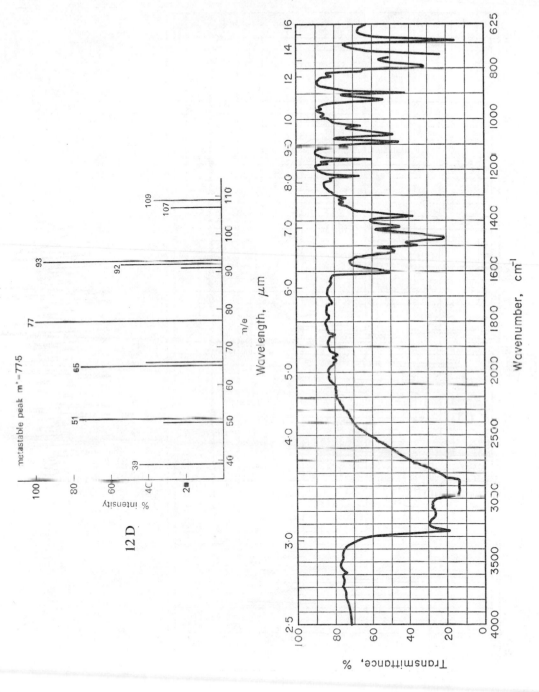

12 D

12 D

PROBLEM 12 (cont.)

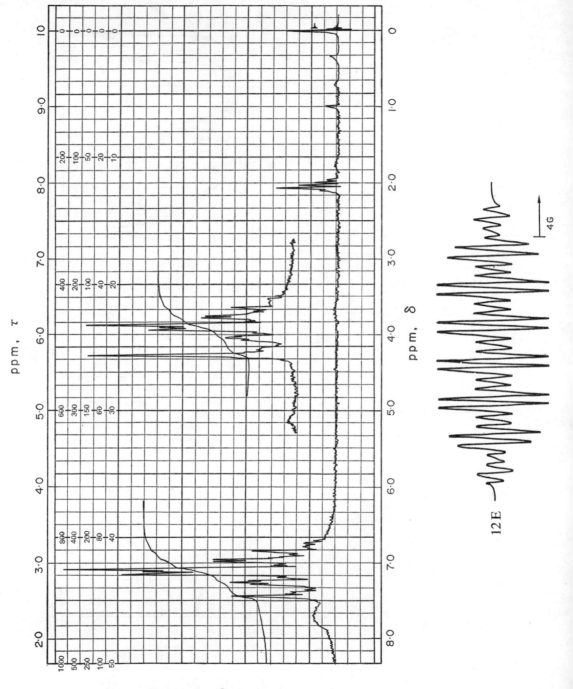

12 D

12 E

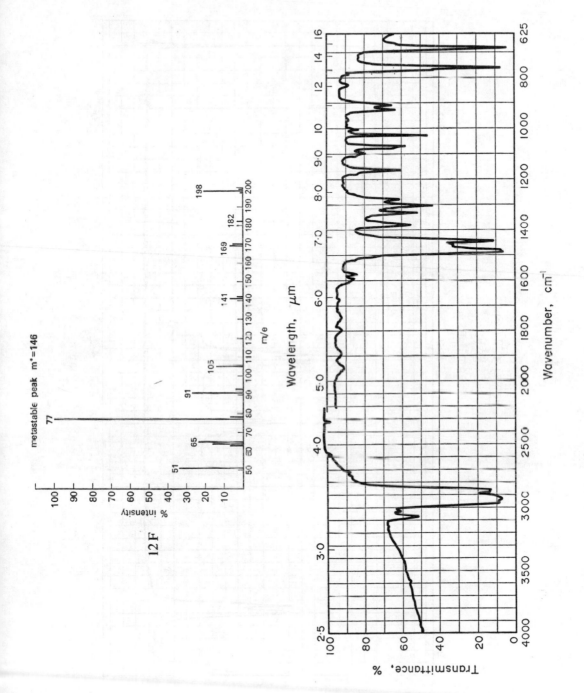

12 F

% intensity

metastable peak m* = 146

m/e

Wavelength, μm

Transmittance, %

Wavenumber, cm⁻¹

12 F

PROBLEM 12 (cont.)

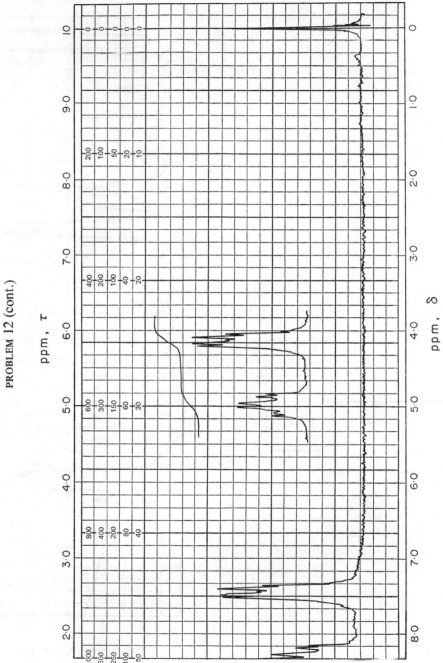

12 F

Answers

ANSWER TO PROBLEM 1—USING A COMBINATION OF SPECTRAL METHODS

The molecular formula $C_3H_2O_2$ indicates the presence of three double-bond equivalents, and unsaturation is confirmed by the ultraviolet spectrum.

The broad peak $v_{(OH)}$ from 3500 to 2700 cm^{-1} taken into consideration with the peak at 1710 cm^{-1} $v_{(C=O)}$ in the i.r. spectrum suggests a carboxylic acid group, and this is confirmed by the peak at $-\tau$ 0·3 in the n.m.r. spectrum which exchanges with D_2O. The sharp peak at 2130 cm^{-1} in the i.r. spectrum indicates a triple bond in the compound, and this agrees with the singlet at τ 6·85 in the n.m.r. spectrum which could arise from an acetylenic hydrogen of a H—C≡C—CO— group (see Vol. 1, n.m.r. Table 4, p. 62). This assignment is supported by a peak at 3300 cm^{-1}, $v_{(≡C—H)}$, in the i.r. spectrum. The i.r. fingerprint region is extremely simple, and the simplest acetylenic carboxylic acid, propiolic acid, agrees with the molecular ion in the mass spectrum at m/e 70. Cleavage α to the carbonyl group of the carboxylic acid gives peaks in the mass spectrum at m/e 53 (loss of .OH) and m/e 25 (loss of .COOH). Breakdown to carbon dioxide (m/e 44) and acetylene (m/e 26) gives rise to very abundant peaks and may be induced thermally. The u.v. spectrum, which shows an absorption λ_{max} 235 nm (ε 2500) is also in accord with this structure.

ANSWER TO PROBLEM 2—USING A COMBINATION OF SPECTRAL METHODS

The molecular formula $C_6H_5NO_3$ indicates five double-bond equivalents and the ultra-violet spectrum shows the presence of conjugated unsaturation.

The mass spectrum gives a molecular ion at m/e 139. The i.r. spectrum is of a Nujol mull, and so it must be remembered that the regions 2850–3000, 1380, and 1450 cm^{-1} are obscured. The peaks at 3090 $v_{(=C—H)}$, 1620 $v_{(C=C)}$, 1590 $v_{(C=C)}$, 855, 760, and 695 cm^{-1} (skeletal vibrations), suggest a *mono-* or *para*-disubstituted benzene (see Vol. 1, i.r. Table 5, p. 173). However, the n.m.r. spectrum immediately confirms the presence of a *p*-disubstituted benzene since the two groups of signals approximating to doublets at τ 1·8 and 2·9 arise from an *ortho*-coupled AA'XX' system. The absorption at 3330 cm^{-1} suggests the presence of an OH or NH proton, and the signal at τ 0·1 in the n.m.r. spectrum which exchanges with D_2O confirms this. The i.r. stretching vibrations in the region of 1500 cm^{-1} [also possibly due to aromatic $v_{(C=C)}$] and 1330 cm^{-1} are consistent with an aromatic nitro group, which is confirmed by the mass spectrum which shows loss of 46 (NO_2) and 30 (NO) mass units to give the peaks at m/e 93 and m/e 109. *p*-Nitrophenol with one nitrogen fits the odd mass number molecular ion. The base peak at m/e 65 corresponds to the elements of the cyclopentadienyl cation formed by loss of carbon monoxide from a phenol cation fragment.

The u.v. spectrum of *p*-nitrophenol in ethanol shows an absorption maximum at 312 nm, which undergoes a bathochromic shift on addition of alkali due to formation of the phen oxide ion.

ANSWER TO PROBLEM 3—USING A COMBINATION OF SPECTRAL METHODS

The molecular formula $C_8H_{10}O$ shows four double-bond equivalents and is in accord with the benzene chromophore shown by the u.v. spectrum. The broad absorption at 3350 cm^{-1} in the i.r. spectrum is characteristic of an intermolecular hydrogen-bonded OH group, and the presence of this is confirmed by the i.r. C=O stretching vibration at 1050 cm^{-1} and the n.m.r. signal at τ 6·85 which is removed with D_2O. The i.r. spectrum also shows the presence of an aromatic ring since peaks are present at 3030, 3070, 3090, 1610, 1500, 750, and 700 cm^{-1}. The skeletal vibrations (750, 700 cm^{-1}) indicate a mono-substituted benzene, and this is confirmed by the n.m.r. spectrum since the signal at τ 2·8 integrates to five aromatic protons (taking the peak at τ 6·85 as 1 H). The i.r. spectrum shows the presence of aliphatic C—H (2880 and 2940 cm^{-1}), and these give rise to two triplets ($J = 7$ Hz) in the n.m.r. spectrum at τ 6·3 and 7·3. This is consistent with two adjacent methylene groups of different chemical shift. From n.m.r. Table 1 (Vol. 1, p. 60) the low-field triplet is due to R—CH$_2$—OH (τ 6·44) and the high-field triplet is due to R—CH$_2$—Ph (τ 7·38). The compound is therefore 2-phenylethanol. The mass spectrum is in complete agreement with this assignment showing the molecular ion at m/e 122, and the two most intense peaks at m/e 91 and 92 due to the following fragmentation. 2-Phenylethanol is consistent with these results, and the u.v. spectrum which shows aromatic absorption at 257 nm (ε 122), as expected, resembles that of toluene.

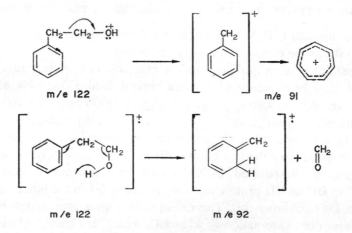

ANSWER TO PROBLEM 4—USING A COMBINATION OF SPECTRAL METHODS

The molecular formula $C_5H_{12}O$ does *not* have a double equivalent and therefore, as expected, the ultraviolet spectrum does not show absorption above 200 nm.

The i.r. spectrum (in CCl$_4$ solution) indicates a fully saturated compound since C—H stretching vibrations appear at $\sim$ 2960, 2900, and 2875 cm^{-1} and peaks around 3050 cm^{-1} are absent. The sharp peak at 3640 cm^{-1} is consistent with a free —OH stretching frequency, and the broad absorption at 3360 cm^{-1} which disappears on dilution is the intermolecular hydrogen-bonded component. These findings are confirmed by the C—O stretching frequencies at 1020 and 1055 cm^{-1} and the broad signal in the 100 MHz n.m.r. spectrum at τ 5·9 which is removed with D_2O. The mass spectrum shows a molecular ion at *m/e* 88 consistent with the saturated alcohol $C_5H_{11}OH$. In the n.m.r. spectrum nine of the protons are equivalent and give rise to a singlet at τ 9·1, the other two appear as a singlet at τ 6·8. This suggests a methylene group adjacent to both an OH group and a fully substituted carbon atom since no coupling is observed. This is confirmed by the mass spectrum with loss of —CH$_2$OH (31 mass units) from the molecular ion giving rise to a base peak at *m/e* 57.

The signal at τ 9·1 devoid of coupling is consistent with three equivalent methyl groups attached to a fully substituted carbon. The above findings are consistent with neopentyl alcohol. A more detailed consideration of the mass spectrum is given in the scheme below.

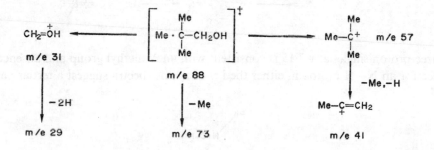

ANSWER TO PROBLEM 5—USING A COMBINATION OF SPECTRAL METHODS

The molecular formula $C_3H_5N_3O_2$ shows three double-bond equivalents and the u.v. spectrum indicates unsaturation.

The i.r. spectrum shows saturated C—H stretching frequencies just below 3000 cm^{-1} and bending frequencies at 1470 and 1372 cm^{-1}, and the n.m.r. spectrum confirms the absence of vinyl or aromatic protons. The triplet at τ 8·67 (3H, $J = 7$ Hz) and quartet τ 5·74 (2H, $J = 7$ Hz) n.m.r. signals are consistent with an ethyl group attached to oxygen (Vol. 1, n.m.r. Table 1 p. 60), and this accounts for all hydrogens in the formula. The i.r. spectrum confirms the presence of an ethoxy group by the C—O—C stretching frequency at 1250 cm^{-1}, and absorption at 1730–1760 cm^{-1} is consistent with an ester carbonyl stretching vibration. The two peaks in the i.r. triple bond region at 2140 and 2190 cm^{-1} are from an azido group, and this accounts for the three nitrogen atoms in the formula, and the compound is, in fact, ethyl azidoformate (Vol. 1, i.r. Table 6, p. 177).

The mass spectrum shows no molecular ion peak (M, 115) but a peak at *m/e* 116 is observed. Since the intensity of this peak increases at high sample pressure, a hydrogen capture process by collision must be involved.[1] Loss of the azido group gives a peak at *m/e* 73 and loss of EtO occurs to give a peak at *m/e* 70. The u.v. spectrum shows absorption at 235 nm (ε 120), which is as reported for an azide conjugated with a carbonyl group;[7] non-conjugated alkyl azides absorb at 285 nm.

ANSWER TO PROBLEM 6—USING A COMBINATION OF SPECTRAL METHODS

Six double-bond equivalents are indicated by the molecular formula $C_{14}H_{19}N$, and the molecular ion peak is at m/e 201 in the mass spectrum. The i.r. spectrum shows absorption bands just above and below 3000 cm^{-1}, which indicate the presence of saturated and unsaturated C—H stretching frequencies. The strong peaks at 1600 and 1510 cm^{-1} and skeletal vibration at 695 and 750 cm^{-1} suggest that a mono-substituted aromatic ring is present, and this is confirmed by the multiplets at τ 2·95 (2H, *meta*-protons) and 3·45 (3H, *ortho*- and *para*-protons) in the n.m.r. spectrum. Relative shielding of the *ortho*- and *para*-protons results from higher electron density at these positions and indicates that the nitrogen atom is adjacent to the aromatic ring.

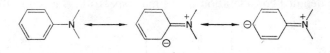

The three proton singlet at τ 7·15 is consistent with an *N*-methyl group and absence of any evidence for an N—H proton in either the i.r. or n.m.r. spectra suggest a tertiary amine of formula

The alkyl group C_7H_{11} must contain two double-bond equivalents since the saturated radical would have the formula C_7H_{15}. Reconsideration of the i.r. spectrum indicates a weak C=C stretching vibration at 1680 cm^{-1}, and since no vinyl proton signals are present in the n.m.r. spectrum, this double bond must be tetra-substituted. A broad singlet at τ 8·5 corresponding to three protons suggests the presence of a methyl group attached to a double bond. The remaining eight aliphatic protons are of similar chemical shift and no first-order splitting is observed. They do, however, form two complex signals of four protons each appearing at τ 8·0 and 8·35. The protons at τ 8·0 suggest the presence of two methylene groups each de-shielded by an adjacent double bond. Since the remaining four protons at τ 8·35 arise from saturated methylene groups, they must be located in a ring.

Taking all the above points into consideration, the group C_7H_{11} contains one double bond, one ring, one allylic methyl, and two allylic methylene groups, consistent with a methylcyclohexene structure. The compound is thus the enamine from *N*-methylaniline and 2-methylcyclohexanone.

The u.v. spectrum supports these findings since the aromatic absorption at 252 nm (ε 20,400) shows similarities to that of *N,N*-dimethylaniline (λ_{max}^{hexane} 250 nm, ε 13,750). The peak at 210 nm is a typical enamine double-bond absorption.[3] The mass spectrum fragmentation is difficult to rationalize completely, but most peaks are consistent with the following scheme:

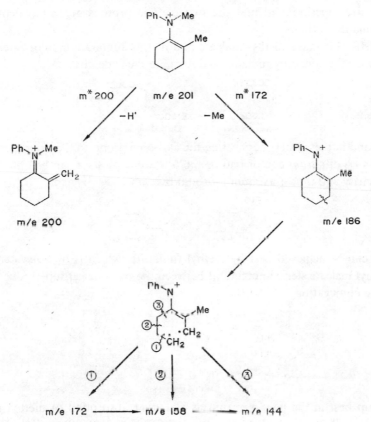

ANSWER TO PROBLEM 7—USING A COMBINATION OF SPECTRAL METHODS

The compounds A and B with molecular formulae $C_8H_{12}O_4$ have three double-bond equivalents and the two isomeric compounds both show a molecular ion peak at *m/e* 172. The i.r. spectra for both show unsaturation from the presence of peaks above 3000 cm^{-1} (A: $\nu_{(C-H)}$ 3080 cm^{-1}; B: $\nu_{(C-H)}$ 3060 cm^{-1}) and double bond stretching absorptions (A: $\nu_{(C=C)}$ 1650 cm^{-1}. B: $\nu_{(C=C)}$ 1645 cm^{-1}). The very strong peaks at 1730 cm^{-1} and in the region 1150–1300 cm^{-1}, indicate an ester group is present in both compounds; however the finger-print regions (625–1300 cm^{-1}) are different. No strong peaks are present at 1600 and 1500 cm^{-1}, which indicates the absence of an aromatic ring, and this is supported by the absence of peaks at about τ 2·0 in the n.m.r. spectrum. The n.m.r. spectrum shows a singlet τ 3·2 for A and 3·8 for B consistent with the presence of de-shielded olefinic protons. The quartet at τ 5·8 and triplet at τ 8·75 in both spectra indicates an ethyl group attached to the oxygen atom of an ester group which is confirmed by the loss of .OEt (45 mass units) and .COOEt (73 mass units) to give the major peaks in both mass spectra at *m/e* 127 and 99 respectively. Since the molecular formula contains four oxygen atoms, two ethyl ester groups are present, and thus two olefinic protons are indicated by comparison of the integral ratios in the n.m.r. spectrum. Furthermore, the signals due to each of the two ester groups are superimposed and the two olefinic protons are equivalent since only a singlet resonance occurs.

The geometrical isomers diethyl maleate and diethyl fumarate are consistent with these observations, and it now only remains to determine their identity.

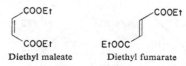

Diethyl maleate   Diethyl fumarate

The single bond in these esters connecting the carbonyl group to the double bond can take up either an s-*cis* or s-*trans* conformation and a low energy state can be obtained by maximum coplanarity and thus maximum *p*-orbital overlap.

s—trans   s—cis

Coplanarity can be achieved best by diethyl fumarate, which is of *trans* stereochemistry, since in diethyl maleate steric interactions between the two ester groups reduce coplanarity and therefore conjugation.

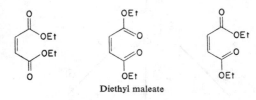

Diethyl maleate

This shows up best in the u.v. spectrum as a weaker absorption of diethyl maleate over diethyl fumarate. Thus A is diethyl fumarate (λ_{max}^{hexane} 223 nm, ε 4100) and B is diethyl maleate (λ_{max}^{hexane} 219 nm, ε 2300).

ANSWER TO PROBLEM 8—USING A COMBINATION OF SPECTRAL METHODS

The mass spectrum shows a molecular ion peak at *m/e* 182 in accord with formula $C_{10}H_{14}O_3$ and four double-bond equivalents. The i.r. spectrum shows a strong and very broad absorption at 1560 cm^{-1}, which suggests the presence of an enol chelate and is confirmed by the u.v. absorptions (231 nm, ε 10,000; 272 nm, ε 10,000) whose intensities are pH dependent.[4] The low field singlet ($-\tau$ 8·3) in the n.m.r. spectrum is from the hydrogen-bonded hydroxyl proton which can be exchanged with D$_2$O.

enol chelate

The i.r. spectrum also shows unsaturated carbonyl absorption at 1670 cm^{-1}, and since the enol-chelate i.r. absorption frequency is unusually low and the hydrogen-bonded hydroxyl proton highly de-shielded it seems this other carbonyl group, which accounts for the third oxygen in the formula, is in conjugation with the enolic system as below.

The n.m.r. spectrum shows a singlet (3H) at τ 7·48, which is consistent with a methyl group on the above system, and the two singlets (2H each) at τ 7·66 and 7·8 are from methylene groups also attached to the above system, and therefore we can write the partial structure:

Thus three double-bond equivalents have been accounted for: two belong to the keto enolate system and one to the ketone group shown in the infrared spectrum at 1670 cm^{-1}. Further n.m.r. analysis shows that the fourth double-bond equivalent is due to a ring system. The remaining peak [(6H) τ 9·1] indicates two equivalent methyl groups on a fully substituted carbon atom and the compound is actually the triketone 2-acetyldimedone.

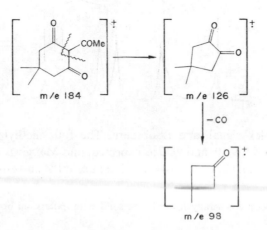

The two major fragmentation peaks in the mass spectrum are probably explained by the following scheme.

ANSWER TO PROBLEM 9 — USING A COMBINATION OF SPECTRAL METHODS

The reaction involves deformylation of the β-ketoaldehyde followed by acid catalysed cyclization to the 3-hydroxycyclohexanone as below.

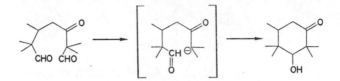

The i.r. spectrum of the product clearly shows a compound with an hydroxyl [$\nu_{(O-H)}^{KBr}$ 3410 cm^{-1}] and carbonyl [$\nu_{(C=O)}^{KBr}$ 1695 cm^{-1}]† function. The low intensity u.v. absorption confirms that the carbonyl is unconjugated (208 nm, ε 207).

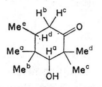

The n.m.r. spectrum shows H^a as a de-shielded singlet at τ 7·1 and the hydroxyl proton gives rise to a singlet at τ 7·6 which exchanges with D$_2$O. Protons H^b and H^c are non-equivalent and so occur at different chemical shift (τ 7·55, 8·1) and suffer geminal split-ting of 14 Hz. H^d subtends a dihedral angle of 60° with H^b and produces little coupling (approaches minimum of Karplus curve), but H^d subtends an angle of up to 180° with H^c and gives rise to a coupling of 5 Hz. This is made clearer by the diagram below.

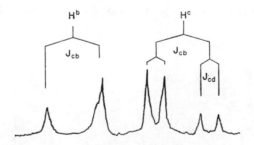

H^d would give a complex signal and is obscure. The four methyl groups Me^a to Me^d give singlets at τ 8·28, 8·95, 9·10, and 9·20 as expected, and Me^e gives a doublet ($J = 7$ Hz) at τ 9·3. The mass spectrum shows a molecular ion at m/e 184 corresponding to the correct

---

† Absorption frequencies are generally lower in the solid state (potassium bromide disc) than in solu-tion.

formula but fragmentation is difficult to rationalize. However, the molecular weight, 184, does show that the action of concentrated hydrochloric acid on 2,2,5,6,6-pentamethyl-3-oxoheptan-1,7-dial (mol. wt. 212) causes the loss of 28 mass units. Accurate mass measurement would show that this corresponds to loss of carbon monoxide and not C_2H_4, but even without this information, loss of carbon monoxide from a β-keto-aldehyde is more reasonable than loss of ethylene.

ANSWER TO PROBLEM 10 — USING A COMBINATION OF SPECTRAL METHODS

The compound is the flavone apigenin:

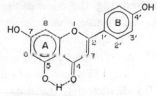

The i.r. spectrum shows a conjugated carbonyl absorption at 1660 cm^{-1}, vinyl ether absorption around 1250 cm^{-1}, and complex aromatic bands around 1600 cm^{-1}. Skeletal vibrations at 750 and 830 cm^{-1} are also present. Broad absorption between 2500 and 3400 cm^{-1} in the i.r. spectrum suggest the presence of phenolic hydroxy groups, and the presence of three are confirmed by the three peaks at τ -0.4, -0.9, and -3.1 in the n.m.r. spectrum.

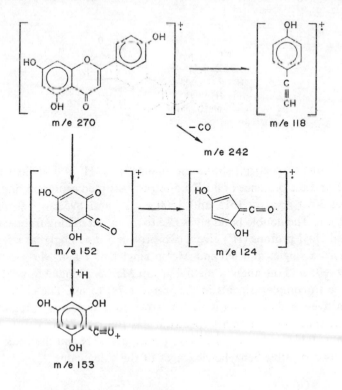

The u.v. spectrum eliminates an isoflavone and neoflavone and is consistent with a 5,7-dihydroxyflavone structure.[4] The n.m.r. spectrum shows the presence of aromatic protons of which two appear as a doublet at τ 2·0 ($J = 8$ Hz) and two as a doublet at τ 2·95 ($J = 8$ Hz). This suggests an AA′BB′ system from a *para*-substituted benzene, and this can only be obtained by the presence of an hydroxyl group at the 4′ position (ring B). The one proton singlet at τ 3·18 is from the proton at C-3, which is at lower field than in a simple unsaturated ketone ($\sim\tau$ 4·0) due to the conjugation of the 2-phenyl substituent. The other two hydroxy groups must be situated on "ring A", and since one hydroxy group gives a signal at τ $-3·1$, this must be at C-5, which is de-shielded by hydrogen bonding to the adjacent carbonyl group with the formation of a favoured six-membered ring. Insertion of the last hydroxy group on "ring A" must leave two aromatic protons which give weakly coupled resonances (τ 3·45 and 3·7) ($J = 2$ Hz) (*meta*-coupling), and so it can only be placed at the C-7 position. The mass spectrum shows a peak at m/e 242 due to loss of carbon monoxide, and a retro-Diels–Alder reaction can produce fragments at m/e 152 and m/e 118 as shown in the scheme on p. 313. The metastable peak at m^* 217·0 indicates that the peak at m/e 242 is derived by direct fragmentation of the molecular ion (calc. m^*, $242^2/270 = 217$).

ANSWER TO PROBLEM 11 — USING A COMBINATION OF SPECTRAL METHODS

The compound is the oestrone derivative:

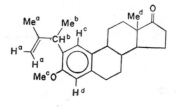

The 100 MHz n.m.r. spectrum shows the signals from H^c and H^d as singlets (1H each) at τ 3·0 and 3·5 and the presence of a 1,2,4,5-tetrasubstituted aromatic ring is also indicated by the peaks at 3090 $\nu_{(=C-H)}$ 1615 and 1500 $\nu_{(C=C)}$ and 895 cm^{-1} (skeletal vibrations) in the i.r. spectrum. The double bond gives rise to an i.r. stretching frequency at 1650 cm^{-1} and the terminal vinyl protons (H^a) give a two-proton n.m.r. singlet at τ 5·2. The methoxyl group (Me^c) gives a singlet at τ 6·3 and Me^a a singlet at τ 8·4. Me^b occurs as a doublet ($J = 7$ Hz) at τ 8·75 and the angular methyl group Me^d as a singlet at τ 9·1. The rest of the protons give rise to complex signals in the region τ 7·0 to 8·8. The carbonyl group which forms part of a five-membered ring is slightly strained and occurs at high frequency (1745 cm^{-1}) in the i.r. spectrum. The mass spectrum has the molecular ion at m/e 352 as base peak indicating a stable structure. The M-15 peak (m/e 337) and the peak at m/e 311 both result from the two possible benzylic cleavages of the side chain.

The peak at *m/e* 173 arises in one step from the peak at *m/e* 337 since a metastable peak is present at m^* 88·7 ($173^2/337 = 88·8$). The *m/e* 173 and *m/e* 187 peaks are possibly due to the species indicated in the scheme below.

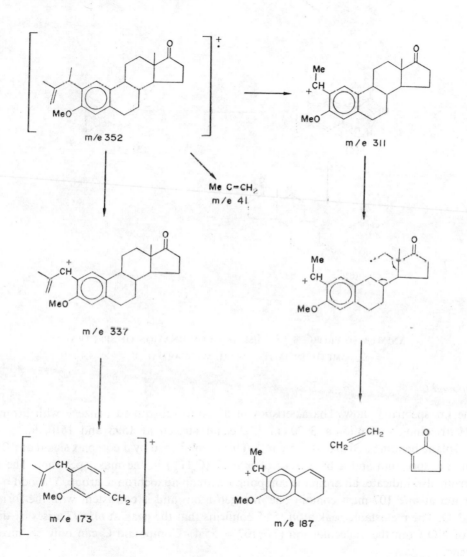

The u.v. spectrum shows phenolic absorption at 285 nm (ε 2,450) and 225 nm (ε 7,950). The compound arises from abnormal Claisen rearrangement of the 3,3-dimethylallyl oestrone ether (2).[5] The expected product (3) undergoes a [1,5]-homosigmatropic hydrogen transfer to the spiroketone (4) which by another sigmatropic hydrogen shift gives the phenol (5). The methyl ether of (5) was used for spectroscopic measurements.

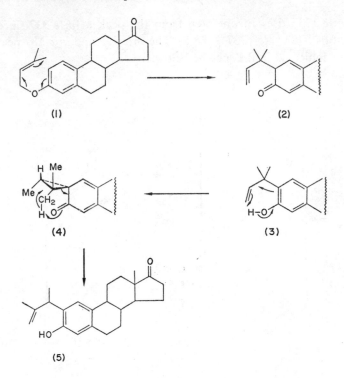

ANSWER TO PROBLEM 12 – USING A COMBINATION OF SPECTRAL
METHODS (WITH DR. H. W. WARDALE)

Compound C

The i.r. spectrum shows characteristics of a mono-substituted benzene with aromatic C—H stretching absorption at 3070 cm^{-1}, skeletal stretch at 1595 and 1510 cm^{-1}, and out-of-plane bending at 690 and 775 cm^{-1}. This is confirmed by a complex signal at τ 2·1 in the n.m.r. spectrum and a base peak of m/e 77 ($C_6H_5^+$) in the mass spectrum. The u.v. spectrum also indicates an aromatic compound with an auxochromic group. The odd molecular ion at m/e 107 must contain one nitrogen atom and is consistent with the formula C_6H_5NO. The metastable peak at m^* 55·5 confirms that the peak at m/e 77 arises by direct loss of NO from the molecular ion ($77^2/107 = 55·4$). Compound C can only be nitroso-benzene.

Compound D

The i.r. spectrum again indicates a mono-substituted benzene with absorptions at 1600, 1500, 690, 750 cm^{-1}, and, in addition, broad peaks at 3250, 3120, and 2700 cm^{-1} suggest the presence of hydrogen atoms as —X—H, where X is oxygen or nitrogen. A carboxylic acid can be ruled out since carbonyl stretching absorptions are absent. An aromatic compound is confirmed by the complex signals at τ 2·9 in the n.m.r. spectrum and a mono-substituted benzene is consistent with the base peak at m/e 77 ($C_6H_5^+$) in the mass spectrum. Exchangeable protons are also indicated by the broad peak at τ 2·3 which is removed when D_2O is added. The odd molecular ion m/e 109 is consistent with the formula C_6H_7NO and the metastable peak m^* 77·5 confirms the loss of 17 mass units (OH) to give a peak at m/e 92 ($92^2/109 = 77·6$). The only mono-substituted benzene with an hydroxyl group and consistent with the formula C_6H_7NO is phenylhydroxylamine. The u.v. absorption spectrum shows some similarities with the data for aniline, λ_{max}^{water} nm (ε) 230 (8,600), 280 (1,430). The mass spectrum shows characteristics of nitrosobenzene (m/e 107), aniline (m/e 93), and the cyclopentadienyl cation (m/e 65). Thus it appears that on electron bombardment phenylhydroxylamine gives the elements of aniline and nitrosobenzene.

Compound F

Compound F, the product of the reaction, shows the presence of a mono-substituted benzene by the i.r. spectrum absorptions at 3080, 1595, 1490, 760, and 680 cm^{-1}, the n.m.r. signals at τ 2·55 and 1·8, and the base peak in the mass spectrum at m/e 77 ($C_6H_5^+$). The u.v. spectrum is also consistent with an aromatic compound with an auxochromic group. In the mass spectrum the molecular ion gives an even peak at m/e 198, and since the product arises from phenylhydroxylamine and nitrosobenzene, it seems reasonable to assume that two nitrogen atoms and two phenyl rings are present. The formula $C_{12}H_{10}N_2O$ consistent with m/e 198 is, in fact, azoxybenzene. Loss of the oxygen atom gives a peak at m/e 182 and confirms an N-oxide. The metastable peak at m^* 146 shows that the peak at m/e 170 arises from rearrangement of azoxybenzene to possibly diphenyl ether in one step ($170^2/198 = 146·0$). The peaks at m/e 105 and m/e 91 are due to PhN_2^+ and PhN^+ respectively. The very strong peak at 1480 cm^{-1} in the i.r. spectrum is probably due to N=N stretching and N—O stretch is observed at 1305 cm^{-1}.

Analysis of Electron Spin Resonance Spectrum Problem 12 E

The most probable free radical intermediates in the reaction of nitrosobenzene and phenylhydroxylamine to give azoxybenzene in strongly basic solution are the nitrosobenzene radical anion (6), the radical anion of the nitrosobenzene dimer (7), and the radical anion of azoxybenzene (8). The e.s.r. spectrum shown in the figure 12 E differentiates between these three possibilities and favours the nitrosobenzene radical anion. The spectrum appears to contain thirty equally spaced lines with a splitting of 1 G between adjacent lines. The spectrum has a centre of symmetry. This is characteristic of radicals in solution and usually indicates that only isotropic hyperfine interactions are being observed. The relative intensity of the strongest lines to the weakest lines is about 6 to 1 and the sum of the intensities of all the lines compared with the intensity of the weakest line is approximately 100 to 1. This indicates that there are about 100 allowed nuclear hyperfine interactions in the free radical. Both (7) and (8) contain ten protons and two ^{14}N nuclei, so that if all hyperfine interactions with these nuclei were resolved we should expect $3^2 \times 2^{10} = 9216$ allowed transitions. The five protons and single ^{14}N nucleus of the nitrosobenzene radical anion (6) produce a total of $2^5 \times 3 = 96$ allowed transitions in good agreement with the observed spectrum. This is a

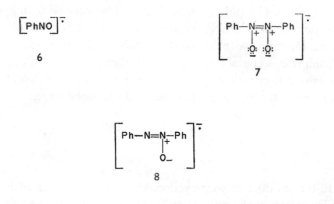

strong indication but not conclusive evidence that the spectrum is of the nitrosobenzene radical anion, for several other possible combinations of nuclei could give rise to about 100 allowed transitions. It is conceivable that not all the nuclear hyperfine interactions for both (7) and (8) are resolvable, and that those that are resolved amount to about 100 transitions. However, if we assume that resolution of the interaction with both nitrogen nuclei always occurs in these two radicals, then the closest we can get to 100 allowed transitions is 72 to 1 for resolution of three protons and two ^{14}N and 144 to 1 for resolution of two ^{14}N and four protons. Both these values are sufficiently far from the experimental value as to make these possibilities unlikely.

Can we now reconcile the splittings and intensities of the observed spectrum with our expectations for the nitrosobenzene radical anion? Clearly the observation of thirty equally spaced lines indicates that there is considerable fortuitous overlapping of transitions and that all the splittings are in relatively simple ratios. Nevertheless, it is instructive to calculate the number and relative intensities of all the lines to be expected in the absence of such overlapping and then to correlate this result with the observed spectrum. It is not unreasonable to assume that there is some equivalence among the five ring protons. The most likely situation is that the two *ortho*-protons are equivalent, the two *meta*-protons are also equivalent and that the *para*-proton is a singleton. On the basis of this equivalence the ninety-six possible transitions can be broken down into a theoretical fifty-four-line spectrum with the relative intensity of the strongest to the weakest lines of only 4 to 1 diagram. In analysis A it is assumed that the nitrogen splitting is largest, the *ortho*-proton splitting next largest, then the *para*-proton splitting, and that the *meta*-proton splitting is the smallest interaction. In analysis B it is assumed that the *para*-proton splitting is largest, the *ortho*-proton splitting next largest, then the nitrogen splitting, and that the *meta*-proton splitting is again the smallest interaction. Many other analyses are possible, but we shall soon see that only these two are of any relevance to the experimental spectrum. The overall width of the analyses and the experimental spectrum is the sum of twice the nitrogen splitting, twice the *ortho*-proton splitting, twice the *meta*-proton splitting, and the *para*-proton splitting.

The smallest resolved splitting in the thirty-line experimental spectrum is 1 G, and hence this must be the size of the smallest hyperfine interaction. We must look in the wings of the spectrum to see the exact nature of this interaction. Here we find a 1 : 2 : 1 triplet so that the smallest interaction is with two equivalent protons. The fourth line of the spectrum is also of intensity 1, so that the 3 G splitting between the first and fourth lines cannot represent the interaction with a second set of two equivalent protons but must be due either to the single *para*-proton or to the nitrogen nucleus. The experimental 1 G splitting is due either to the two *ortho*-protons or to the two *meta*-protons. Thus so far the observed spectrum agrees with both analysis A and analysis B. If the situation shown in analysis B is correct and the observed 3 G splitting is due to the nitrogen interaction, then the first and fourth lines in the experimental spectrum are the first two components of a 1 : 1 : 1 triplet and the third component should then occur as part of the seventh line of the observed spectrum. If analyses A is correct and the 3 G splitting is due to the single *para*-proton, then there will be no such contribution to the seventh line. In either case the intensities of the fifth and sixth lines should be 2 and 1 respectively. The observed intensities of these lines are 4 and 5. This can

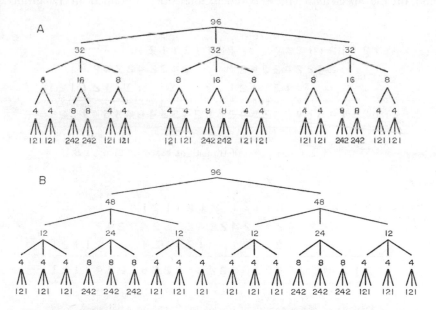

DIAGRAM 1. Hypothetical analysis A and B for splitting and intensities in the e. s. r. spectrum of the nitrosobenzene radical anion assuming there is no overlap of transitions.

only result from the addition of the first two components of the first 2:4:2 triplet precisely on top of the expected lines. The splitting, 4 G, between the outermost line and the first component of the 2 : 4 : 2 triplet, represents a second interaction with two equivalent protons.

Having now extracted the splittings for the two sets of two equivalent protons, we may use the equation for the overall width to partially resolve the ambiguity of the origin of the 3 G splitting. The overall width of the experimental spectrum is 29 G. Thus using the equality previously mentioned

$$(2\times4)+(2\times1)+(2\times{}^{14}N)+(1\times p\text{-}^{1}H) = 29.$$

If the *para*-proton splitting is 3 G, then the nitrogen splitting must be 8 G, but if the nitrogen splitting is 3 G, then the *para*-proton splitting has the improbable value of 13 G. Such a *para*-proton splitting would indicate a spin density of almost 0·6 on the adjacent carbon atom. Examination of the seventh line of the spectrum confirms that the 3 G splitting is due to the *para*-proton. This line is of intensity 2 and arises from the third component of the first 2:4:2 triplet. There is no extra intensity due to the third component of the nitrogen triplet. The two possibilities are shown below. Diagram 2 shows how the overlapping of the transitions based on analysis B occurs. The effects of the large *para*-proton doublet are not observed until the fourteenth line. Diagram 3 shows a similar overlapping based on analysis A. The second components of the nitrogen triplet begin to add in at the ninth line of the spectrum. The theoretical intensities obtained in Diagram 3 agree

```
    1 2 1 1 2 1 1 2 1         1 2 1 1 2 1 1 2 1
        2 4 2 2 4 2 2 4 2         2 4 2 2 4 2 2 4 2
            1 2 1 1 2 1 1 2 1         1 2 1 1 2 1 1 2 1
_____
1 2 1 1 4 5 3 4 6 4 3 5 4 2 3 3 2 4 5 3 4 6 4 3 5 4 1 1 2 1
```

DIAGRAM 2. Overlapping of transitions based on analysis B.

```
    1 2 1 1 2 1      1 2 1 1 2 1      1 2 1 1 2 1
        2 4 2 2 4 2      2 4 2 2 4 2      2 4 2 2 4 2
            1 2 1 1 2 1      1 2 1 1 2 1      1 2 1 1 2 1
_____
1 2 1 1 4 5 2 2 6 6 2 2 6 6 2 2 6 6 2 2 6 6 2 2 5 4 1 1 2 1
```

DIAGRAM 3. Overlapping of transitions based on Analysis A.

remarkably well with the experimental spectrum right through to the centre of symmetry between the fifteenth and sixteenth lines, and the observed intensity ratio of 6:1 for the strongest to the weakest line is also satisfied. Only further experiments can make our analysis more complete. We have extracted one nitrogen and three proton coupling constants and accounted for the observation of thirty equally spaced lines with the correct relative intensities. Diagram 4 below shows the full relationship between the hypothetical analysis A and the observed spectrum.

Selected deuterium substitution has been used to confirm that the smallest splitting is due to the two equivalent *meta*-protons. Other results of these experiments are somewhat more

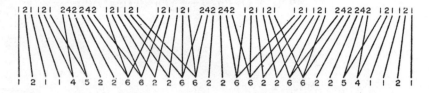

DIAGRAM 4. Relationship between hypothetical analysis A and the observed e. s. r. spectrum of the nitrosobenzene radical anion.

surprising. The 3 G splitting is actually due to one of the *ortho*-protons and the 4 G splitting is the result of the fortuitous equivalence of the other *ortho*-proton with the *para*-proton. The inequivalence of the two *ortho*-protons suggests that the nitroso group does not lie in the plane perpendicular to the ring passing through carbons 1 and 4 but is pushed to one side by a lone pair on the nitrogen atom (9). The π-bonding between the nitroso group and

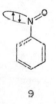

9

the ring carbons is sufficiently strong to prevent rapid free rotation about the carbon to nitrogen bond. A similar situation, observed in aromatic aldehyde and ketone radical anions, was discussed in section 2 (p. 178) of the chapter devoted to electron spin resonance spectroscopy.

In summary, the e.s.r. spectrum may be analysed in terms of four coupling constants:

(i) $a_{iso}(^{14}N) = 8$ G;
(ii) $a_{iso}(^{1}H) = 4$ G (two equivalent);
(iii) $a_{iso}(^{1}H) = 3$ G;
(iv) $a_{iso}(^{1}H) = 1$ G (two equivalent).

These values suggest that the radical intermediate present in the reaction between nitroso-benzene and phenylhydroxylamine in strongly basic solutions is the nitrosobenzene radical anion. The mechanism for the reaction is shown in the following scheme.[6]

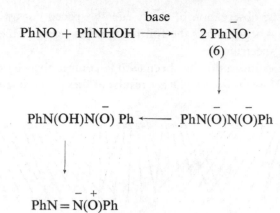

$$\text{PhNO} + \text{PhNHOH} \xrightarrow{\text{base}} 2 \text{ PhNO}^{\cdot -}$$

(6)

$$\text{PhN(OH)N(\overline{O}) Ph} \longleftarrow \text{PhN(\overline{O})N(\overline{O})Ph}$$

$$\text{PhN} = \overset{+}{\underset{}{\text{N}}}(\text{O})\text{Ph}$$

References

1. K. BIEMANN, *Mass Spectrometry, Organic Chemical Applications*, McGraw-Hill, New York, 1962, p. 46.
2. Cf. YU. N. SHEINKER, *Dokl. Akad. Nauk SSSR*, 1951, **79**, 1043; *Chem. Abs.*, 1951, **45**, 6927c.
3. G. OPITZ, H. HELLMANN, and H. W. SCHUBERT, *Annalen.*, 1959, **623**, 112.
4. A. I. SCOTT, *Interpretation of the Ultraviolet Spectra of Natural Products*, Pergamon Press, Oxford, 1964, p. 37.
5. A. JEFFERSON and F. SCHEINMANN, *J. Chem. Soc.* (C), 1969, 243.
6. G. A. RUSSELL, *Science*, 1968, **161**, 433.

Documentation of Molecular Spectra

R. W. A. OLIVER

Department of Chemistry and Applied Chemistry, University of Salford

THE previous chapters in this monograph have given an outline of the theory, practice, and some of the typical applications of spectroscopy to organic chemistry. If the organic chemist is to utilize fully the wealth of spectroscopic data which his experiments will generate, he must devise an efficient documentation system. Such a system must provide both for the storage and retrieval of individual sets of data and also for a system of group classification so that related spectra may be compared.

The ideal system would be storage and searching by computer on a national or international scale. However, such systems are extremely expensive to set up and maintain, so that to date only a few companies offer such facilities to their chemists, and it is therefore necessary for the individual to devise and maintain an efficient documentation system of his own. It is the purpose of the present author to advise on this problem and to draw attention to the existence of the generally available aids on which an individual system should be based. Before proceeding to this discussion, a note of caution concerning the reliability of spectral data is in order.

With the increasing cost and complexity of spectroscopic instruments it is becoming more widespread for departments to employ special technicians to operate them and to institute a spectroscopic service for the chemist rather than allow him to make his own measurements. This system has many advantages, but unless care is taken in the supervision of the service by qualified spectroscopists, incorrect data may be generated. Further, due to the limitations of instrument design, it is still unfortunately possible, even with skilled operators, to find that spectral measurements of the same substance are not identical even though made on two similar machines and using the same sampling techniques. One should, therefore, become acquainted thoroughly with the limitations as well as the capabilities of the instruments providing the data. The best way to obtain this knowledge is to read carefully the manufacturer's specifications which are to be found in the instrument manuals. Whilst good spectroscopic service departments provide estimates of the accuracy of their measurements, others do not. In a personal documentation system, the first set of data to be stored and indexed must be the instrument calibration curves obtained on agreed standards under the conditions to be applied to future work. In the author's own research group the spectroscopic standards given in Table 1 are used.

TABLE 1
SPECTROSCOPIC STANDARDS

Ultra violet and visible spectra	Wavelength scale	Glass filters provided by the instrument makers
	Absorbance scale	Alkaline potassium chromate solutions (*J.Res.Nat.Bur.Stand.*, **48**, 414, 1952)
Infrared spectra	Wave number, or wavelength scale	Low resolution work; polystyrene film, provided by the instrument makers; high resolution work; various gases (*Calibration Tables for I.R. Spectrometers*, IUPAC, Butterworths, London)
	Transmission scale	No agreed standard; generally use the strongest band as internal standard
Proton magnetic resonance spectra	Magnetic field scale	A60: use tetramethyl silane standard (*Anal.Chem.*, **35**, 1985, 1963)
	Frequency scale	HA100: use tetramethyl silane in conjunction with frequency counter
Electron spin resonance spectra	Frequency	Electronic counter as provided by maker
	Field	PMR probe in conjunction with electronic counter; alternatively use standard, e.g. charred dextrose or DPPH
	Concentration	Comparison with standard sample, e.g irradiated glycine, galvinoxyl
Mass spectra	*m/e* scale	Low *m/e* values and low-resolution work; atmospheric O_2 and N_2 used; high resolution work; fluorokerosene standards provided by instrument makers
	Intensity scale	No agreed standard; generally use the strongest ion peak as the internal standard

For those chemists who have to make their own spectral measurements, the following texts giving practical details and tips on techniques and sources of error are recommended:

D. J. PASTO and C. R. JOHNSON, *Organic Structure Determination*, Prentice-Hall, 1969. This book contains information on the preparation and handling of samples for the various spectral measurements.

G. H. BEAVEN, E. A. JOHNSON, H. A. WILLIS, and R. G. MILLER, *Molecular Spectroscopy: Methods and Applications in Chemistry*, Heywood, 1962. This book gives an excellent introduction to practical ultraviolet, visible, and infrared spectroscopy.

A. D. CROSS, *Introduction to Practical Infrared Spectroscopy*, Butterworths, 1969. This text contains a set of standard reference spectra suitable for checking infrared instruments. Reference spectra are also given by G. EGLINTON in Volume 1, p. 123.

R. H. BIBLE, *The Interpretation of NMR Spectra: An Empirical Approach*, Plenum Press, 1965. This book contains one chapter on practical methods and an appendix giving detailed instructions for the use of a Varian A60 spectrometer.

R. S. ALGER, *Electron Paramagnetic Resonance: Methods and Applications*, Wiley–Interscience, 1968.

To summarize the comments made in this section, care must be taken to ensure the validity of one's spectroscopic data; there is little point in attempting to interpret and store incorrect data.

Since the majority of spectrometers yield spectra in graphical form of various sizes (mostly large), the second task of the experimenter is to reduce the physical bulk of the data. Currently there are two methods of doing this: either one may digitize all or some of the peaks in the spectra manually or a slave recorder may be coupled to the output device of the spectrometer so as to reduce the spectra to a more convenient size, e.g. a file card. If the first method is adopted, it should be realized that one is likely to be discarding some of the available information since it is difficult to digitize the shape of absorption bands; if the second method is used, some digitization will have to be done in order to retrieve the spectra efficiently. Because of the very great differences between the mode of presentation of the same spectra from different instruments and because of the differences between the various kinds of spectra, it is not possible to recommend a uniform treatment for this initial data reduction process. Rather, the author recommends that the reader adopts an identical data reduction process for his own spectra to that found in the most suitable commercially available collection of similar spectra. The advantages of following this recommendation are that one reaps the benefit of the great care and thought which has been put into these published collections and that one's own collection may be easily compared with the larger collection. The problem which faces the reader is therefore one of discovering the existence and location of a suitable collection of like spectra *before* attempting the documentation process. In order to solve this problem recourse should be made to Tables 2–7 which contain summaries of generally available collections of molecular spectra. These tables have been designed to be self-explanatory, but the final column has been left blank so that it may be completed by the user.[1]

Should the reader be unable to find any collection of spectra which match his own from the 225,000 spectra summarized in Tables 2–7, recourse should be had to the original literature. A tremendous amount of work has been performed by a number of authors in order to assist in this searching process, and summaries of the various bibliographies of molecular spectral data which contain 297,000 literature references are given in Tables 8–11. Again, these tables are devised to be self-explanatory in use, and the final column is left blank so that the reader may complete it once a convenient location of the required bibliography has been found.

In conclusion, the need for providing a suitable index for one's collection of molecular spectra cannot be too strongly recommended. There is no doubt that the key index is the molecular formula index, since not only is this simple to prepare but also easy to use. It is again recommended that the user should model his index on that of the published collection which contains the most similar spectra. It should be emphasized that the time spent in data reduction and classification and indexing of one's own spectral data, is time well spent, since a personal collection will in time become the most valuable of all.

Finally the author would like to acknowledge the help of Mrs. M. I. Lomax of the library of this University, in the compilation of tables.

Reference

1. R. W. A. OLIVER and M. I. LOMAX, *A Guide to the Published Collections and Bibliographies of Molecular Spectra*. Perkin Elmer Ltd., Beaconsfield, Bucks, 1971.

TABLE 2

SUMMARY OF THE PUBLISHED COLLECTIONS OF ELECTRONIC ABSORPTION SPECTRA

Ref. No.	Name and format of collection	Number of spectra	Type of compound	Molecular formula index	Alphabetical name index	Other indexes	Location
1	Sadtler spectra collections; large quarto loose leaf sheets						
	(a) Standard ultraviolet spectra: Vols. 1–60	28,000	Organic compounds	Yes	Yes		
	(b) Dyes, pigments, and stains	1,600	Dyes, pigments, and stains	Yes	Yes	Ultraviolet locator available for these separate collections	
	(c) Pharmaceuticals	1,500	Pharmaceuticals	Yes	Yes		
	(d) Biochemicals	650	Biochemicals	Yes	Yes		
	(e) Agricultural chemicals	300	Agricultural chemicals	Yes	Yes		
2	Lang spectra collection; quarto loose-leaf sheets; Vols. I–XVII	3,000	Organic, biochemical inorganic, and organometallic compounds	Yes	Yes	None	
3	API/TRC collection C; Quarto loose-leaf sheets	1,280	Hydrocarbons, organic and, some inorganic compounds	Yes	Yes	None	
4	Mabry collection; bound monograph	1,200	Flavonoids	Yes	Yes	None	
5	DMS ultraviolet atlas; folio loose-leaf sheets; Vols. I–V	1,000	Organic, biochemical, and inorganic compounds; solvents and spectroscopic standards	Yes	No	(i) Chromophoric group index. (ii) BPPS numerical code: ref. 5a	
6	Bolshakov collection; bound monograph	811	N, O₂, S, Si, P₄ heterocyclic compounds	No	No	Chemical class	

(Continued overleaf)

TABLE 2 (*continued*)

Ref. No.	Name and format of collection	Number of spectra	Type of compound	Molecular formula index	Alphabetical name index	Other indexes	Location
7	Holubek and Strouf collection; quarto loose-leaf sheets: Vols. I–III	600	Alkaloids	No	Yes	Botanic species index	
8	Friedel and Orchin collection; quarto loose-leaf sheets: Vol. I	579	Aromatic compounds	Yes	Yes	None	
9	N. Neuss collection: Vols. I, II, and supplements	350	Indoles and dihydroindole alkaloids	No	Yes	None	
10	M. M. Kusakov collection; bound monograph.	268	Aromatic hydrocarbons	No	No	Chemical class	
11	Fikhtengal'ts collection; bound monograph	141	Elastomers and rubber	No	No	Chemical class	
12	Venkstern and Baev collection	65	Nucleic acid compounds	No	No	General index	
13	Yamaguchi collection: Vol. I	Not given	Natural products	No	Yes	Botanic species index	

References

1. The complete Sadtler ultraviolet spectral collections are available from Heyden & Son Ltd., London. Published periodically.
2. L. LANG, *Absorption Spectra in the Ultraviolet and Visible Region*, Akadémiai Kiadió, Budapest and Academic Press, New York. Published periodically.
3. *API/TRC Collection C. Ultraviolet Spectral Data*, prepared and published by the Thermodynamics Research Center, Texas A & M University, in collaboration with the American Petroleum Institute. Published periodically.
4. T. J. MABRY, K. R. MARKHAM, and M. B. THOMAS, *The Systematic Identification of Flavonoids*, Springer, 1970.
5. *Documentation of Molecular Spectroscopy. Ultraviolet Atlas of Organic Compounds*, Butterworths, London.
5a. *Angewandte Chemie (International Edition)*, (1965) **4**, 516–18.
6. G. F. BOLSHAKOV, *Ultraviolet Spectra of Hetero-organic Compounds*. Khemia Publishing Co., Moscow, 1969.
7. J. HOLUBEK and O. STROUF, *Spectral Data and Physical Constants of Alkaloids*, Czechoslovak Academy of Sciences, Prague, and Heyden & Son Ltd., London. Published periodically.
8. R. A. FRIEDEL and M. ORCHIN, *Ultraviolet Spectra of Aromatic Compounds*, Wiley, New York, and Chapman & Hall, London, 1951.
9. N. NEUSS, *Physical Data of Indole and Dihydro Indole Alkaloids*, Available from Lilly Research Laboratories, Eli Lilly & Co., Indianapolis 6, Indiana, USA. Published periodically.
10. M. M. KUSAKOV, N. A. SIMANKO, and M. V. SIMINRA, *Ultraviolet spectra of aromatic hydrocarbons*, Akad. Nauk USSR Moscow, 1963.
11. V. S. FIKHTENGAL'TS, R. V. ZOLOTAREVA, and YU. A. LVOV, *Ultraviolet Spectra of Elastomers and Rubber Chemicals* Plenum Press, Data Division, New York, 1966.
12. T. V. VENKSTERN and A. A. BAEV, *Spectra of Nucleic Acid Compounds*, IFI/Plenum Data Corp., New York, 1968.
13. K. YAMAGUCHI, *Spectra Data of Natural Products*, Elsevier, 1970.

TABLE 3

SUMMARY OF THE PUBLISHED MAJOR COLLECTIONS OF INFRARED ABSORPTION SPECTRA

Ref. No.	Name and format of collection	Number of spectra	Type of compound	Molecular formula index	Alphabetical name index	Other indexes	Location
1	Sadtler spectra collections; large quarto loose-leaf sheets:						
	(a) Standard prism infrared spectra: Vols. 1–39	39,000	Simple aliphatic and heterocyclic compounds	Yes	Yes	Spec-finder	
	(b) Standard grating infrared spectra: Vols. 1–19	19,000	Simple aliphatic and heterocyclic compounds	Yes	Yes	Spec-finder	
	(c) Coblentz Soc. spectra: Vols. 1–6	7,000	Organic	Yes	Yes	Chemical class	
	(d) Biochemicals	2,000	Biochemicals	Yes	Yes	Spec-finder	
	(e) Inorganic	1,000	Inorganics	Yes	Yes	None	
	(f) Organometallics	900	Organometallics	Yes	Yes	None	
	(g) Pharmaceuticals	850	Pharmaceuticals	Yes	Yes	Spec-finder	
	(h) Steroids	750	Steroids	Yes	Yes	None	
2	Sadtler commercial spectra collections; large quarto loose-leaf sheets:						
	(a) Prism spectra; 22 different collections; one composite index	21,280	Includes monomers and polymers, surface active agents, dyes, pigments and stains, fats and waxes, solvents	Yes	Yes	Spec-finder	
	(b) Grating spectra; 18 different collections; one composite index	12,700	Includes monomers and polymers, surface active agents, dyes, pigments and stains, fats and waxes, solvents	Yes	Yes	None	
3	DMS collection; rim-punched cards, $3\frac{3}{4}'' \times 8\frac{1}{4}''$ in box files	16,000	Organic compounds	Yes	Yes	Can be searched mechanically	

4	Aldrich Library of infrared spectra; bound monograph	8,000	Hydrocarbons, heterocyclic, and organometallic compounds	Yes	No	Chemical class
5	API/TRC collection B; quarto loose-leaf sheets	3,416	Hydrocarbons, organic, and some inorganic compounds	Yes	Yes	None
6	NRC-NBS collection; quarto loose-leaf sheets	2,500	Organic compounds	Yes	Yes	None
7	Mecke and Langenbucher collection; lose-leaf sheets; Vols. 1–8	1,880	Organic compounds	Yes	Yes	Chemical class

References

1, 2. The complete Sadtler infrared spectra collections are available from Heyden & Son Ltd., London. Published periodically.

3. *Documentation of Molecular Spectroscopy. Infrared Absorption Spectra*, Butterworths, London. Published periodically.

4. C. J. POUCHERT, *The Aldrich Library of Infrared Spectra*, Aldrich Chemical Co., Inc., Milwaukee, 1970.

5. *API/TRC Collection B. Selected Infrared Spectral Data*, prepared and published by the Thermodynamics Research Center, Texas A & M University, in collaboration with the American Petroleum Institute. Published periodically.

6. *TRC/NBS Collection of Infrared Spectral Data, 1952–1960*, National Bureau of Standards, Washington, USA.

7. R. MECKE and F. LANGENBUCHER, *Infrared Spectra of Selected Chemical Compounds*, Heyden & Son Ltd, London, 1965.

TABLE 4

SUMMARY OF THE PUBLISHED MINOR COLLECTIONS OF INFRARED ABSORPTION SPECTRA

Ref. No.	Name and format of collection	Number of spectra	Type of compound	Molecular formula index	Alphabetical name index	Other indexes	Location
1	Hummel and Scholl Atlas; bound monograph	1,758	Polymers, resins, and additives	No	Yes	Chemical class index (pp. 5-7)	
2	Bentley, Smithson, and Rozek collection; quarto bound monograph	1,566 Far infrared spectra	Organic, organometallic, and inorganic compounds	Yes	No	Chemical class index	
3	Neudert and Ropke, Atlas of steroid spectra; bound monograph	900	Steroids	Yes	Yes	Chemical structure index devised by authors (see p. 5); coding given for a peek-a-boo search system	
4	Dobriner, Katzellenbogen, Roberts, Gallagher and Jones, Atlas of steroid spectra; bound monographs: Vol. 1, 1953 Vol. 2, 1958	309 451	Steroids Steroids	No No	Yes Yes	None Functional group index	
5	Holubek and Strouf collection; quarto loose-leaf sheets: Vols. I-III	600	Alkaloids	No	Yes	Botanic species index	
6	Hummel, Atlas of surface active agents; bound monograph	466	Surface active agents	No	Yes	None	

	Collection	No.	Compound type			Index
7	Pliva collection; quarto loose-leaf sheets: Vol. I Vol. II	288 221	Sesquiterpenes Monoterpenes	No	Yes	Chemical group
8	Neus collection; quarto loose-leaf sheets: Vol. 1, 1961 Vol. 2, 1964 Supplement 1966 Supplement 1968	151 155 35 17	Indoles and dihydro indole alkaloids	No	Yes	None
9	Maenke collection: loose-leaf sheets	355	Minerals	No	Yes	Chemical group index
10	Bolshakov collection; bound monograph	344	Hetero-organic compounds	No	No	Chemical group index
11	Velti collection; bound monograph	300	Volatile organic compounds	Yes	Yes	Includes index to 2000 ASTM reference
12	Haslam and Willis collection; bound monograph	300	Plastics and resins	No	Yes	None
13	Bellanato and Hildago collection; bound monograph	214	Essential oils	No	Yes	Chemical class index
14	Chicago Society for Paint Technology collection; bound monograph	195	Paint, coatings, and pigments	No	Yes	Chemical class index
15	Nyquist collection: bound monograph	125	Plastics and resins	No	No	Chemical class index
16	Yamaguchi collection: Vol. 1	Not given	Natural products	No	Yes	Botanic species index

References

1. D. O. HUMMEL and F. SCHOLL, *Infrared Analysis of Polymers, Resins and Additives: An Atlas*, Vol. 1, Part 1, Text; Part 2 Spectra, Wiley–Interscience, 1969.
2. F. F. BENTLEY, L. D. SMITHSON, and A. L. ROZEK, *Infrared Spectra and Characteristic Frequencies 700–300* cm^{-1}, Wiley–Interscience, 1968.
3. W. NEUDERT and H. ROPKE, *Atlas of Steroid Spectra*, translated by J. B. LEANE, Springer-Verlag, 1965.
4. K. DOBRINER, E. R. KATZELLENBOGEN, and R. N. JONES, *An Atlas of Infrared Spectra Steroids*, Vol. 1. 1953, Vol. 2, Wiley–Interscience, 1958.
5. J. HOLUBEK and O. STROUF, *Spectral Data and Physical Constants of Alkaloids*, Czechoslovak Academy of Sciences, Prague, and Heyden & Sons Ltd., London. Published periodically.
6. D. O. HUMMEL, *Identification and Analysis of Surface Active Agents by Infrared and Chemical Methods*, Vol. 1, Part 1, Text and Alphabetical Name Index; Part 2, Atlas of Infrared Spectra, Wiley–Interscience, 1962.
7. J. PLIVA (ed.), *Die Terpene, Sammlung der Spektren und physikalischen Konstanten*, Vol. I, 1960; Vol. II, 1963, Akademie Verlag, Berlin.
8. N. NEUSS, *Physical Data of Indole and Dihydroindole Alkaloids*, Available from Lilly Research Laboratories, Eli Lilly & Co., Indianapolis 6, Indiana, USA. Published periodically.
9. H. MOENKE, *Spektralanalyse von Mineralien und Gesteiner: eine Anleitung zur Emission und Absorptionspektroskopie*, Akademie Verlag, Berlin, 1962.
10. G. F. BOLSHAKOV, E. A. GLEBOVSKAIA, and Z. G. KAPLAN, *Infrared-Spectra and X-ray Diffraction of Hetero-organic Compounds*, Khemia Publishing Co., Moscow, 1967.
11. D. WELTI, *Infrared Vapour Spectra. Group Frequency Correlations, Sample Handling and the Examination of Gas Chromatographic Fractions*, Heyden & Son Ltd., London, 1970.
12. J. HASLAM and H. A. WILLIS, *Identification and Analysis of Plastics*, Iliffe Books Ltd., London, van Nostrand, New Jersey, 1965.
13. J. BELLANATO and A. HILDAGO, *Infrared Analysis of Essential Oils*, Heyden & Son Ltd., London, 1970.
14. CHICAGO SOCIETY FOR PAINT TECHNOLOGY. *Infrared Spectroscopy*, 1961.
15. R. A. NYQUIST, *Infrared Spectra of Plastics and Resins*, 2nd edn. Dow Chemical Company, Midland, Michigan, USA, 1961.
16. K. YAMAGUCHI, *Spectral Data of Natural Products*, Elsevier, 1970.

TABLE 5

SUMMARY OF THE PUBLISHED DIGITIZED COLLECTIONS OF ELECTRON IMPACT MASS SPECTRA

Ref. No.	Name and format of collection	Number of spectra	Type of compound	Molecular formula index	Alphabetical name index	Other indexes	Location
1	ICI/MSDC eight-peak index of mass spectra	17,124	Organic compounds	Yes	No	Eight most abundant	
2	ASTM index of mass spectral data (AMD 11); bound monograph, 1969	8,000	All classes of compounds	Yes	No	Molecular ion; six strongest fragment ion peaks	
3	Stenhagen atlas of mass spectral data: bound monographs: Vols. I–III, 1969	6,000	Organic compounds	Yes: exact molecular weight index	No	All; five strongest ion peaks abstracted	
4	Cornu and Massot, Compilation of mass spectral data: bound monographs: Main Vol. 1966 / Supplement 1967	5,000 / 1,000	All classes of compounds	Yes	No	Molecular ion; ten/strongest fragment ion peaks	
5	MSDC collection A4; loose-leaf sheets, 1970	3,000	All classes of compounds	No: Molecular weight index	No	All	
6	API/TRC collection E; quarto loose-leaf sheets.	2,784	Hydrocarbons and related organic compounds	Yes	No	All	
7	Yamaguchi collection: Vol. 1	Not given	Natural products	No	Yes	All or the most prominent ones	

References

1. *Eight-peak Index of Mass Spectra*, compiled by ICI Ltd. (Dyestuffs Division) in collaboration with the Mass Spectrometry Data Centre, HMSO, 1970.
2. AMERICAN SOCIETY FOR TESTING AND MATERIALS, *Index of Mass Spectral Data (AMD 11)*, *1969;* available from Heyden & Son Ltd., London.
3. E. STENHAGEN, S. ABRAHAMSSON, and F. W. MCLAFFERTY, *Atlas of Mass Spectral Data*, Vols. I – III, 1969, Interscience.
4. A. CORNU and R. MASSOT, *Compilation of Mass Spectral Data*, main volume, 1966; first supplement 1967; second supplement in press. Presses Uniersitaires de France; available from Heyden & Son Ltd., London.
5. *Mass Spectrometry Data Centre Collection*, published periodically by HMSO; available directly from the Data Centre at AWRE, Aldermaston, Berkshire.
6. *API/TRC Collection E. Mass Spectral Data*, prepared and published by the Thermodynamics Research Center, Texas A & M University, in collaboration with the American Petroleum Institute. Published periodically.
7. K. YAMAGUCHI, *Spectral Data of Natural Products*, Elsevier, 1970.

TABLE 6

SUMMARY OF THE PUBLISHED COLLECTIONS OF NUCLEAR MAGNETIC RESONANCE SPECTRA

Ref. No.	Name and format of collection	Number of spectra	Type of compound	Molecular formula index	Alphabetical name index	Other indexes	Location
1	Sadtler standard n.m.r. collections; large quarto loose-leaf sheets: (a) 60 MHz spectra (b) 100 MHz spectra	10,000 300	Organic compounds	Yes	Yes	Chemical shift index	
2	Saciler commercial n.m.r. collection; large quarto loose-leaf sheets: (i) Lubricants (60 MHz) (ii) Polymers (60 MHz) (iii) Polymers (100 MHz) (iv) Pharmaceuticals (100 MHz) (v) Plasticizers (60 MHz) (vi) Solvents (60 MHz) (vii) Surface active agents (60 MHz)	300 300 300 300 300 300 300	Lubricants Polymers Polymers Pharmaceuticals Plasticizers Solvents Surface active agents	Yes	Yes	Chemical shift index	
3	Bovey n.m.r. data tables; bound monograph	4,230	Organic and biochemical compounds	Yes	No	None	
4	Brügel collection; bound monograph	over 3,000	Organic, fluorine, boron and phosphorus compounds	No	Yes	None	
5	API/TRC collection F; quarto loose-leaf sheets: 40 MHz, 60 MHz, and 100 MHz spectra	1,592	Hydrocarbons and related organic compounds	Yes	Yes	None	

(Continued overleaf)

TABLE 6 (*continued*)

Ref. No.	Name and format of collection	Number of spectra	Type of compound	Molecular formula index	Alphabetical name index	Other indexes	Location
6	Varian n.m.r. spectra catalogues; large quarto loose-leaf sheets: 2 vols. 60 MHz and 100 MHz spectra	700	Organic compounds	No	Yes	Chemical shift and functional group indexes	
7	JEOL n.m.r. spectra collection; large quarto loose-leaf sheets: 60 MHz and 100 MHz spectra	225	Organic and fluoro compounds, steroids, Polymers	Yes	Yes	Chemical shift and chemical class	
8	Mabry collection; bound monograph	128	Flavonoids	Yes	Yes	None	
9	Neudert and Ropke, Atlas of steroid spectra; bound monograph	95	Steroids	No	No	None	
10	Yamaguchi collection: Vol. 1	Not given	Natural products	No	No	Botanic species index	

References

1, 2. The complete Sadtler nuclear magnetic resonance spectra collections are available from Heyden & Son Ltd., London. Published periodically.

3. F. A. BOVEY, *N.M.R. Data Tables for Organic Compounds*, Vol. 1, Abstracts data from the literature up to and including 1962, Wiley–Interscience, 1967.

4. W. BRÜGEL, *Nuclear Magnetic Resonance Spectra and Chemical Structure*, Vol. 1, *The Spectral N.M.R. Parameters of Compounds with Analysed Spectra*, translated by P. Haugh, Academic Press, 1967.

5. *API/TRC Collection F. Nuclear Magnetic Resonance Spectral Data*, prepared and published by the Thermodynamics Research Center, Texas A & M University, in collaboration with the American Petroleum Institute. Published periodically.

6. *Varian N. M. R. Spectra Catalogues*, Vol. 1, by N. S. BHACCA, L. F. JOHNSON, and J. N. SHOLLERY; Vol. II by N. S. BHACCA, D. P. HOLLIS, L. F. JOHNSON, and E. A. PIER, Published by Varian Associates.

7. *JEOL. Nuclear Magnetic Resonance Spectra*, Japan Electron Optics Laboratory collection, indexed and edited by Sadtler Research Laboratories, Heyden & Son Ltd., London, 1967.

8. T. J. MABRY, K. R. MARKHAM, and M. B. THOMAS, *The Systematic Identification of Flavonoids*, Springer, 1970.

9. W. NEUDERT and H. ROPKE, *Atlas of Steroid Spectra*, Chapter 3 by P. HIGHAM and J. B. LEANE. Springer-Verlag, Berlin, 1965.

10. K. Y. YAMAGUCHI, *Spectral Data of Natural Products*, Elsevier, 1970.

TABLE 7

SUMMARY OF THE PUBLISHED COLLECTIONS OF ELECTRON SPIN RESONANCE SPECTRAL DATA

Ref. No.	Name and Format	Number of spectra	Radical formula index	Alphabetical name index	Other indexes	Location
1	Koenig collection, Magnetic properties of co-ordination and organometallic metal compounds; bound monograph	Approx. 800	Yes, within a chemical class index	No	None	
2	Fischer collection, Magnetic properties of free radicals; bound monograph	Approx. 600	Yes, within a chemical class index	No	None	
3	Bielski and Gebicki, Atlas of e.s.r. spectra; bound monograph	600	No	Yes	Author index	

References

1. E. KOENIG, *Magnetic Properties of Co-ordination and Organometallic Metal Compounds*, Landolt-Börnstein, Numerical Data and Functional Relationships in Science and Technology, New Series, Group II, Vol. 2, Springer-Verlag, Berlin, 1966.
2. H. FISCHER, *Magnetic Properties of Free Radicals*, Landolt-Börnstein, Numerical Data and Functional Relationships in Science and Technology, New Series, Group II, Vol. 1, Springer-Verlag, Berlin, 1965.
3. B. H. J. BIELSKI and J. M. GEBICKI, *Atlas of Electron Spin Resonance Spectra*, Academic Press, New York, 1967.

TABLE 8

SUMMARY OF THE PUBLISHED BIBLIOGRAPHIES OF ELECTRONIC SPECTRAL DATA

Ref. No.	Name and date of period covered	Number of spectra or entries	Type of compound	Molecular formula index	Alphabetical name index	Other indexes	Location
1	Organic electronic spectral data: Vol. I 1946–52 Vol. II 1953–5 Vol. III 1956–7 Vol. IV 1958–9 Vol. V 1960–61	λ_{max} and log ε_{max} given for 100,000 spectra	Organic compounds	Yes	No	None	
2	Ultraviolet and visible absorption spectra, H. M. Hershenson: Index 1930–54 Index 1955–9 Index 1960–3	73,000 literature references	All classes of compound	No	Yes	None	
3	ASTM, Molecular formula list of compounds: names and references to published ultraviolet and visible spectra (AMD 41)	Literature references given for spectra of 25,000 compounds	All classes of compound	Yes	No	None	

4	Handbook of ultraviolet and visible absorption spectra of organic compounds, Kenzo Hirayama	λ_{max} and log ε_{max} given for 8443 compounds	Organic and organo-metallic compounds	No	No	(i) Chemical structure index based on chromophoric group devised by the author (see p. 4) (ii) Index of λ_{max} leading to chromophoric systems
5	Atlas of steroid spectra, W. Neudert and H. Ropke	λ_{max} and ε_{max} given for 900 steroids	Steroids	Yes	Yes	Spectral type code devised by authors
6	International tables of selected constants, Vol. 15	λ_{max} and log ε_{max} given for 195 compounds	Sesquiterpenoids	No	Yes	Author index

References

1. *Organic Electronic Spectral Data: Catalogue of Ultraviolet and Visible Spectra*, Interscience, Vol. I (edited by J. M. Kamlet), 1961; Vol. II (edited by H. E. Ugnade), 1961; Vol. III, (edited by C. H. Wheeler and L. A. Kaplan), 1966; Vol. IV (edited by J. Phillips and F. C. Nachod, 1963; Vol. V (edited by R. E. Lyle and P. R. Jones), 1969.

2. H. M. HERSHENSON, *Ultraviolet and Visible Absorption Spectra*, Academic Press, 1956, 1961, and 1966.

3. AMERICAN SOCIETY for TESTING and MATERIALS, *Molecular Formula List of Compounds, Names, and References to Published Ultraviolet and Visible Spectra*, to be published late 1970. Available from Heyden and Son Ltd., London.

4. KENZO HIRAYAMA, *Handbook of Ultraviolet and Visible Absorption Spectra of Organic Compounds*, Plenum Press, 1967.

5. W NEUDERT and H. ROPKE, *Atlas of Steroid Spectra*. Translated by J. B. Leane, Springer-Verlag, Berlin, 1965.

6. G. OURISSON, S. MUNAVALLI, and G. EHRET, *International Tables of Selected Constants*, Vol. 15, *Data Relative to Sesquiterpenoids*, Pergamon Press, 1966.

TABLE 9

SUMMARY OF THE PUBLISHED BIBLIOGRAPHIES OF INFRARED SPECTRAL DATA

Ref. No.	Name and date of period covered	Number of spectra or entries	Type of compound	Molecular formula index	Alphabetical name index	Other indexes	Location
1	ASTM, Molecular formula list of compounds, names and references to published Infrared Spectra, Vol. I (AMD 31)	92,000	All classes of compound. Also includes literature references to near and far infrared spectra	Yes	No, for organics. Yes, for inorganics	Serial number index	
2	Infrared absorption spectra, H. M. Hershenson: Index 1945–57 Index 1958–62	27,000 literature references	All classes of compounds	No	Yes	None	
3	Index of published Infrared Spectra, M. B. B. Thomas, 2 vols.: covers period up to 1957	10,000 literature references	All classes of compounds	No	Yes	None	

References

1. AMERICAN SOCIETY FOR TESTING AND MATERIALS, *Molecular formula List of Compounds, Names, and References to Published Infrared Spectra,* Vol. 1 (AMD 31), 1969. Available from Heyden & Son Limited, London

2. H. M. HERSHENSON, *Infrared Absorption Spectral; Index for 1945–57;* published 1959, *Index for 1958–1962;* published 1964, Academic Press.

3. M. B. B. THOMAS, *An Index of Published Infrared Spectra,* 2 vols., HMSO, 1961.

TABLE 10

SUMMARY OF THE PUBLISHED BIBLIOGRAPHIES OF MASS SPECTRAL DATA

Ref. No.	Name and date of period covered	Number of spectra or entries	Type of compound	Molecular formula index	Other indexes	Location
1	Mass spectrometry bulletin, annual cumulative index: Vol. 1, 1967 Vol. 2, 1968 Vol. 3, 1969	5265 3322 2951	All classes of compounds	No	Molecular weight index Author index	
2	Structure Indexed literature of organic mass spectra: Vol. 1, 1966 Vol. 2, 1967	2616	General organic, organometallic and naturally occurring organic compounds	Yes	Index based on chemical structure	
3	Index and bibliography of mass spectrometry (1963–5), McLafferty and Pinzelik	6000 literature references	All classes of compounds	No	Key word-in-context index (KWIC)	
4	Cumulative chemical compound index to organic mass spectrometry: Vol. 1, 1968; Vol. 2, 1969 (published together)	2429	All classes of compounds	No	Molecular weight index	
5	Bibliography on mass spectrometry (1938–57), AEI	565 literature references	All classes of compounds	No	Author index	

References

1. *Mass Spectrometry Bulletin*, Annual Indexes, published periodically by HMSO. Available directly from the Data Centre at AWRE, Aldermaston, Berkshire.

2. *Structure Indexed Literature of Organic Mass Spectra* (edited by the Society of Mass Spectroscopy of Japan), Academic Press of Japan: Vol. 1, 1966; Vol. 2, 1967.

3. F. W. McLafferty and J. Pinzelik, *Index and Bibliography of Mass Spectrometry, 1863–1965*, Wiley, 1967.

4. *Cumulative Chemical Compound Index to the Journal of Organic Mass Spectrometry*, Vol. 1 and Vol. 2 published together, 1970. Heyden & Son Ltd., London.

5. *Bibliography on Mass Spectrometry, 1938–1967*. Pergamon Press, for AEI, Manchester, 1951.

TABLE 11

SUMMARY OF THE PUBLISHED BIBLIOGRAPHIES OF NUCLEAR MAGNETIC RESONANCE, ELECTRON SPIN RESONANCE (AND NUCLEAR QUADRUPOLE RESONANCE)

Ref. No.	Name and date of period covered	Number of literature references	Type of compound	Molecular formula index	Alphabetical name index	Other indexes	Location
1	DMS literature list n.m.r., n.q.r., and e.p.r. loose-leaf sheets: Vol. I, 1953–65 Vol. II, 1965–6 Vol. III, 1967–8	4799 5199 5199	All classes of compounds	No	No	Peek-a-Boo search system Author index, Vols. II and III	
2	Formula index to n.m.r. literature data, Howell Kende, and Webb; bound monographs: Vol. 1, Prior to 1961 Vol. 2, 1961–2	2500 5700	General, organic, heterocyclic, phosphorus compounds	Yes	No	None	
3	n.m.r. and e.s.r. spectra, Hershenson; bound monograph: 1963–5	over 8000	All classes of compounds	No	Yes	None	

References

1. DOCUMENTATION OF MOLECULAR SPECTROSCOPY, *Current Literature List NMR, N.Q.R. and E.P.R.*, Vols. I–III, 1964–8, Butterworth, & London.
2. M. GERTRUDE HOWELL, A. S. KENDE, and JOHN S. WEBB, *Formula Index to N.M.R. Literature Data*. Plenum Press, 2 vols., Vol. 1, 1966; Vol. 2, 1965.
3. H. M. HERSHENSON, *Nuclear Magnetic Resonance and Electron Spin Resonance Spectra. Index for 1958–1963*, Academic Press, 1965.

Index

E.s.r. radicals and specific compounds appear in **bold type**. The following abbreviations have been used:

c.s.m.	= data from a combination of spectral methods.
(e.s.r.)	= e.s.r. data.
f.	= fragmentation in mass spectrometry.
(i.r.)	= i.r. data.
m.s.	= mass spectrum.
(n.m.r.)	= n.m.r. data.
(u.v.)	= u.v. data.